Diving Physiology of Marine Mammals and Seabirds

Analyzing the physiological adaptations of marine mammals and seabirds, this book provides a comprehensive overview of what allows these species to overcome the challenges of diving to depth on a single breath of air. Through comparative reviews of texts on diving physiology and behavior from the last 75 years, Ponganis combines this research into one succinct volume.

Investigating the diving performance of marine mammals and seabirds, this book illustrates how physiological processes to extreme hypoxia and pressure are relevant to the advancement of our understanding of basic cellular processes and human pathologies. This book underscores the biomedical and ecological relevance of the anatomical, physiological, and molecular/biophysical adaptations of these animals to enable further research in this area.

An important resource for students and researchers, this text not only provides an essential overview of recent research in the field, but will also stimulate further research into the behavior and physiology of diving.

Paul J. Ponganis is a Research Marine Biologist and Marine Physiologist at the Scripps Institution of Oceanography, University of California, San Diego. A leading expert in the field and also an anesthesiologist, his primary clinical interests are in cardiac anesthesia, which he has practiced for the last 30 years in conjunction with his research at the Scripps Institution of Oceanography. His research has focused on the diving physiology of marine mammals and penguins at field sites around the world. In recognition of their Antarctic research, the Ponganis Icefall on Coulman Island was named after him and his wife.

Diving Physiology of Marine Mammals and Seabirds

PAUL J. PONGANIS
Scripps Institution of Oceanography, University of California, San Diego

CAMBRIDGE
UNIVERSITY PRESS

CAMBRIDGE
UNIVERSITY PRESS

University Printing House, Cambridge CB2 8BS, United Kingdom

One Liberty Plaza, 20th Floor, New York, NY 10006, USA

477 Williamstown Road, Port Melbourne, VIC 3207, Australia

4843/24, 2nd Floor, Ansari Road, Daryaganj, Delhi - 110002, India

79 Anson Road, #06-04/06, Singapore 079906

Cambridge University Press is part of the University of Cambridge.

It furthers the University's mission by disseminating knowledge in the pursuit of education, learning and research at the highest international levels of excellence.

www.cambridge.org
Information on this title: www.cambridge.org/9780521765558

First published 2015

A catalogue record for this publication is available from the British Library

Library of Congress Cataloging in Publication data
Ponganis, Paul J., 1950–
Diving physiology of marine mammals and seabirds / Paul J. Ponganis, Scripps Institution of Oceanography,
University of California, San Diego.
pages cm
Includes bibliographical references and index.
ISBN 978-0-521-76555-8 (Hardback : alk. paper)
1. Marine mammals–Physiology. 2. Water birds–Physiology. 3. Deep diving–Physiological aspects. 4. Underwater physiology. I. Title.
QL713.2.P66 2015
599.5–dc23 2015020038

ISBN 978-0-521-76555-8 Hardback

To Katherine, my love and inspiration.

Contents

Preface

The goal of this book is to provide students and researchers with a reference resource for the diving physiology of marine mammals and seabirds. To make progress in the future, it is essential to know what has been achieved in the past. It is my hope that this book will serve that purpose, and stimulate further research into both how these animals function in their environment and how their extreme adaptations may provide insight into basic physiology and pathophysiological processes.

It has been 25 years since the publication of Kooyman's *Diverse Divers*. Since that time, advances in biomedical technology and the advent of electronic backpack recorders have greatly expanded the field. Prior to that period, the pioneering work of Irving and Scholander in the 1930s and 1940s began the "modern" era of diving investigations. Consequently, more than 75 years of research, ranging from anatomical/physiological studies to biochemical/molecular investigations, are reviewed in this book. In my view, these topics are all parts of the animal's "physiology," and are relevant to the questions of both field biologists and laboratory investigators. In addition to future advances in electronic behavioral recorders, genomics, proteomics, and computer modeling, I want to encourage the development of physiological investigations to see how an animal functions and to determine what is actually happening within the body. To develop the tools and not interfere with natural behaviors is the challenge of the future.

As regards the book's layout, after initial chapters on diving behavior and the physiological challenges of diving, the chapters are primarily arranged along physiological themes, with a concluding chapter on biomedical applications. Cardiorespiratory physiology, oxygen store management, and hypoxemic/pressure tolerance receive the most emphasis, reflecting my primary interests.

I have tried to be as comprehensive as possible, but as with any book there will undoubtedly be newer papers published even as the book goes to press. In addition, if there are any significant publications omitted, especially in areas outside my expertise, I am responsible and attribute that to a lack of time and a demanding clinical schedule. Lastly, the inclusion of both marine mammals and seabirds into one book has necessitated division of many chapters into separate marine mammal and bird sections. As an aid to students, I have included the scientific name of a given species the first time it is used in every chapter as well as in each figure legend and table.

My perspectives on diving physiology stem from those of both a biologist and physician. I have been fortunate to be able to work as a biologist at Scripps Institution of Oceanography as well as to have an active clinical practice in anesthesiology at Sharp Memorial Hospital in San Diego. It has been a unique experience: to collaborate with Jerry Kooyman, the foremost diving physiologist of my era, and to conduct an exciting clinical practice at what I consider has been San Diego's premier heart transplant and ventricular assist device center over the past 30 years. I still remember driving into the hospital the night we performed San Diego's first heart transplant in 1985.

To be able to pursue careers as both a biologist and anesthesiologist, I want to acknowledge K. S. Norris, my graduate student sponsor, and R. W. Pierce. Dick Pierce advised and supported my graduate work, taught the basics of diving physiology to two young Norris graduate students (D. Costa and myself), and was the force behind the scenes in the building of the UCSC Long Marine Lab. Similarly, my medical education and training at Stanford University were outstanding. And, of course, there is Jerry Kooyman, my long-time colleague and good friend at Scripps.

Much of my work has also benefited from the advice and assistance of SeaWorld's outstanding veterinary and animal care staff, and from consultations with Sam Ridgway, Red Howard, and the National Marine Mammal Foundation. What a luxury to have such expertise closely available. Anesthesia Service Medical Group has provided an outstanding clinical anesthesia practice opportunity with the flexibility to devote time to research and remote expeditions. My anesthesia partners and other physician colleagues at Sharp have been entirely supportive of this work. Again, I do not know of other medical practices in which such opportunities are available. To all, I am indebted.

I have also benefited from long-time associations and friendships with Roger Gentry, Phil Thorson, Mike Castellini, Randy Davis, Terrie Williams, Dan Costa, Markus Horning, Fritz Trillmich, Yvon Le Maho, Katsu Sato, Scott Eckert, Greg Marshall, Judy St. Leger, Tom Jue, and E. (Zhenya) Baranov. And in the era of microprocessor recorders, I cannot forget both my brother, Ed Ponganis, who developed our early electronic recorders in the 1980s, and the late Harve Hanish and his staff at UFI, who developed multiple custom physiological recorders for my more recent research. Many other collaborators have made significant contributions to my work including L. Winter, L. Welch, O. Matthieu-Costello, M. Costello, M. Scadeng, R. Spragg, D. Houser, T. Zenteno-Savin, S. Barber-Meyer, J. Heil, C. Champagne, H. Goforth, Y. Habara, and K. Shiomi. Lastly, there have also been my graduate students and fellows, all of whom have been outstanding, including P. Jobsis, D. Levenson, J. Meir, C. Williams, M. Tift, A. Wright, R. van Dam, T. Stockard, B. McDonald, J. Goldbogen, T. Welch, my US Navy anesthesia residents from Balboa Hospital, San Diego, and the many students Dan Costa allows me to "borrow" for field research.

I appreciate the patience and support of Ilaria Tassistro, Victoria Parrin, and Cambridge University Press in the preparation of this book, and thank M. Piscitelli and A.W. Vogl for use of photographs, P. Miller, J. Goldbogen, K. Goetz, D. Costa, and K. Sato for graphics data from their research and publications, P. Jobsis for the use of figures, and J. St. Leger and E. Nielsen for necropsy observations.

Lastly, I want to thank and acknowledge my collaborator and wife, Dr. Katherine V. Ponganis. It is she who has brought Jerry Kooyman and me kicking and screaming into the computer age. At the same time, she pursued research in cosmochemistry. It is also her programming skills that have allowed us to decipher the reams of data collected by our physiological recorders. Many a graduate student has benefited from these programs. And she has had the patience and understanding to allow me to be both a biologist and physician, as well as to write this book.

Paul Ponganis MD, PhD
Scholander Hall
Center for Marine Biotechnology & Biomedicine, Scripps Institution of Oceanography

1 Diving behavior

Diving physiology is best interpreted in light of behavior. Therefore, this first chapter is an overview of the diving behavior of marine mammals and seabirds. Prior to Kooyman's development of the first time–depth recorder in the 1960s (Kooyman, 1965, 1968), reviews of diving behavior primarily cited (a) dive depths and durations of harpooned whales, (b) depths at which carcasses were found entangled in nets or cables, and (c) even average time until last movement during forced submersion (Andersen, 1966, Harrison and Kooyman, 1968, Kooyman and Andersen, 1968, Piatt and Nettleship, 1985, Schorger, 1947). However, the development of microprocessor-based data loggers and satellite transmitters over the past three decades has now allowed documentation of remarkable diving behaviors in many marine species. Amazingly, at the time of this writing, a depth recorder has been deployed on every pinniped species. The number of studies is now so extensive that, even with today's internet search engines, it is difficult to be sure that every investigation has been found, especially in regard to maximum dive depths and durations. One excellent source of such information for readers is on the internet, Penguiness Book (http://penguinessbook.scarmarbin.be/index.php), a website created and maintained by Drs. Yan Rupert-Coudert and Akiko Kato.

The tables in this chapter list both common and maximum dive depths and durations that had been published at the time of writing. As future studies will increase the size of the data base as well as the range of diving environments encountered, these values, especially the maximum dive durations and depths, will undoubtedly change. Nonetheless, these data, especially the common dive durations and depths, provide a background to (a) appreciate differences in the dive performance, breath-hold capacities, and ecology of various species; (b) understand the role of physiological adaptations underlying such dive behaviors; and (c) evaluate differences between dives and physiological responses under experimental conditions versus those in the wild.

In addition to review of the depths and durations of dives in marine mammals and seabirds, this chapter will also highlight recent advances in documentation of underwater behaviors with the advent of more sophisticated dive recorders. We are only beginning to glimpse how these animals function and thrive in an underwater world hidden from our view. From a biologist's perspective, these are indeed exciting times.

When reviewing these dive behaviors, readers should remember that physiology ultimately underlies the dive capacity and foraging capabilities of a given species. All marine mammals and diving birds are potentially faced with alterations in prey availability and prey distribution due to overfishing, pollution, and climate change. Consequently, how an animal dives and how close it pushes itself to its physiological limits during foraging activity are important topics. While it is true that understanding diving physiology does not directly "save" a species, it does help provide a rational basis for arguments to (a) regulate the fishing industry and pollution, (b) create marine sanctuaries, and (c) limit greenhouse gas emissions to decrease global warming.

Systematics, evolution, and foraging ecology of marine mammals and seabirds are not included in this chapter. Those topics could easily constitute another book. For such details, readers are referred to several texts and reviews (Berta *et al.*, 2006, Castellini and Mellish, 2015, Croxall, 1987, Davis and Darby, 1990, Perrin *et al.*, 2009, Reynolds and Rommel, 1999, Ridgway and Harrison, 1981–1998, Schreiber and Burger, 2002, Sibley and Monroe Jr., 1990).

1.1 Marine mammals

Marine mammals are composed of representatives from four mammalian orders, Pinnipedia, Cetacea, Sirenia, and Carnivora. The pinnipeds, which has also been classified as a suborder of the Carnivora, consist of the phocids (true or earless seals), otariids (eared seals – the fur seals and sea lions), and odobenids (walruses). The cetaceans are divided into the odontocetes (toothed whales) and mysticetes (baleen whales), while the sirenians include the manatees (three species) and dugongs (one species). Marine carnivore species include the polar bear (*Ursus maritimus*), sea otter (*Enhydra lutris*), and several other marine otters. After review of dive behaviors in each of these mammalian orders, this section on marine mammals will conclude with brief mention of the diving abilities of several aquatic mammals.

1.1.1 Pinnipeds: phocids

Although routine dive durations and depths of many phocid seals are less than 10–15 min and 100 m, almost all of these seals are capable of occasional exceptional dives of 25–30-min duration and greater than 300-m depth (Table 1.1). Both the routine and extreme dives are quite a physiological accomplishment in terms of breath-hold duration and pressure tolerance. However, it is the large phocid seals that are the longest-duration and deepest divers among the pinnipeds (Table 1.1). These include Weddell seals (*Leptonychotes weddellii*), hooded seals (*Cystophora cristata*), and northern and southern elephant seals (*Mirounga angustirostris, M. leonina*). Routine dive durations are up to 30 min, and common depths are 400–600 m. The southern elephant seal currently holds the record for the deepest (2388 m) and longest (120 min) dives of any pinniped (Hindell *et al.*, 1992, Costa et al., 2010).

Table 1.1 Dive characteristics of phocid pinnipeds.

Species	Duration (min)		Depth (m)		Reference
	Common	Max	Common	Max	
Harbor seal *Phoca vitulina*	<10	35	5–100	481	A
Spotted seal *P. largha*	<10	–	<100	–	B
Harp seal *P. groenlandica*	2–15	>15	50–300	568	C
Ringed seal *P. hispida*	<10	>50	20–100	500	D
Baikal seal *P. sibirica*	2–6	>40	5–50	324	E
Caspian seal *P. caspica*	1	<4	<50	>200	F
Hawaiian monk seal *Monachus schauinslandi*	5–15	>20	20–450	>500	G
Mediterranean monk seal *M. monachus*	5–10	18	10–80	123	H
Grey seal *Halichoerus grypus*	1–8	32	10–120	436	I
Ribbon seal *Histriophoca fasciata*	–	–	200–600	>600	J
Bearded seal *Erignathus barbatus*	1–6	19	10–60	480	K
Hooded seal *Cystophora cristata*	5–25	>52	100–600	>1016	L
Crabeater seal *Lobodon carcinophagus*	5	24	90	713	M
Ross seal *Ommatophoca rossii*	5–15	>20	100–300	792	N
Leopard seal *Hydrurga leptonyx*	<5	15	10–50	304	O
Weddell seal *Leptonychotes weddellii*	10–15	96	150–400	904	P
Northern elephant seal *Mirounga angustirostris*	15–30	119	200–600	1735	Q
Southern elephant seal *Mirounga leonina*	20–29	120	269–552	2388	R

References: A: Bowen *et al.*, 1999, Eguchi and Harvey, 2005, Lesage *et al.*, 1999; B: Lowry *et al.*, 1994; C: Folkow *et al.*, 2004, Lydersen and Kovacs, 1993; D: Born *et al.*, 2004, Gjertz *et al.*, 2000a, Kelly and Wartzok, 1996;, E: Stewart *et al.*, 1996, Watanabe *et al.*, 2004, 2006; F: Miyazaki, 2001; G: Parrish *et al.*, 2002; H: Dendrinos *et al.*, 2007, Gazo and Acuilar, 2005, Kirac *et al.*, 2002; I: Beck *et al.*, 2003, Goulet *et al.*, 2001, Thompson and Fedak, 1993, Thompson *et al.*, 1991; J: London *et al.*, 2014; K: Gjertz *et al.*, 2000b, Krafft *et al.*, 2000; L: Folkow and Blix, 1999; M: Burns *et al.*, 2004; N: Blix and Nordoy, 2007; O: Nordoy and Blix, 2009; P: Castellini *et al.*, 1992a, Heerah *et al.*, 2013, Schreer and Testa, 1996; Q: Le Boeuf *et al.*, 1988, Robinson *et al.*, 2012, Stewart and DeLong, 1995; R: Bennett *et al.*, 2001, Costa *et al.*, 2010; Hindell *et al.*, 1991).

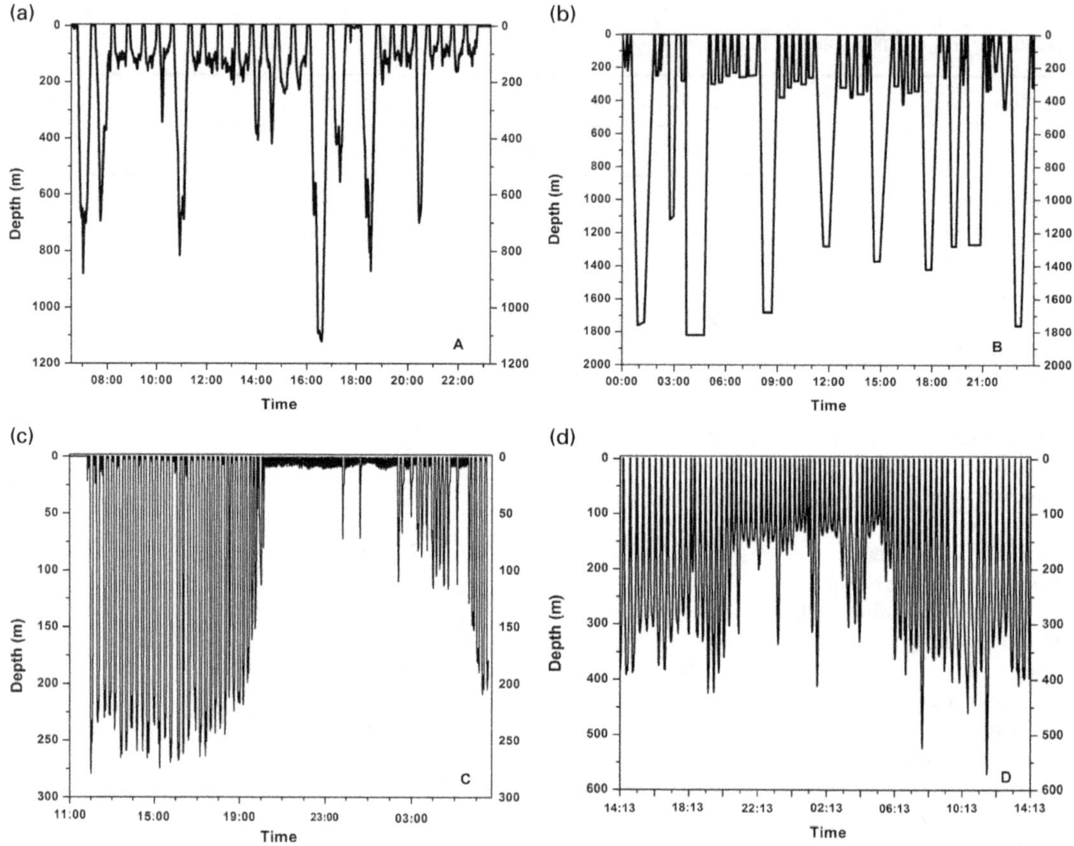

Figure 1.1 Representative 15–24-hour dive activity profiles of marine mammals and seabirds. **(a)** Sperm whale (*Physeter macrocephalus*) in the Bleik Canyon near Andenes, Norway, 2011 (courtesy of P.O. Miller, unpublished data). **(b)** Cuvier's beaked whale (*Ziphius cavirostris*), adapted from Schorr *et al.*, 2014. **(c)** Fin whale (*Balaenoptera physalus*), adapted from Goldbogen *et al.*, 2006, data courtesy of J. Goldbogen. **(d)** Juvenile northern elephant seal (*Mirounga angustirostris*), adapted from data of Meir *et al.*, 2009.

Weddell seals dive in bouts, with long rest periods on the sea ice between bouts (Kooyman, 1981, Kooyman *et al.*, 1980). Harbor seals (*Phoca vitulina*) and many other phocid seals also haul out regularly between foraging trips (Thompson *et al.*, 1989, 1991). This interrupted pattern of diving contrasts with that in northern/southern elephant seals and hooded seals (see Fig. 1.1). During the several-month-long trips of these three species to sea, the animals are underwater 80–90% of the time (Folkow and Blix, 1999, Hindell *et al.*, 1991, Le Boeuf *et al.*, 1988). Surface intervals are less than a few minutes. Accordingly, these species have been termed "surfacers," in contrast to the Weddell seal, which has been considered a "diver" (Kramer, 1988). Although this surfacer–diver concept was developed in regard to foraging ecology, the distinction between surfacers and divers also has implications for diving physiology.

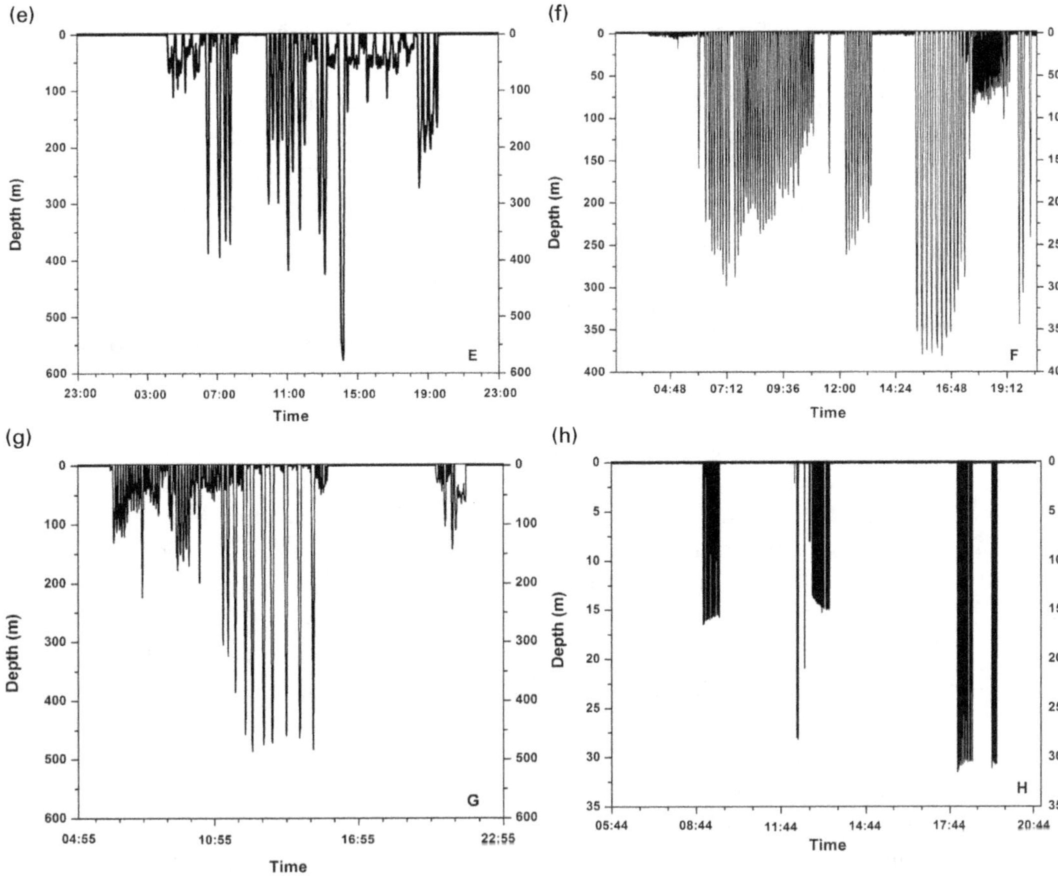

Figure 1.1 *(cont.)* **(e)** Weddell seal (Leptonychotes weddellii) in the Ross Sea, 2011 (courtesy of K. Goetz and D. Costa, unpublished data). **(f)** California sea lion (Zalophus californianus), adapted from data of McDonald and Ponganis, 2014. **(g)** Emperor penguin (*Aptenodytes forsteri*), adapted from data of Sato *et al.*, 2011. Last dive is the longest reported dive of an emperor penguin. **(h)** European shag (*Phalacrocorax aristotelis*), adapted from Sato *et al.*, 2008, data courtesy of K. Sato.

A surfacer such as the elephant seal, which goes to sea for 2–7 months, at times gaining an average of 1 kg of body mass per day (Le Boeuf *et al.*, 1988, Robinson *et al.*, 2012), must forage, process food, build up fat reserves, recover from long dives, and probably sleep all while holding its breath at depth underwater. It is remarkable that even after a two-hour dive, a southern elephant seal continued to make 30-min dives with short surface intervals for five hours, and then resumed serial one-hour dives after that (Hindell *et al.*, 1992). There was no prolonged surface recovery period. In contrast, after a 45-min dive, a Weddell seal spent nearly 70 min of the next two hours at the surface (Kooyman *et al.*, 1980). And, even without long dives, Weddell seals typically dive for only about 12 hours of the day (Castellini *et al.*, 1992a, Kooyman *et al.*, 1980).

Partitioning of the metabolic demands of travel, foraging, and digestion into different dive types may play a significant role in the continuous dive behavior of a surfacer such as the elephant seal. In the gray seal (*Halichoerus grypus*), a diver which routinely hauls out (Thompson *et al.*, 1991), the cost of digestion can be delayed until rest or haul-out periods (Sparling *et al.*, 2007). That option to process food at the surface does not occur in elephant seals. However, intensive research on the dive behavior of elephant seals has revealed several distinct dive profiles, including V-shaped dives (transit), active flat-bottom dives (benthic feeding), active bottom dives (pelagic feeding), and drift dives (dives with prolonged periods with no flipper strokes) (Crocker *et al.*, 1997, Hindell *et al.*, 1991, Le Boeuf *et al.*, 1993, Mitani *et al.*, 2010, Robinson *et al.*, 2012). Drift drives, considered to be rest dives because of the lack of stroking activity, have received much attention and have been proposed as periods of sleep as well as food processing. Changes in drift rates in such dives throughout a trip to sea have been used to identify the times and locations of increases in body buoyancy (fat deposition) and hence successful foraging by elephant seals (Biuw *et al.*, 2003, Robinson *et al.*, 2010). Partitioning of energy demands in different dive types so that dives are energetically similar in cost in elephant seals is supported by similar rates of blood oxygen depletion during those different dive types (Meir *et al.*, 2013).

Further documentation of foraging behavior, prey ingestion, and prey identification during foraging dives are being developed with use of stomach temperature sensors, jaw motion sensors, three-dimensional dive profile reconstructions, and digital/video cameras (Davis *et al.*, 1999, Horsburgh *et al.*, 2008, Kuhn *et al.*, 2009, Liebsch *et al.*, 2007, Naito *et al.*, 2013, Parrish *et al.*, 2002, Suzuki *et al.*, 2009, Ydesen et al., 2014). Such studies should yield insight into the time partitioning of energy demands of surfacers and, hence, into their ability to dive continuously.

1.1.2 Pinnipeds: otariids

Investigations of diving behavior in the otariids have primarily been conducted on lactating females during maternal foraging trips to sea. Such studies take advantage of a natural behavior – the regular return of lactating females from foraging trips to nurse their pups. Observation of this behavior in northern fur seals (*Callorhinus ursinus*) by Gentry in the early 1970s led to his collaboration with Kooyman to develop a longer-duration time–depth recorder and to their eventual documentation of dive behavior in six otariid species distributed from Alaska to the Antarctic and from the Galapagos to South Africa (Gentry and Kooyman, 1986, Kooyman *et al.*, 1976b). Since that time, every otariid species has been studied (Table 1.2).

In general, the fur seals and sea lions have not been considered as good divers as the phocid seals. The dives of otariids are shorter and shallower than those of phocid seals, with most dives less than four-min duration and less than 100 m in depth. However, at least some otariids do regularly perform deep dives. Even in the earliest

Table 1.2 Dive characteristics of otariid pinnipeds.

Species	Duration (min)		Depth (m)		Reference
	Common	Max	Common	Max	
Northern fur seal *Callorhinus ursinus*	2	10	65	256	A
Antarctic fur seal *Arctocephalus gazella*	<2	11	30	240	B
Sub-Antarctic fur seal *A. tropicalis*	<2	7	10–30	208	C
South African fur seal *A. pusillus pusillus*	2	8	45	204	D
Australian fur seal *A. pusillus doriferus*	2–4	9	65–85	164	E
New Zealand fur seal *A. forsteri*	2–3	11	30–75	274	F
Galapagos fur seal *A. galapagoensis*	<2	5	26	115	D
Juan Fernandez fur seal *A. philippii*	2–4	–	<10	90	G
Guadalupe fur seal *A. townsendi*	2–3	18	10–20	130	H
South American fur seal *A. australis*	2–4	7	20–60	170	I
California sea lion *Zalophus californianus*	2	16	62	>482	J
New Zealand sea lion *Phocarctos hookeri*	4	20	123	597	K
Australian sea lion *Neophoca cinerea*	3–4	9	60	200	L
Southern sea lion *Otaria flavescens*	2–4	8	20–40	243	M
Steller sea lion *Eumetopias jubata*	<2	8	9–24	452	N
Galapagos sea lion *Zalophus wollebaeki*	3–6	11	50–150	387	O

References: A: Gentry *et al.*, 1986, Ponganis *et al.*, 1992a, Sterling and Ream, 2004; B: Boyd and Croxall, 1992, Kooyman *et al.*, 1986; C: Georges *et al.*, 2000a, 2000b; D: Horning and Trillmich, 1997, Kooyman and Trillmich, 1986; E: Arnould and Hindell, 2001; F: Mattlin *et al.*, 1998; G: Francis *et al.*, 1998; H: Gallo-Reynoso *et al.*, 2008, Lander *et al.*, 2000; I: Trillmich *et al.*, 1986; J: Feldkamp *et al.*, 1989, Melin *et al.*, 2008; K: Chilvers *et al.*, 2006; Gales and Mattlin, 1997; L: Costa *et al.*, 2001, Fowler *et al.*, 2006; M: Thompson *et al.*, 1998, Werner and Campagna, 1995; N: Merrick and Loughlin, 1997, Pitcher *et al.*, 2004; O: Villegas-Amtmann and Costa, 2010, Villegas-Amtmann *et al.*, 2008).

studies of northern fur seals, dive depths clustered at 50–60 m and 175 m, and individual fur seals were classified as shallow, deep, or mixed divers, dependent on their dive profiles (Gentry *et al.*, 1986). And although initial studies found that most dives of California sea lions (*Zalophus californianus*) and Galapagos sea lions

(*Zalophus wollebaeki*) were shallow (<100 m) (Feldkamp *et al.*, 1989, Kooyman and Trillmich, 1986), later investigations of these species in other locations have documented dives in the 400-m depth range (McDonald and Ponganis, 2013, Melin *et al.*, 2008, Villegas-Amtmann and Costa, 2010). Although such deep dives in Galapagos and California sea lions are a small percentage of all dives, performance of such dives, especially serial deep dives as in the California sea lion, demonstrate the physiological capacity of these animals. The deepest diving otariid is the benthic-feeding New Zealand sea lion (*Phocarctos hookeri*), which can dive as deep as 600 m and as long as 20 min (Chilvers, 2008). In regard to the surface–diver classification, otariids are divers, often not diving for about 50% of their time at sea, and, if some individuals do continuously dive while in the water, they regularly haul out onto land during their foraging trips (Thompson *et al.*, 1998).

1.1.3 Pinnipeds: odobenids

The walrus (*Odobenus rosmarus*), the lone member of the Odobenidae, is difficult to study due to both size and accessibility. Typically, these animals dive for 4–6 min, with a maximum of 24 min during underwater territorial displays and during foraging activity (Born and Knutsen, 1997, Gjertz *et al.*, 2001, Jay *et al.*, 2001, Nowicki *et al.*, 1997, Wiig *et al.*, 1993). Depths of dives are 30–70 m in range.

1.1.4 Cetaceans

The ability to document cetacean diving behavior began with the pioneering radio telemetry research of Evans on dolphins, and Watkins and Schevill on whales (Evans, 1971, Watkins, 1978, 1979, Watkins and Schevill, 1977, Watkins and Tyack, 1991). Miniaturization of recorders/transmitters and refinement of attachment techniques, including suction cups and percutaneous darts, have since allowed application to a wide variety of species (Andrews *et al.*, 2008, Balmer *et al.*, 2014, Baird, 1998, Hooker and Baird, 2001, Johnson and Tyack, 2003, Mate *et al.*, 2007). In particular, Johnson and Tyack's development of the DTAG, an archival recorder specifically designed to examine behavioral responses of marine mammals to sound, has been a significant breakthrough in at-sea cetacean behavioral research. Data from DTAGs have allowed detailed analyses of depth profiles, orientation, stroking, echolocation signals, prey capture, and foraging efficiency in sperm whales (*Physeter macrocephalus*) and beaked whales (family: Ziphiidae) (Johnson *et al.*, 2004, 2006, Madsen *et al.*, 2002, 2005, 2007, Miller *et al.*, 2004b, 2004c, Tyack *et al.*, 2006, Watwood *et al.*, 2006, Zimmer *et al.*, 2003).

Among the cetaceans, the most notable dive depths and durations are those of the large toothed whales, specifically sperm whales and beaked whales (Baird *et al.*, 2008, Hooker and Baird, 1999, Schorr *et al.*, 2014, Tyack *et al.*, 2006, Watkins *et al.*, 1993) (Table 1.3). Routine depths and durations range from 400 to 800 m and 40 to 60 min, respectively, with maximum depths and durations greater than 2000 m and 120 min. Currently, in 2015, Cuvier's beaked whale (*Ziphius cavirostris*) holds the record for any mammal for the

Table 1.3 Dive characteristics of some odontocete cetaceans.

Species	Duration (min)		Depth (m)		Reference
	Common	Max	Common	Max	
Harbor porpoise *Phocoena phocoena*	1	5	14–40	226	A
Finless porpoise *Neophocoena phocoenoides*	2	–	<25	–	B
Dall's porpoise *Phocoenoides dalli*	2–4	–	<70	94	C
Bottlenose dolphin *Tursiops truncatus*	1	8	20	>500	D
Spotted dolphin *Stenella attenuata, S. frontalis*	1–2	5	20–60	213	E
Spinner dolphin *S. longirostris*	<4	–	–	–	F
Common dolphin *Delphinus delphis*	–	5	30–60	280	G
Dusky dolphin *Lagenorhynchus obscurus*	–	–	50–65	130	H
Atlantic white-sided dolphin *L. acutus*	<1	1	–	–	I
Pacific white-sided dolphin *L. obliquidens*	–	6	–	215	J
Risso's dolphin *Grampus griseus*	<50	>400	–	–	K
Pilot whale *Globicephala* sp.	5–15	21	100–800	1019	L
Narwhal (*Monodon monoceros*)	1–15	26	150–500	>1000	M
Beluga whale *Delphinapterus leucas*	9–16	23	50–350	647	N
False killer whale *Pseudoorca crassidens*	5–12	15	50–650	650	O
Killer whale *Orcinus orca*	2–5	–	–	265	P
N. bottlenosed whale *Hyperoodon ampullatus*	40	70	800	1483	Q
Cuvier's beaked whale *Ziphius cavirostris*	58–70	138	1070–1334	2992	R
Baird's beaked whale *Berardius bairdii*	10–45	65	100–1500	1777	S
Arnoux's beaked whale *Berardius arnuxii*	10–45	>70	–	–	T
Blainville's beaked whale *Mesoplodon densirostris*	47–55	84	835–1099	1599	U

Table 1.3 *(cont.)*

Species	Duration (min)		Depth (m)		Reference
	Common	Max	Common	Max	
Sperm whale *Physeter macrocephalus*	40–60	138	400–900	2250	V

References: A: Otani *et al.*, 1998, Westgate *et al.*, 1995; B: Akamatsu *et al.*, 2010; C: Hanson and Baird, 1998, Jefferson, 1987; D: Klatsky *et al.*, 2007, Mate *et al.*, 1995, Ridgway, 1986; E: Baird *et al.*, 2001, Davis *et al.*, 1996, Scott and Chivers, 2009; F: Norris and Dohl, 1980; G: Evans, 1971, Ridgway, 1986; H: Benoit-Bird *et al.*, 2004; I: Mate *et al.*, 1994; J: Hall, 1970, Ridgway, 1986; K: Wells *et al.*, 2009; L: Aguilar de Soto *et al.*, 2008, Baird *et al.*, 2002, Heide-Jorgensen *et al.*, 2002, Nawojchik *et al.*, 2003; M: Heide-Jorgensen and Dietz, 1995, Laidre *et al.*, 2002; N: Martin and Smith, 1992, 1999, Ridgway *et al.*, 1984; O: Minamikawa *et al.*, 2013; P: Baird *et al.*, 2006a; Q: Hooker and Baird, 1999; R: Baird *et al.*, 2008, Schorr *et al.*, 2014, Tyack *et al.*, 2006; S: Minamikawa *et al.*, 2007; T: Hobson and Martin, 1996, Ponganis *et al.*, 1995; U: Baird *et al.*, 2008, Tyack *et al.*, 2006; V: Jaquet *et al.*, 2000, Norris and Harvey, 1972, Papastavrou *et al.*, 1989, Watkins *et al.*, 1993, Watwood *et al.*, 2006.

Table 1.4 Dive characteristics of some mysticete whales.

Species	Duration (min)		Depth (m)		Reference
	Common	Max	Common	Max	
Humpback whale *Megaptera novaeanglieae*	4–8	–	23–118	~160	A
Blue whale *Balaenoptera musculus*	5–10	–	180–200	–	B
Fin whale *Balaenoptera physalus*	3–8	12	<100	470	C
Minke whale *Balaenoptera acutorostrata, B. bonaerensis*	<5	9	<50	105	D
Bryde's whale *Balaenoptera brydei*	5	9	40–200	292	E
Right whale *Eubalaena glacialis*	12	–	120	–	F
Bowhead whale *Balaena mysticetus*	≤1–15	63	<16–100	487	G
Gray whale *Eschrictius robustus*	2–5	13	<30	–	H

References: A: Witteveen *et al.*, 2008; B: Croll *et al.*, 1998; C: Panigada *et al.*, 1999; D: Friedlaender *et al.*, 2014, Stockin *et al.*, 2001; E: Alves *et al.*, 2010; F: Baumgartner and Mate, 2003; G: Krutzikowsky and Mate, 2000, Laidre *et al.*, 2007; H: Stelle *et al.*, 2008, Stewart *et al.*, 2001, Woodward and Winn, 2006, Würsig *et al.*, 1986).

longest (137.5 min) and deepest (2992 m) dives (Schorr *et al.*, 2014). Pilot whales (*Globicephala* sp.), narwhals (*Monodon monceros*), beluga whales (*Delphinapterus leucas*), and false killer whales (*Pseudoorca crassidens*) are also deep divers, although dive durations are usually less than 20 min (Table 1.3).

The well-known bottlenose dolphin (*Tursiops truncatus*) was considered to typically dive for fewer than 5 min to shallow depths <20 m (Mate *et al.*, 1995). But, again, more recent studies of bottlenose dolphins in other locations have revealed regular dives to near 500-m depth (Klatsky *et al.*, 2007). Such deep dives are consistent with trained, open-water dives of Atlantic bottlenose dolphins (*Tursiops truncatus*), and Pacific bottlenose dolphins (*Tursiops gillii*) to 390 m and 535 m, respectively, with durations of 7.5 to 8 min (Ridgway, 1986). Dives of pelagic spotted dolphins (*Stenella attenuata*) are up to five min in duration, and can be as deep as 200 m (Mate *et al.*, 1995, Scott and Chivers, 2009). Documentation of the dive behavior of other dolphins and porpoises, as well as the mysticete whales, has been limited due to difficulties in capture or recorder deployment/attachment. In some cases, as for river dolphins and gray whales (*Eschrictius robustus*), only visual observations are primarily available for estimation of dive durations. River dolphins usually dive for fewer than 2 min, although dives as long as 8 min have been reported in Ganges River dolphins (*Platanista gangetica*) and Irawaddy dolphins (*Orcaella brevirostris*) (Bashir *et al.*, 2013, Edwards and Schnell, 2001, Renjun *et al.*, 1994, Stacey and Hvenegaard, 2002). In general, the baleen whales have dive durations under 10 min; they too, however, can reach depths of 200 m (Croll *et al.*, 1998, Friedlaender *et al.*, 2014, Witteveen *et al.*, 2008).

Classification of cetaceans as surfacers or divers depends on the species. Data are limited due to lack of long-term deployments of dive recorders, but currently available evidence indicates that sperm whales and beaked whales should be considered surfacers. Similar to the elephant seal, 80–90% of the day they are submerged below 10-m depth (Hooker *et al.*, 2012, Watkins *et al.*, 1999). Surface periods of Cuvier's beaked whale averaged less than two minutes (Schorr *et al.*, 2014). However, killer whales (*Orcinus orca*) and pilot whales should be classified as divers as they have been reported to spend almost 70–80% of their time within 10 m of the surface (Hooker *et al.*, 2012). Hanging or logging at the surface has been observed in some baleen whales and beluga whales, but the percentage time at the surface is unknown (Lyamin *et al.*, 2000, Sjare and Smith, 1986). Similarly, the percentage of time that some dolphin species may swim apparently asleep in shallow water is also not known (Lyamin *et al.*, 2008, Norris and Dohl, 1980, Wursig and Wursig, 1980).

1.1.5 Sirenians

Manatees (*Trichechus* sp.) and dugongs (*Dugong dugon*) usually dive for 2–3 min (Chilvers *et al.*, 2004, Gallivan and Best, 1980, Gallivan *et al.*, 1986, Marsh *et al.*, 1978, Reynolds III, 1981, Reynolds III and Odell, 1991). Maximum dive depths of these herbivorous mammals are about 12 m.

1.1.6 Marine carnivores

The polar bear, perhaps the most famous and certainly the largest of marine carnivores, is an excellent swimmer (Stirling, 1974). Most dives are less than 0.5 min; the longest reported dive is 3.2 min (Dyck and Romberg, 2007, Stirling and van Meurs, 2015). Most dives of sea otters are 1–3 min in duration and less than 30 m in depth (Ralls *et al.*, 1995, Tinker *et al.*, 2007, Yeates *et al.*, 2007). Dive durations of other otters such as the Eurasian otter (*Lutra lutra*) are usually less than 1 min (Conroy and Jenkins, 1986, Nolet *et al.*, 1993).

1.1.7 Aquatic mammals

Among aquatic animals, the muskrat (*Ondatra zibethicus*) has been perhaps the most frequent subject of physiological investigations. Dive durations of the muskrat are usually less than 2 min, although exploratory and alarm dives can be as long as 4 min (MacArthur *et al.*, 2001). Beavers (*Castor canadensis*), another research subject, usually dive for less than 2 min, but have been observed to stay underwater as long as 15 min at rest (Irving and Orr, 1935, Tevis, 1950). The Eurasian beaver performs dives as long as 4.9 min and as deep 4.2 m (Graf *et al.*, 2014) The platypus (*Ornithorhynchus anatinus*) usually dives for less than 2 min duration and to depths of less than 10 m (Bethge *et al.*, 2003). One of the smallest mammalian divers, the star-nosed mole (40–60 g, *Condylura cristata*) makes 10-sec dives with a reported maximum dive duration of 47 sec (McIntyre *et al.*, 2002). The capybara (*Hydrochaeris hydrochaeris*), a South American aquatic rodent, usually dives for less than 30 sec, although it has been reported to dive for several minutes (Creed, 2004).

1.2 Seabirds

Diving seabirds are composed of members from several avian orders, Procellariiformes, Charadriiformes, Pelecaniformes, and Sphenisciformes. The Procellariiformes (tube noses) include diving petrels, storm petrels, fulmars, prions, shearwaters, other petrels, and albatrosses. The family, Alcidae, comprises the diving seabirds of the Charadiformes, an order which also includes gulls, skuas, skimmers, and terns. The alcids include murres, little auks, razorbills, true guillemots, two groups of murrelets, puffins, auklets, and the extinct great auk (*Pinguinus impennis*). Cormorants (shags), gannets, and boobies are the accomplished divers of the order, Pelicaniformes; this order also includes pelicans, frigatebirds, and tropicbirds. Penguins comprise the Sphenisciformes.

Documentation of the diving activities of many seabirds has been limited by recorder size, flight requirements, and body size. Dive durations of many of the smaller alcids have only been determined by observation or surface radio transmissions, while diving

Table 1.5 Dive characteristics of the Procellariiformes.

Species	Duration (min)		Depth (m)		Reference
	Common	Max	Common	Max	
Common diving petrel *Pelecanoides urinatrix*	<1	–	<50	64	A
Peruvian diving petrel *Pelecanoides garnotii*	–	–	<40	83	B
S. Georgian diving petrel *Pelecanoides georgicus*	–	–	<40	49	C
Antarctic prion *Pachyptila desolata*	–	–	<5	7	D
Slender-billed petrel (thin-billed prion) *Pachyptila belcheri*	–	–	5	7	E
Blue petrel *Halobaena caerulea*	–	–	5	6	E
Bulwer's petrel *Bulweria bulwerii*	–	–	3	5	F
Northern fulmar *Fulmarus glacialis*	<0.1	0.1	<1	<3	G
Westland petrel *Procellaria westlandica*	–	–	3	<8	H
White-chinned petrel *Procellaria aequincotialis*	<0.1	0.1	<10	13	I
Monteiro's storm petrel *Oceanodroma monteiroi*	–	–	<1	<2	J
Audubon's shearwater *Puffinus lherminieri*	–	–	<30	35	K
Balearic shearwater *Puffinus mauretanicus*	<0.5	1.1	<10	26	L
Black-vented shearwater *Puffinus opisthomelas*	–	–	<40	52	M
Flesh-footed shearwater *Puffinus carneipes*	–	–	<10	67	N
Short-tailed shearwater *Puffinus tenuirostris*	–	–	<60	71	O
Sooty shearwater *Puffinus griseus*	–	–	<40	70	P
Wedge-tailed shearwater *Puffinus pacificus*	–	–	<50	66	K
Barolo shearwater *Puffinus baroli*			<20	23	Q
Streaked shearwater *Calonectris leucomelas*	–	–	–	5	R
Black-browed albatross *Thalassarche melanophris*	<0.1	0.2	<4	4.5	S
Grey-headed albatross *Thalassarche chrysostoma*	<0.1	0.2	<5	6.5	T

Table 1.5 *(cont.)*

Species	Duration (min)		Depth (m)		Reference
	Common	Max	Common	Max	
Shy albatross *Thalassarche cauta*	<0.2	0.3	<5	7.4	U
Light-mantled albatross *Phoebetria palpebrata*	–	–	<8	12.4	V
Wandering albatross *Diomedea exulans*	–	–	<0.5	0.6	V

References: A: Bocher *et al.*, 2000a, 2000b, Chastel, 1994, Ryan and Nel, 1999; B: Zavlaga and Jahncke, 1997; C: Bocher *et al.*, 2000a, Prince and Jones, 1992; D: Cherel *et al.*, 2002; E: Chastel and Bried, 1996; F: Mougin and Mougin, 2000; G: Garthe and Furness, 2001; H: Freeman *et al.*, 1997; I: Huin, 1994; J: Bried, 2005; K: Burger, 2001; L: Aguilar *et al.*, 2003; M: Keitt *et al.*, 2001; N: Rayner *et al.*, 2011; O: Weimerskirch and Cherel, 1998; P: Shaffer *et al.*, 2006, 2009, Weimerskirch and Sagar, 1995; Q: Neves *et al.*, 2012; R: Oka, 1994; S: Prince *et al.*, 1994, Sakamoto *et al.*, 2009; T: Huin and Prince, 1997, Prince *et al.*, 1994; U: Hedd *et al.*, 1997; V: Prince *et al.*, 1994.

depths of most procellariforms have been determined only with capillary depth gauges. Consequently, there are only limited observations on many of the species in Tables 1.5 (Procellariiformes) and 1.10 (other aquatic birds).

The miniaturization of electronic dive recorders, and conduct of studies on many seabird species, have been pioneered by Naito and colleagues at Japan's National Institute of Polar Research and the University of Tokyo (Croxall et al., 1991, 1993, Kato *et al.*, 1992, 2003, Kuroki *et al.*, 2003, Ropert-Coudert *et al.*, 2004a, Watanuki *et al.*, 1996, Williams *et al.*, 1992b). Their international collaborations have provided much of the data in Tables 1.6–1.9. Despite size limitations, more advanced recorders have also been developed for analysis of behavior and activity during diving of seabirds. These recorders and techniques include velocity meters, accelerometers, three-dimensional dive profile reconstructions, stomach and esophageal temperature recorders, beak angle detectors, and digital/video cameras (Charrassin *et al.*, 2001, Ponganis *et al.*, 2000, Sato *et al.*, 2002, 2007, Shiomi *et al.*, 2012, Takahashi *et al.*, 2004a, 2004b, 2008, Watanabe *et al.*, 2011, Wilson *et al.*, 1992a, 1995, 2002b). In addition to documenting behavior during the dive, such applications have also provided insight into foraging ecology, biomechanics, and even diving physiology (Sato *et al.*, 2002, 2009, 2011, Watanabe and Takahashi, 2013, Watanabe *et al.*, 2012, Watanuki *et al.*, 2006, 2008, Wilson *et al.*, 2010).

1.2.1 Procellariiform seabirds

Among Procellariiformes (Table 1.5), the diving petrels and shearwaters appear to be the best divers, reaching depths of 60–70 m. These birds plunge dive and use winged

Table 1.6 Dive characteristics of alcids.

Species	Duration (min)		Depth (m)		Reference
	Common	Max	Common	Max	
Atlantic puffin *Fratercula arctica*	<1	1.9	10–40	68	A
Rhinoceros auklet *Cerorhinca monocerata*	<1	2.5	10–30	60	B
Cassin's auklet *Ptychoramphus aleuticus*	<1	1.2	10–20	43	C
Common murre (common guillemot) *Uria aalge*	<2	3.4	20–50	138	D
Thick-billed murre (Brunnich's guillemot) *Uria lomvia*	<2	4.1	10–60	210	E
Little auk (Dovekie) *Alle alle*	<1	1.5	8–12	27	F
Razorbill *Alca torda*	<1	–	15–30	140	G
Pigeon guillemot *Cepphus columba*	<2	2.4	10–40	45	H
Xantu's murrelet *Synthliboramphus hypoleucus*	<1	–	–	–	I
Marbled murrelet *Brachyramphus marmoratus*	<1	1.9	–	–	J

References: A: Burger and Simpson, 1986, Wanless *et al.*, 1988; B: Burger *et al.*, 1993, Kuroki *et al.*, 2003; C: Ainley *et al.*, 1990, Burger and Powell, 1990; D: Burger and Simpson, 1986, Tremblay *et al.*, 2003, Wanless *et al.*, 1988; E: Croll *et al.*, 1992a, Elliott *et al.*, 2007; F: Brown *et al.*, 2012, Falk *et al.*, 2000, Harding *et al.*, 2009; G: Benvenuti *et al.*, 2001; Jury, 1986; H: Clowater and Burger, 1994; I: Hamilton *et al.*, 2005; J: Henkel *et al.*, 2004, Jodice and Collopy, 1999, Thoresen, 1989.

propulsion, although shearwaters can also use their legs for underwater locomotion (Brown *et al.*, 1978). The depths are impressive, given that diving petrels weigh only 100–150 g (Chastel, 1994, Prince and Jones, 1992). Prions and other petrels appear confined to shallow depths (<10 m) while the occasional dives of albatrosses are also shallow. The southern giant petrel (*Macronectes giganticus*) has also been observed to dive and feed on submerged carrion in shallow water (Van Den Hoff and Newbery, 2006).

1.2.2 Charadriiform seabirds

The diving behaviors of the alcids have been more extensively studied than those of the Procellariiformes. Dive durations of the wing-propelled alcids are usually less than 1–2 min and less than 60 m in depth (Table 1.6). The 1-kg murres are the best divers

Table 1.7 Dive characteristics of cormorants and shags.

Species	Duration (min)		Depth (m)		Reference
	Common	Max	Common	Max	
Antarctic shag *Phalacrocorax bransfieldensis*	<2	5.4	10–70	113	A
Cape cormorant *Phalacrocorax capensis*	<1	1.2	10–20	34	B
Crozet shag *Phalacrocorax melanogenis*	1–2	6.2	< 40	145	C
Great cormorant *Phalacrocorax carbo*	<1	2.5	<10	33	D
Japanese cormorant *Phalacrocorax capillatus*	<1	2.4	<20	45	E
King cormorant *Phalacrocorax albiventer*	1–3	5.1	<40	109	F
European shag *Phalacrocorax aristotelis*	<1	2.7	<40	61	G
Blue-eyed shag *Phalacrocorax atriceps*	2–4	6.3	10–80	125	H
Flightless cormorant *Phalacrocorax harrisi*	<2	3.3	<15	73	I

References: A: Casaux, 2004, Casaux and Barrera-Oro, 2006, Casaux *et al.*, 2001; B: Ryan *et al.*, 2010; C: Tremblay *et al.*, 2005; D: Grémillet *et al.*, 1999, Kato *et al.*, 2006; E: Watanuki *et al.*, 1996; F: Kato *et al.*, 2000; G: Wanless *et al.*, 1993, 1997, 1999; H: Croxall et al., 1991, Wanless and Harris, 1993, Wanless *et al.*, 1992; I: Wilson et al., 2008).

among the group, with remarkable maximal dive durations and depths near 3–4 min and 138–210 m. Puffins, auklets, and the razor bill (*Alca torda*) have slightly shallower routine depths and shorter dive durations (Table 1.6).

1.2.3 Pelecaniform seabirds

Cormorants and shags of the Pelecaniformes order are foot-propelled divers with some maximum dive depths and durations that are equally or perhaps more impressive than those of murres (Table 1.7). Both the Crozet shag (*Phalacrocorax melanogenis*) and blue-eyed shag (*Phalacrocorax atriceps*), about 2–2.5 kg in weight, have maximum dive durations greater than 6 min and maximum depths greater than 125 m (see Table 1.7 for references). The diving performances of the Antarctic shag (*Phalacrocorax bransfieldensis*) and king cormorant (*Phalacrocorax albiventer*) are only slightly less.

Boobies and gannets are short-duration (<1 min) plunge divers that primarily reach depths less than 20 m (Table 1.8). Although not studied extensively, pelicans appear limited to near-surface activities. Darters or anhingas are foot-propelled,

Table 1.8 Dive characteristics of the Pelecaniformes, excluding cormorants and shags (Phalacrocoracidae). All are marine birds except the African darter.

Species	Duration (min)		Depth (m)		Reference
	Common	Max	Common	Max	
Peruvian pelican *Pelecanus thagus*	<0.1	–	–	–	A
Blue-footed booby *Sula nebouxii*	<0.3	0.6	<8	22	B
Peruvian booby *Sula variegata*	0.1	<0.3	<5	10	C
Red-footed booby *Sula sula*	<0.1	<0.2	<7	10	D
Cape gannet *Morus capensis*	0.1	0.5	<5	13	E
Northern gannet *Morus bassanus*	<0.3	0.7	<15	34	F
Red-tailed tropic bird *Phaethon rubricauda*	–	–	<7	13	G
African Darter *Anhinga rufa*	<1.5	1.8	–	–	H

References: A: Duffy, 1983; B: Zavalaga *et al.*, 2007; C: Duffy, 1983, Ludynia *et al.*, 2010; D: LeCorre, 1997, Weimerskirch *et al.*, 2005; E: Grémillet *et al.*, 2004, Ropert-Coudert *et al.*, 2004a, 2004b; F: Brierley and Fernandes, 2001, Lewis *et al.*, 2002, Ropert-Coudert *et al.*, 2009; G: LeCorre, 1997; H: Ryan, 2007.

shallow-diving, freshwater members of the Pelecaniformes, but are in a separate family (Anhingidae) from the cormorants and shags (Phalacrocoracidae).

1.2.4 Penguins

Penguins (order: Sphenisciformes) have a broad range of diving behaviors (Table 1.9). Routine dive depths and durations of many penguin species with smaller body masses are less than 60 m and 2 min, respectively. These values and even the maximum values of those penguin species are equivalent to the dive data for murres and the deeper-diving cormorant species. The gentoo penguin (*Pygoscelis papua*) appears to be the best diver among the pygoscelid and other smaller-bodied penguin species, with routine dive depths greater than 80 m and regular dive durations of three or more minutes (see Table 1.9). The largest-bodied penguins, king (*Aptenodytes pagagonicus*) and emperor (*A. forsteri*) penguins, are the most accomplished avian divers. The 12-kg king penguin routinely makes dives of 3–6-min duration, with most maximum depths near 300 m, while 25-kg emperor penguins have regular dives as long as 10 min and as deep as 400 and even 500 m (see Table 1.9). Although 400–500-m deep dives of emperor penguins form a small percentage of all the dives during a

Table 1.9 Dive characteristics of penguins.

Species	Duration (min)		Depth (m)		Reference
	Common	Max	Common	Max	
Galapagos penguin *Spheniscus mendiculus*	<1	3.2	<6	52	A
Magellanic penguin *S. magellanicus*	1–2	4.6	<60	97	B
Humboldt penguin *S. humboldti*	<1	2.8	<20	54	C
African penguin *S. demersus*	1–2	2.3	20–80	130	D
Little penguin *Eudyptula minor*	<1–1.5	1.5	<10	69	E
Yellow-eyed penguin *Megadyptes antipodes*	1–2	–	10–30	56	F
Royal penguin *Eudyptes schlegeli*	1–3	7.5	<10–60	226	G
N. rockhopper penguin *E. chrysocome moseleyi*	<1–1.5	3.2	<10–40	168	H
S. rockhopper penguin *E. chrysocome filholi*	1–2	11	<10–60	113	I
Macaroni penguin *E. chrysolophus*	1–2	4	20–35	113	J
Adélie penguin *Pygoscelis adeliae*	1–2	5.9	10–40	180	K
Chinstrap penguin *P. anarcticus*	1–2	3.6	10–40	179	L
Gentoo Penguin *P. papua*	1–3	9.1	10–100	225	M
King penguin *Aptenodytes patagonicus*	2–6	9.2	<100–250	343	N
Emperor penguin *A. forsteri*	<5–10	27.6	<100–400	564	O

References: A: Boersma, 1976, Mills, 2000, Steinfurth *et al.*, 2008; B: Peters *et al.*, 1998, Walker and Boersma, 2003; C: Culik *et al.*, 2000, Luna-Jorquera and Culik, 1999; D: Petersen *et al.*, 2006, Wilson, 1985, Wilson and Wilson, 1990; E: Bethge *et al.*, 1997, Gales *et al.*, 1990, Montague, 1985, Ropert-Coudert *et al.*, 2003, 2006; F: Seddon and Vanheezik, 1990; G: Hull, 2000; H: Cherel *et al.*, 1999, Tremblay and Cherel, 2003; Tremblay *et al.*, 1997; I: Hull, 2000, Schiavani and Rey, 2004, Tremblay and Cherel, 2000, Wilson et al., 1997a; J: Boyd and Croxall, 1996, Croxall *et al.*, 1993, Green *et al.*, 1998, 2003; K: Chappell *et al.*, 1993, Norman and Ward, 1993, Watanuki *et al.*, 1997; L: Bengtson *et al.*, 1993, Miller and Trivelpiece, 2008, Takahashi *et al.*, 2003, Wilson and Peters, 1999; M: Lescroel and Bost, 2005, Robinson and Hindell, 1996, Williams *et al.*, 1992b; N: Charrassin and Bost, 2001, Kooyman *et al.*, 1992a, Pütz and Cherel, 2005, Pütz *et al.*, 1998; O: Kirkwood and Robertson, 1997, Kooyman and Kooyman, 1995, Rodary *et al.*, 2000, Sato *et al.*, 2011, Wienecke *et al.*, 2007.

foraging trip to sea, the emperor penguins in the Ross Sea do regularly perform such deep dives (Kooyman and Kooyman, 1995, Sato *et al.*, 2011). The frequency of these dives is much lower for emperor penguins in other areas of Antarctica (Wienecke *et al.*, 2007).

Table 1.10 Dive characteristics of other aquatic birds.

Species	Duration (min)		Depth (m)		Reference
	Common	Max	Common	Max	
Common loon *Gavia immer*	<1.5	2	–	55	A
Australasian grebe *Tachybaptus novaehollandiae*	<0.5	–	–	–	B
Least grebe *Tachybaptus dominicus*	<0.5	–	–	–	C
Great crested grebe *Podiceps cristatus*	<1	–	–	–	D
Red-necked grebe *Podiceps grisegena*	<0.6	–	–	–	E
Horned grebe *Podiceps auritus*	<0.6	–	–	–	F
Hoary-headed grebe *Poliocephalus poliocephalus*	<0.5	–	–	–	G
New Zealand dabchick *Poliocephalus rufopectus*	<0.5	<1.0	–	–	H
Pie-billed grebe *Podilymbus podiceps*	<0.5	–	–	–	C
Western grebe *Aechmophorus occidentalis*	<0.5	<1.1	–	–	I
Common goldeneye *Bucephala clangula*	<0.5	–	–	–	J
Common scoter *Melanitta nigra*	<1.0	–	–	–	K
Long-tailed duck *Clangula hyemilis*	–	–	–	60	L
Common eider *Somateria mollissima*	<1.0	1.3	<6	9	M
Common pochard *Aythya ferina*	<0.3	–	–	–	N
Tufted duck *Aythya fuligula*	<0.5	0.8	–	–	O
Lesser scaup *Aythya affinis*	<0.5	–	–	–	P
Red-headed duck *Aythya americana*	<0.5	–	–	–	Q
Canvasback duck *Aythya valisineria*	<0.5	–	–	–	R

Table 1.10 *(cont.)*

Species	Duration (min)		Depth (m)		Reference
	Common	Max	Common	Max	
Mallard (pekin) duck *Anas platyrhynchos*	<0.1	<0.2	–	–	S
American coot *Fulica americana*	<0.1	–	–	–	T
Dipper *Cinclus mexicantis*	<0.5	0.5	<1.0	6.0	U

References: A: Nocera and Burgess, 2002, Schorger, 1947; B: Ropert-Coudert and Kato, 2009; C: Jenni and Gambs, 1974; D: Ulenaers *et al.*, 1992; E: Simmins, 1969; F: Dow, 1964, Ladhams, 1968; G: Ropert-Coudert and Kato, 2009; H: Edgar, 1962; I: Forbes and Sealy, 1988; J: Heintzelman, 1963; K: Kaiser *et al.*, 2006; L: Schorger, 1947; M: Guillemette *et al.*, 2004, Ron and Guillemette, 1991; N: Butler and Woakes, 1979; O: Butler and Woakes, 1979, 1982, Stephenson *et al.*, 1989b; P: Stephenson, 1994; Q: Furilla and Jones, 1987b; R: Woodin and Stephenson, 1998; S: Furilla and Jones, 1987a; T: Batulis and Bongiorno, 1972; U: Murrish, 1970.

In terms of the surfacer–diver classification, penguins are best considered divers. King penguins, for example, spend only about 40% of time at sea diving during their 5–10-day foraging trips (Pütz and Cherel, 2005, Pütz *et al.*, 1998). In contrast to the continuous diving activity of a surfacer such as the elephant seal, after the longest reported dive (27.6 min) of an emperor penguin, the bird required 6 min to stand up from the prone position on the sea ice, another 20 min to begin to walk, and a total of 8.4 hours before it began to dive (Sato *et al.*, 2011). During foraging trips to sea during the chick-nurturing period, emperor penguins spend 20–40% of their time out of the water on the sea ice (Watanabe *et al.*, 2012).

1.2.5 Other aquatic birds

The last table in this chapter reviews known diving behavior of other aquatic birds. Although many of these birds are freshwater rather than marine species, they are included both for comparison and because many diving physiology studies have been conducted on ducks. Although they are members of different avian orders, the loons, grebes, and ducks are all foot-propelled swimmers. Eiders also use their wings. As of this writing, the diving of grebes and loons has been little studied. Loons, however, have been caught in nets as deep as 55 m and observed to dive for as long as 2 min (see Table 1.10). Such diving performance is probably comparable to those of diving petrels and shearwaters, boobies and gannets, some of the smaller alcids, and even the smallest-bodied penguin species.

Examination of the diving behaviors of different duck species has been primarily limited to observation, often in experimental aquaria (Table 1.10). In general, the best divers among the ducks are considered to be the diving ducks (pochards, and scaups including the genus *Aythya*, and stiff-tailed ducks), and the sea ducks (multiple species including eiders, scoters, goldeneyes, mergansers, smews, the harlequin duck (*Histrionicus histrionicus*), and the long-tailed duck (*Clangula hyemilis*)). The long-tailed duck has been caught in nets at 60 m depth (Schorger, 1947). Otherwise, the known diving depths and durations of diving ducks and sea ducks are shallow and short ($<$1 min). The dabbling ducks (primarily members of the genus *Anas*, including teals, pintails, mallards, and shovelers) are not considered divers.

The distinction between diving and dabbling ducks is important because diving physiology studies have been conducted on both types of ducks and some physiological responses to breath-holding are distinct between the two groups (Furilla and Jones, 1987a). Consequently, in reviewing physiological findings in ducks, it is important to always remember whether the subject species is a diver or a dabbler.

2 Challenges of the breath hold and the environment

The physiological responses and adaptations that underlie the dives of a marine mammal or seabird are just as remarkable as the dive depths and durations described in Chapter 1. To put these into perspective and to prepare for further discussion in later chapters, the physiological challenges of human diving and the physical characteristics of the underwater environment at depth must first be reviewed. In addition, this chapter will conclude with brief comments on the sensory adaptations that allow these animals to navigate, communicate, and forage at depth. Although too complex to have the space to adequately review in this book, these sensory adaptations deserve note as they are just as essential to the success of a dive as the cardiovascular and respiratory responses are to its performance.

2.1 Challenges in human breath-hold diving

Human breath-hold dives are relatively shallow and of short duration in comparison to those of marine mammals and penguins. Most humans probably cannot match the famous diving women of Japan and Korea (the ama), who typically dive to routine depths of 20 m in dives of about 1-min duration during their daily work (Lindholm and Lundgren, 2009). However, even the dives of unassisted, competitive breath-hold divers pale in comparison to the routine deep dives of 25-kg emperor penguins (80 m, 3–4 min duration for humans vs. >500 m, 8–10 min duration for emperor penguins (Ferrigno and Lundgren, 2003), Chapter 1). As of 2009, the depth record for a human diver pulled down by a weight was 214 m during a dive of 4.4-min duration (Lindholm and Lundgren, 2009). The world record for a human breath hold was 10.2 min in 2009; this was accomplished by a swimmer at rest with the head immersed in a pool (Lindholm and Lundgren, 2009). Although these impressive records are far less than the "unassisted" dive performance of many seals, they do provide the framework to consider the physiological challenges and complications facing the human breath-hold diver.

2.1.1 Breath-hold duration "break point"

The duration of a breath hold in a human diver is considered to be determined by a "break point" at which the inspiratory muscles begin to involuntarily contract, and, after which, the urge to breathe eventually results in inspiration (Ferrigno and Lundgren, 2003). The break point is associated with a threshold level of the partial pressure of carbon dioxide (P_{CO2}), and is reportedly dependent on four factors: (1) tolerance to hypercapnia and hypoxia; (2) O_2 and CO_2 storage capacity; (3) metabolic rate; and (4) willingness to tolerate the experience (Ferrigno and Lundgren, 2003). Alveolar P_{CO2} values typically ranged between 44 and 54 mm Hg (5.8–7.1 kPa) at termination of out-of-water breath holds. A reduced CO_2 sensitivity is one mechanism to prolong breath-hold capacity, and there is some evidence that training can promote greater CO_2 tolerance as well as hypoxemic tolerance (Ferrigno and Lundgren, 2003). Another way to prolong dive duration is to start with a lower P_{CO2} through pre-dive hyperventilation. A lower initial P_{CO2} increases the time to reach the break point P_{CO2} threshold, and can thus prolong the duration of a breath hold.

2.1.2 Shallow water black out

However, pre-dive hyperventilation is a high-risk maneuver in a human breath-hold diver as it increases the risk of shallow water black out during ascent from depth. Shallow water black out is secondary to hypoxia and occurs when the alveolar/arterial partial pressure of oxygen (P_{O2}) reaches a threshold near 30 mm Hg (4.0 kPa) (Ferrigno and Lundgren, 2003, Lindholm and Lundgren, 2009). Pre-dive hyperventilation prolongs time at depth, during which continued oxygen consumption lowers the concentration of oxygen in the lungs. Although the P_{O2} in the lungs at depth may be above the threshold for unconsciousness, the P_{O2} during ascent to the surface may decrease below that threshold due to the decrease in ambient pressure as the diver returns to the surface. This is a consequence of the fact that the partial pressure of a gas is the product of the gas's concentration and the ambient pressure (Henry's Law). While a given O_2 fraction in the lung may result in an adequate P_{O2} at depth, the decrease in ambient pressure during ascent may well result in a P_{O2} below the threshold for unconsciousness. The Korean ama do not hyperventilate prior to dives, and their mean end-of-dive arterial P_{O2} and P_{CO2} values are 60 and 50 mm Hg (8.0 and 6.7 kPa), respectively (Qvist *et al.*, 1993). In competitive breath-hold divers on the verge of unconsciousness, end tidal P_{O2}s have been as low as 19–22 mm Hg (near 3.0 kPa) (Lindholm and Lundgren, 1996, Overgaard *et al.*, 2006). Such episodes in competitive divers have been accompanied by elevations of biochemical markers for brain damage (Andersson *et al.*, 2009b, Liner and Andersson, 2009).

2.1.3 Oxygen storage

Total O_2 storage in a 70-kg human has been calculated to be near two liters, with 41% in the lungs, 44% in blood, 12% in muscle, and the remainder dissolved in solution in tissue. As reviewed in Ferrigno and Lundgren (2003), greater O_2 storage and longer dive durations in trained divers may be afforded through larger total lung capacity secondary to increased chest wall compliance. In contrast, the blood O_2 store, as reflected by the concentration of hemoglobin (Hb, the O_2 transport protein in blood) is not elevated at rest in professional divers such as the Korean ama (Hurford *et al.*, 1990, Qvist *et al.*, 1993). However, during diving activity of the Korean ama, the magnitude of splenic contraction and subsequent elevation in blood Hb concentration appears to be greater in the ama than in non-diver controls (Hurford *et al.*, 1990). A 20% decrease in splenic volume in the ama was associated with a 10% increase in blood Hb content. Larger spleens and greater splenic contraction during breath holds have also been reported in elite competitive divers (Schagatay, 2012). Lastly, anaerobic metabolism as evidenced by blood lactate accumulation may also contribute to the diver's breath hold capacity (Ferretti *et al.*, 1991, Scholander *et al.*, 1962).

2.1.4 Carbon dioxide storage

Carbon dioxide storage in humans can be divided between a static store in bone (123 liters) and a labile component (16.6 liters), which is primarily in the lungs, blood, and muscle (Farhi and Rahn, 1960, Ferrigno and Lundgren, 2003). It is in that labile component in which CO_2 accumulates during a breath hold. It is notable that CO_2 storage capacity can be elevated two-fold in elite human divers (Ferretti *et al.*, 1991). Seventy percent of that CO_2 was estimated to be stored in the lungs and blood. Thus, an increased CO_2 storage capacity in trained or elite divers can also delay onset of the "breaking point" and so can prolong dive duration.

2.1.5 Metabolic rate and the dive response

Metabolic rate controls the rates at which oxygen stores are consumed and carbon dioxide produced during a dive. Consequently, because of the effects of P_{CO2} and P_{O2} on dive duration, metabolic rate is a significant factor in how long a diver can stay underwater on a breath of air. As will be reviewed in later chapters, heart rate is a prime determinant of metabolic rate because the oxygen consumption of many tissues is perfusion dependent. Declines in heart rate during breath holds were first noted in ducks by the French physiologist Paul Bert in the late 1800s (Bert, 1870). In 1935, Irving and colleagues suggested that this dive response (a low heart rate (bradycardia) and peripheral vasoconstriction) served to conserve oxygen for essential tissues (Irving *et al.*, 1935a).

In humans, a diving bradycardia was first reported in 1940 (Irving, 1963, Irving *et al.*, 1940). The subject was a professional underwater alligator wrestler, and his heart rate was as low as 30 beats per minute (bpm). Bradycardias of 40–50 bpm have been recorded in both Korean ama and Australian pearl divers (Hong *et al.*, 1967, Scholander *et al.*, 1962). During simulated dives of three elite divers to 50 m in a pressure chamber, heart rates and cardiac outputs declined to 20–30 bpm and 3 l min^{-1}, respectively (Ferrigno *et al.*, 1997). Evidence for peripheral vasoconstriction in human divers included post-dive elevations in blood lactate concentrations, maintenance and even elevation of blood pressure, decreased limb blood flow, and splenic contraction (Ferrigno *et al.*, 1997, Hurford *et al.*, 1990, Joulia *et al.*, 2009, Scholander *et al.*, 1962). The end result of these cardiovascular responses is conservation of blood oxygen; this has been demonstrated in many studies by decreases in arterial oxygen desaturation rates, and decreased pulmonary gas exchange (Andersson and Evaggelidis, 2009, Andersson *et al.*, 2002, 2008, Ferrigno and Lundgren, 2003, Joulia *et al.*, 2009). Thus, it is advantageous to have a slow heart rate during dives, and many competitive divers train and even use meditation techniques to relax and slow their heart rates during breath holds (Elsner, 2015)

2.1.6 Pressure: gas laws

The other major challenge faced by human breath-hold divers is depth (hydrostatic pressure). The primary physiological effects of pressure are two: gas compression and elevated blood nitrogen levels (P_{N2}). Both of these effects are due to the effects of pressure on gases, and are explained by well-known gas laws. Compression effects are due to Boyle's Law, which states that the product of pressure times volume is constant. Elevated P_{N2} is secondary to Henry's Law, which, as discussed above, states that the partial pressure of a gas is the product of its concentration and ambient pressure. Ambient pressure, of course, increases as the diver descends to depth.

2.1.7 Pressure: barotrauma

Increased ambient pressure can lead to tissue damage (barotrauma) when air volume within a relatively rigid body cavity is compressed to a lesser volume than that of the cavity (Ferrigno and Lundgren, 2003). Most divers are familiar with the need to equilibrate pressure within the chambers of the bony nasal sinuses and middle ear in order to avoid barotrauma to those structures (mucosal bleeding and ear drum rupture). The same constraint is also true of the thoracic cavity. Theoretically, the chest is at its lowest volume when the lung is at residual volume (RV), the lung volume after a maximal exhalation. In humans, this value is about 1 L or approximately 20–25% of total lung capacity (TLC). On the basis of the TLC/RV ratio, humans should not be able to breath-hold dive beyond a four- to fivefold compression of the lungs, i.e. about 40 m depth. Otherwise, the lung would be at less than RV, and the diver would suffer from "chest squeeze."

Clearly, human breath-hold divers are not limited to 40 m depth. There are several explanations. First, as described above, trained divers may have a more compliant chest wall, and lower effective RV. Second, exceptional divers may have larger than normal TLCs due to body habitus (Schagatay, 2012). Third, due to compression of the body, blood is redistributed from the extremities into the thoracic cavity. This can be as much as 1–2 L of blood (Craig, 1968, Ferrigno and Lundgren, 2003). This increased intrathoracic blood volume thus occupies space as the lung is compressed below residual volume. However, intrathoracic pooling of blood may also distend the pulmonary vasculature and heart, predisposing deep divers to pulmonary edema (Liner and Andersson, 2008), pulmonary hemorrhage (Lindholm *et al.*, 2008), and cardiac arrhythmias (Hansel *et al.*, 2009, Scholander *et al.*, 1962). Fourth, there are various pre-dive maneuvers which may increase the start-of-dive air volume in the chest (Ferrigno and

Lundgren, 2003). Both the ama and Tuamotuan pearl divers whistle as they exhale; this is thought to transiently increase intrathoracic pressure, decrease intrathoracic blood volume, and thereby allow for a larger final inspiration of air prior to a dive. And, then, there is "lung packing" (Lindholm and Lundgren, 2009). Here, the diver inhales to total lung capacity, and then gulps in air forcing it through an open glottis with a swallowing-like maneuver. This has been reported to increase TLC as much as 39%, thus creating a much larger TLC/RV ratio. This technique, however, elevates airway pressures, which may lead to lung rupture, and which may also decrease venous return to the heart, lowering blood pressure and predisposing the diver to loss of consciousness (Andersson *et al.*, 2009a).

2.1.8 Pressure: ambient pressure and heart rate

The potential contribution of hydrostatic pressure to regulation of heart rate during breath-hold dives has not been fully evaluated. In breath-hold dives of elite human divers to depths as deep as 50 m, the degree of bradycardia was similar to that during breath-holds of the same subjects at the surface (Ferrigno *et al.*, 1997). It was concluded that depth, per se, did not have an effect on heart rate. However, exposure of human subjects to 3–5 ATA ambient pressure (20–40 m depth) during saturation dives of several days' duration in hyperbaric chambers was associated with a 14–23% reduction in heart rate (Eckenhoff and Knight, 1984). In 2.5% of ECG records in this study, there were isolated supraventricular arrhythmias (atrioventricular nodal escape rhythms and wandering atrial pacemakers); these were associated with the slowest heart rates, and always resolved during exercise. Rare premature ventricular beats also occurred. In experimental studies, extreme increases in ambient pressure (150 ATA, 1490 m depth) did cause a 5–30% decrease in the spontaneous discharge rate of sinoatrial node pacemaker cells and 40% decrease in conduction velocity in the atria and ventricles (Daniels and Grossman, 2003).

2.1.9 Pressure: nitrogen, decompression sickness and nitrogen narcosis

The second major challenge due to pressure in human breath-hold divers is excess nitrogen absorption. Elevated blood nitrogen levels can lead to decompression sickness (DCS) as well as nitrogen narcosis. As the partial pressure of N_2 (P_{N2}) in the lung increases due to the increase in ambient pressure at depth, there will be greater uptake of nitrogen into the blood and tissues of divers. As will be discussed further in Chapter 12, DCS develops after return to the surface secondary to bubble formation due to an increased ratio of blood or tissue P_{N2} to ambient pressure. DCS most commonly occurs in SCUBA (self-contained underwater breathing apparatus) divers because there is a continuous air supply at high ambient pressure (and high P_{N2}) from the SCUBA tank. For a breath-hold human diver, the risk of excess nitrogen uptake from the lungs is probably mitigated by (1) a limited amount of nitrogen (one lung volume); (2) the relatively short duration of time at depth; (3) compression of the lungs and decreased alveolar gas exchange secondary to the development of ventilation–perfusion mismatch; and (4) decreased uptake of nitrogen due to a slower heart rate and lower cardiac output at depth (Ferrigno and Lundgren, 2003).

Nonetheless, DCS-like symptoms have been reported in breath-hold divers. Pearl divers from the Tuamotu Archipelago reported vertigo, syncope, partial paralysis, and even death after diving (Cross, 1965). The syndrome was called "taravana." Such episodes were associated with six-hour work periods of 1.5-min dives to 30-m depth with short surface intervals. Paulev also reported his own DCS experience after 60 1-min submarine escape dives from depths of 15–20 m over a five-hour period (Paulev, 1965). Most breath-hold divers, however, do not make such frequent, repetitive dives. DCS symptoms have not been reported in the Korean ama, presumably due to their relatively shallow, short-duration dives. In a study of Korean women diving 3–6 m deep, venous P_{N2} increased mildly from a baseline of 584 torr to 635 torr (77.9 kPa to 84.7 kPa); at the end of a three-hour work period, P_{N2} declined with a half-time of 36 min and reached baseline in 3–4 hours (Radermacher *et al.*, 1992).

The other potential complication of elevated nitrogen pressure is nitrogen narcosis. This depressant effect of nitrogen on the central nervous system occurs at P_{N2}s equivalent to 30-m depth (corresponding to a P_{N2} of 2400 mm Hg (320 kPa)) (Halsey, 1982). Although common in SCUBA divers, given the constraints described above for breath-hold divers and DCS, it is unlikely that nitrogen narcosis will occur in most human breath-hold divers.

2.1.10 Pressure: high pressure nervous syndrome

Another potential physiological complication due to pressure is high pressure nervous syndrome (HPNS, see Chapter 12 for discussion). Its onset in humans is at a depth of about 190 m (Halsey, 1982). HPNS has been observed during SCUBA diving and pressure chamber dives. Its occurrence in most breath-hold divers is unlikely given its depth threshold. However, competitive divers, as described above, have now reached this depth threshold. As depth limits are further pushed, HPNS may become a serious threat to the safety of such activities.

2.2 Challenges of the environment

The primary environmental physiological challenges to an animal descending into the ocean depths are pressure, the thermal conductivity and specific heat of water, and the absorption of light.

2.2.1 Challenges of the environment: pressure

Pressure increases by one atmosphere (ATM) every 10 m of depth. One atmosphere is equivalent to 760 mm Hg or 101.3 kPa. Some of the physiological consequences of pressure have already been alluded to above, and will be further discussed in Chapter 12. One other frequently used measure of pressure is the term "atmospheres absolute"

(ATA). At the surface, pressure is considered one ATM, and is equivalent to one ATA. At 10 m depth, the pressure has increased by one ATM and is at two ATA. In diving physiology, the use of the term ATA is thus an index of the ambient pressure at depth relative to that at the surface.

2.2.2 Challenges of the environment: temperature and heat loss

Heat loss of a human is greater in water than in air due to both the specific heat and thermal conductivity of water (Mekjavic *et al.*, 2003). Specific heat is the energy required to raise the temperature of 1 kg of a substance by 1 K (degree kelvin). The specific heat of water at 30 °C is 4180 joule kg^{-1} $°K^{-1}$; this value is more than four times greater than that of air. And, on a volume basis, the specific heat of water is even greater, at 3500 times that of air. Thermal conductivity is the quantity of heat that passes in a unit of time through a unit area of a substance with thickness equal to unity when its opposite face differs in temperature by one degree. The thermal conductivity of water at 27 °C is 6096 W cm^{-1} $°K^{-1}$, about 23 times that in air. In addition, heat loss due to convection is maximized in air with air speeds greater than 4.2 m s^{-1}, whereas in water it is maximized at 0.5 m s^{-1}. As a result, core temperature in humans decreases 2–5 times faster during immersion in water than during exposure to air at the same temperature.

As will be reviewed in Chapter 8, immersion in cold water elicits a range of physiological responses to prevent hypothermia. In humans, peripheral vasoconstriction and increased heat generation through muscular exercise or shivering serve to preserve core temperatures. However, additional insulation is needed in cold water, and has been achieved with the use of wet suits and dry suits. In animals, insulation is provided by fur, feather, and fat insulation (see Chapter 8). One additional physiological consequence of peripheral vasoconstriction and body compression at depth is the intrathoracic pooling of blood and distention of cardiopulmonary baroreceptors. Baroreceptor stimulation results in increased diuresis, which may predispose divers working in cold water over long periods to dehydration.

2.2.3 Challenges of the environment: light

The third environmental challenge faced by divers is the absorption of light by water (Duntley, 1963, Wozniak and Dera, 2007). Although less than 5% of sunlight is reflected by water, the penetration of light to depth is limited by scattering and by its absorption by water molecules, pigments, and various particulates. Red, yellow, and orange wavelengths are absorbed most quickly. By 1-m depth, 60% of incoming light has been absorbed, and by 10 m, 85% of light – and this is in clear water. The upper 100–200 m of the ocean are considered the photic zone, and by 150-m depth, 99% of light has been absorbed. Amazingly, blue-green light, the least absorbed, has been detected photoelectrically as deep as 600 m. Given the absorption of light by water and the dive depths of many of the species reviewed in Chapter 1, sensory adaptations are required in these animals for successful detection and capture of prey.

2.3 Sensory adaptations

Investigations of sensory adaptions have been primarily conducted in marine mammals. This research has focused on vision, tactile vibrissae, and sound production/hearing.

2.3.1 Sensory adaptations: olfaction

Olfactory adaptations have not been extensively studied, but it should be noted that olfactory structures in the central nervous system including cranial nerve I, the olfactory bulb, and olfactory tract are absent in all toothed whales examined to date (Oelschlager and Oeschlager, 2008). In mysticetes, a few observations suggested the presence of such structures, and, recently, anatomical examinations in bowhead whales (*Balaena mysticetus*) have confirmed their presence (Thewissen *et al.*, 2011). The authors suggested that olfaction may play a role in the detection of krill, on which these whales feed. Olfactory brain structures are present in pinnipeds, but whether olfaction participates in prey detection is unknown (Thewissen *et al.*, 2011, Watkins and Wartzok, 1985). Olfaction may play a significant role in location of prey by procellariiform seabirds. Dimethyl sulfide, a compound associated with phytoplankton, has been shown to attract petrels and prions at sea (Nevitt *et al.*, 1995). Behavioral investigations have also demonstrated that Humboldt penguins (*Spheniscus humboldti*) and South African penguins (*S. demersis*) are able to detect dimethyl sulfide (Culik, 2001, Wright *et al.*, 2011).

2.3.2 Sensory adaptations: vision

Successful detection of underwater prey is dependent on vision in many diving species. Visual adaptations have occurred at both the anatomical and biochemical levels. Because of the similarity of the refractive indices of water (1.34) and corneal tissue (1.35), the cornea does not participate in light refraction, and, as a consequence, is flattened in cetaceans, pinnipeds, penguins, and other seabirds (Martin, 1998, 1999, Martin and Brooke, 1991, Mass and Supin, 2009). In contrast, because of the refractive index of air (1.0), corneas in terrestrial mammals are convex, and the anterior chambers of eyes are larger than in marine mammals. In both marine mammals and diving birds, the lens tends to be spherical as it is primarily responsible for light refraction (Mass and Supin, 2009, Suburo and Scolaro, 1990).

Because of the absorption of light by water and decreased illumination at depth, the size of the eye and that of the fully dilated pupil are large, especially in deep divers (Levenson and Schusterman, 1997, 1999, Martin, 1999). It has also been noted that the range of maximum pupillary dilatation in the elephant seal (*Mirounga angustirostris*) and king penguin (*Aptenodytes patagonicus*), is extreme (469- and 300-fold, respectively) (Levenson and Schusterman, 1999, Martin, 1999). Those authors suggest that a pinpoint, constricted pupil at the surface limits incoming light sufficiently that the retina is pre-adapted for darkness (the retinal rod cells are not “bleached out”). Such dark pre-adaptation, as well as rapid pupillary dilation, during descent would allow optimal retinal function even during the relatively short descent times of these animals. Evidence for this has been found in the elephant seal's dark adaptation curve; adaptation from daylight to maximum sensitivity requires only 6 min, similar to its descent times to routine foraging depths (Levenson and Schusterman, 1999).

In addition to a large eye, adaptations for low-light vision include a predominance of rod photoreceptors (98–99% of photoreceptors in pinnipeds and cetaceans) and a well-developed tapetum lucidum, the reflective layer which reflects light back through the retina and increases light sensitivity (Mass and Supin, 2009, Walls, 1942). There are no short-wave (S) cone cells in all cetacean and pinniped species examined to date (Levenson and Dizon, 2003, Peichl *et al.*, 2001). Peak sensitivity of the rod pigments of bottlenose dolphins is 488 nm, and is considered blue-shifted in relation to the 500–550 nm peak sensitivities of rods from terrestrial mammals and most pinnipeds (Levenson *et al.*, 2006, Mass and Supin, 2009). Only the rod pigments of the deep-diving elephant seals are as blue-shifted as in the dolphin (Levenson *et al.*, 2006, Lythgoe and Dartnall, 1970). The middle- and long-wavelength (M/L) cone pigments of pinnipeds have maximum sensitivities near 550 nm, again similar to terrestrial mammals but greater than that in the dolphin (525 nm) (Levenson *et al.*, 2006, Lythgoe and Dartnall, 1970). Overall, therefore, it appears that both the rods and cones of the completely aquatic cetaceans as well as the rods of the deep-diving elephant seals are blue-shifted in relation to terrestrial mammals. And, in both cetaceans and pinnipeds, there is an absence of S cones.

2.3.3 Sensory adaptations: touch

Tactile sensation may contribute significantly to successful underwater foraging. In manatees, modified vibrissae, perioral bristles, have a functional role in prey manipulation and ingestion (Marshall, 2009, Marshall *et al.*, 1998). The dense innervation of the elaborate vibrissae of ring seals and bearded seals is about ten-fold greater than in terrestrial mammals (Hyvärinen, 1989, Marshall *et al.*, 2006). It has been suggested that such vibrissae are adaptations for benthic foraging. In addition to a role in prey detection and ingestion, the vibrissae of some pinnipeds, may contribute to the long-range detection and tracking of prey. The whiskers of harbor seals (*Phoca vitulina*) are capable of detection of movement in water (Dehnhardt *et al.*, 1998). Blindfolded seals and sea lions can both successfully follow hydrodynamic trails (Dehnhardt *et al.*, 1998, 2001, Gläser *et al.*, 2011, Hanke *et al.*, 2010, Schulte-Pelkum *et al.*, 2007, Wieskotten *et al.*, 2011a, 2011b).

2.3.4 Sensory adaptations: sound production

The sounds created by prey, predators, and the environment undoubtedly contribute to the underwater behavior of both marine mammals and seabirds (Van Opzeeland *et al.*, 2010). Although relatively little is known about the role of sound in the activities of diving birds, sound production and hearing have been extensively studied in marine mammals, especially odontocetes, which use echolocation to find prey. Underwater sound production in pinnipeds and sirenians is thought to involve shunting of air between the trachea and pharyngeal/nasal structures, and may involve vibration of tracheal membranes (Frankel, 2009). In mysticetes, sound production is probably accomplished through the vibration of a laryngeal U-shaped tissue fold, which, in contrast to the vocal cords of terrestrial mammals, is parallel to the flow of air (Reidenberg and Laitman, 2007). Sound frequencies range from as low as 7 Hz in the long-distance calls of blue whales (*Balaenoptera musculus*) to 24 kHz in singing humpback whales (*Megaptera novaeangliae*) (Frankel, 2009). Sound production in odontocetes, however, involves unique structures in the nasal region of the head. The "monkey lips"/dorsal bursae complex, located beneath the blowhole, is the source of clicks and whistles in odontocetes (Cranford *et al.*, 1996, 1997, Madsen *et al.*, 2013). Sound frequencies produced by dolphins and porpoises can be as high as 120 kHz or greater (Frankel, 2009). Readers are referred to detailed descriptions of the anatomy and pathway of sound transmission in these animals (Cranford *et al.*, 1996, 2008, Madsen *et al.*, 2013). In regard to physiology and compression due to pressure, the important point here is that the nasal passages, especially in deep divers, must have air available from the respiratory system for sound production at depth. For example, beaked whales do not begin to echolocate until they reach depths of 400–500 m (Johnson *et al.*, 2006, Tyack *et al.*, 2006).

2.3.5 Sensory adaptations: hearing

Adaptations for the underwater detection of sound by the ear are most extreme among the cetaceans. In pinnipeds, hearing in air is accomplished by sound transmission from the tympanic membrane (ear drum) to the middle ear ossicles and then to the cochlea. Underwater sound transmission in pinnipeds is thought to be accomplished through bone conduction (Nummela, 2009).

In sirenians, it is unclear how sound reaches the cochlea. The external auditory meatus ends in a blind sac. The middle ear bones are large and have the highest bone density of any mammal. It has been proposed that underwater sounds are transmitted via the zygomatic process of the squamous bone via bone conduction (Chapla *et al.*, 2007, Nummela, 2009).

In cetaceans, the external auditory meatus is narrow, and the tympanic membrane does not function in sound reception (Nummela *et al.*, 2007). Sound transmission to the inner ear of odontocetes occurs via fat pads extending from the mandibular canal (Cranford *et al.*, 2008, 2010, Norris, 1968). In mysticetes, the primary sound path to the inner ear probably involves the transmission of skull vibrations (Cranford and Krysl, 2015). Another mechanism may involve pressure transmissions through soft tissues. Fats pads adjacent to the inner ear have been recently documented (Nummela, 2009, Yamato *et al.*, 2012). Relevant to diving physiology and gas compression are the presence of air sinuses around the tympanic bulla, which isolate the inner ear and prevent bone conduction. Maintenance of this air layer even at depth is essential in odontocetes to allow for directional hearing (Nummela, 2009). As that air volume is compressed, "ear squeeze" is avoided by engorgement of large venous plexuses within the sinuses (Cranford *et al.*, 2010). Similar venous plexuses are also found in the middle ear of pinnipeds (Odend'hal and Poulter, 1966, Welsch and Riedel-Sheimer, 1997).

3 Respiratory gas exchange

Rapid replenishment of depleted oxygen stores and removal of accumulated carbon dioxide are essential during the surface intervals of breath-hold divers. Respiratory anatomy, lung volumes, and respiratory mechanics underlie pulmonary gas exchange and are the primary topics of this chapter. Marine mammals and seabirds will be reviewed separately because of the differences between the structure and function of mammalian and avian respiratory systems (Maina, 2006, Maina *et al.*, 2010).

In mammals, airflow in the lung is bidirectional, the lung is compliant, and gas exchange occurs in the alveoli, the sac-like terminations of the distal airways. In contrast, the avian respiratory system includes not only the lungs but also the air sacs. In birds, the lung is relatively rigid, gas exchange occurs in the air capillary (a tubular structure), and airflow is unidirectional and crosscurrent to the flow of blood (Powell, 2000, Scheid, 1979). Although avian lung volumes are about one-quarter less than those of mammals on a mass-specific basis, combined air sac and lung volumes are 3–5 times the lung volumes of similarly sized mammals (Maina, 2006). In addition, on average, the avian respiratory surface areas are about 15% greater than in similarly sized mammals, the blood–gas barrier in the lung about 62% thinner, and the pulmonary capillary blood volume 22% greater (Maina, 2006). The anatomical differences in the respiratory systems of birds and mammals thus potentially affect divers in regard to the size of respiratory oxygen stores, air volume/buoyancy, the magnitude of gas exchange at depth, and the pressure tolerance of a compliant versus rigid lung.

In this chapter, respiratory anatomy, respiratory mechanics, and ventilation will be reviewed first for marine mammals and then for seabirds. A recent review of cetacean lung morphometry and mechanics is also highly recommended to readers for an excellent and thorough overview of cetacean lung mechanics and pulmonary function (Piscitelli *et al.*, 2013). The effects of pressure on the respiratory system and gas exchange in these animals are examined in Chapter 12.

3.1 Marine mammal respiratory anatomy and function

3.1.1 Airway and lung anatomy in marine mammals

Although histological investigations of airway and lung structure had begun in the 1920s–1930s (for a review, see Piscitelli *et al.* 2013), modern research on respiratory anatomy and respiratory mechanics in diving mammals began with Scholander's investigations of whales on board whaling ships. There, he measured and dissected lungs from whales with specimens probably similar to those illustrated in Fig. 3.1. His observations of cartilaginous distal airways in whales led him to hypothesize that more rigid airways would allow (1) movement of air into those airways during compression of the lungs at depth; (2) collapse of alveoli; (3) cessation of gas exchange at depth; and (4) the avoidance of excess nitrogen absorption during dives (Scholander, 1940) (see Chapter 12).

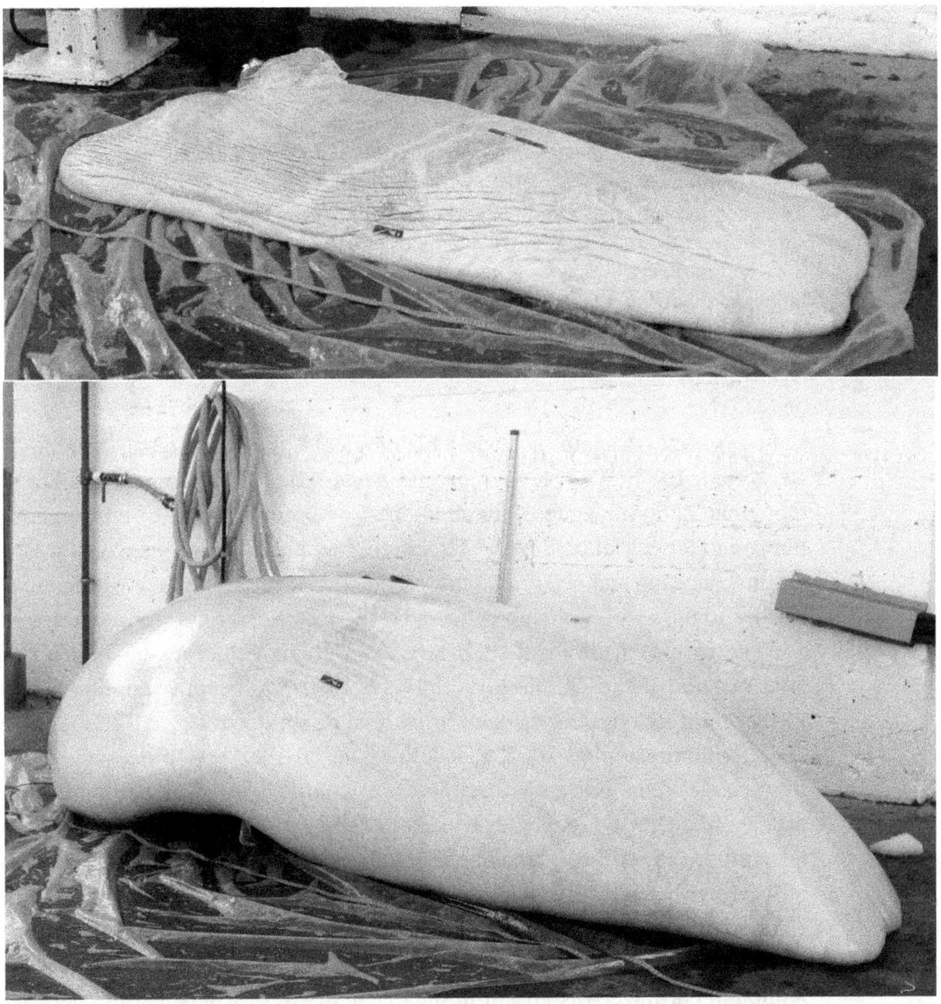

Figure 3.1 Deflated and inflated lung of a fin whale, *Balaenoptera physalus*. Such observations of lungs on a whaling vessel probably led Scholander to the hypothesis of lung collapse at depth as a mechanism for the prevention of decompression sickness in deep-diving marine mammals. Photographs with permission and courtesy from W. Vogl and M. Piscitelli.

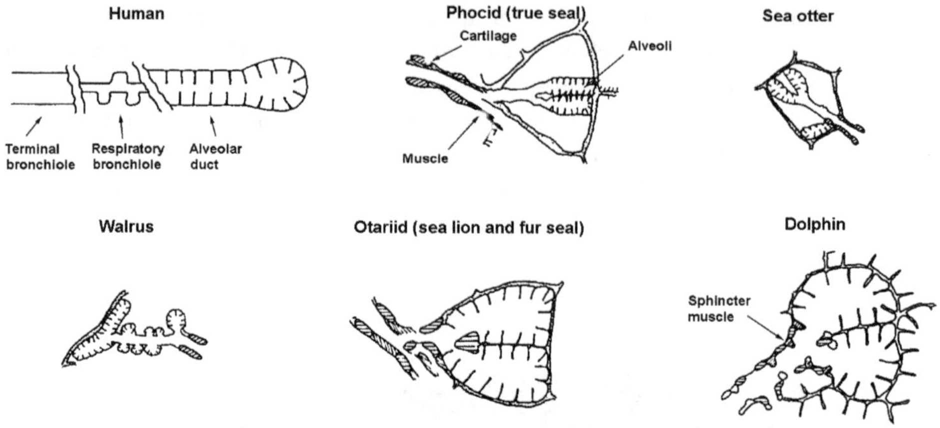

Figure 3.2 Comparison of terminal airway and alveolar diagrams of humans and marine mammals. **(a)** In humans, the terminal bronchiole extends to the respiratory bronchiole and then into the alveolar duct. **(b)** In phocid seals, the distal end of the terminal bronchiole lacks cartilage but is lined by thick smooth muscle that extends into the respiratory bronchiole. **(c)** In sea otters, some distal airways are supported by cartilage all the way to the alveolar sacs, while others lack such support over the final 1 mm. **(d)** Walrus distal airways are similar to those of the otter. **(e)** In sea lions and fur seals, airways are reinforced with cartilage all the way to the alveoli. **(f)** In dolphins, all airways have cartilaginous support. Terminal segments are marked by a series of muscular sphincters. Adapted from Kooyman, 1973.

In a series of anatomical and histological studies (Denison and Kooyman, 1973, Kooyman, 1972, 1973, Kooyman and Andersen, 1968, Tarasoff and Kooyman, 1973), Kooyman and colleagues examined airway reinforcement of diving mammals in comparison to terrestrial mammals. This strengthening of the air passages was most prominent in cetaceans and sea lions with cartilaginous reinforcement of the airways from the trachea to the level of the alveolar sac (Fig. 3.2). Respiratory bronchioles were absent.

More recent histological studies of deep-diving cetaceans have also revealed extensive vascular plexuses along the airways that may become engorged during dives to further reinforce the trachea and bronchi at depth (Cozzi *et al.*, 2005, Davenport *et al.*, 2013, Ninomiya *et al.*, 2005, Piscitelli *et al.*, 2013). These plexuses, composed primarily of large veins but also arterioles, are well developed and extend into the terminal bronchi in deep-diving whales (Piscitelli *et al.* 2013). Other possible functions that have been proposed for these vascular structures have included warming of air, dampening of airway pressures, oxygen storage, and prevention of barotrauma.

In Kooyman's examinations of phocid seals, tracheal reinforcement was minor and ranged from flexible cartilaginous rings to ventral bars of cartilage in some species. Although cartilage was absent in the distal airways of phocids, the presence of oblique muscle fibers in the bronchial walls was thought to reinforce these segments (Fig. 3.2). In walruses and sea otters, distal airways were reinforced with a mix of cartilage or muscle elements. Conceptually, more rigid airways should allow alveoli to empty and collapse more fully (no gas trapping) during compression.

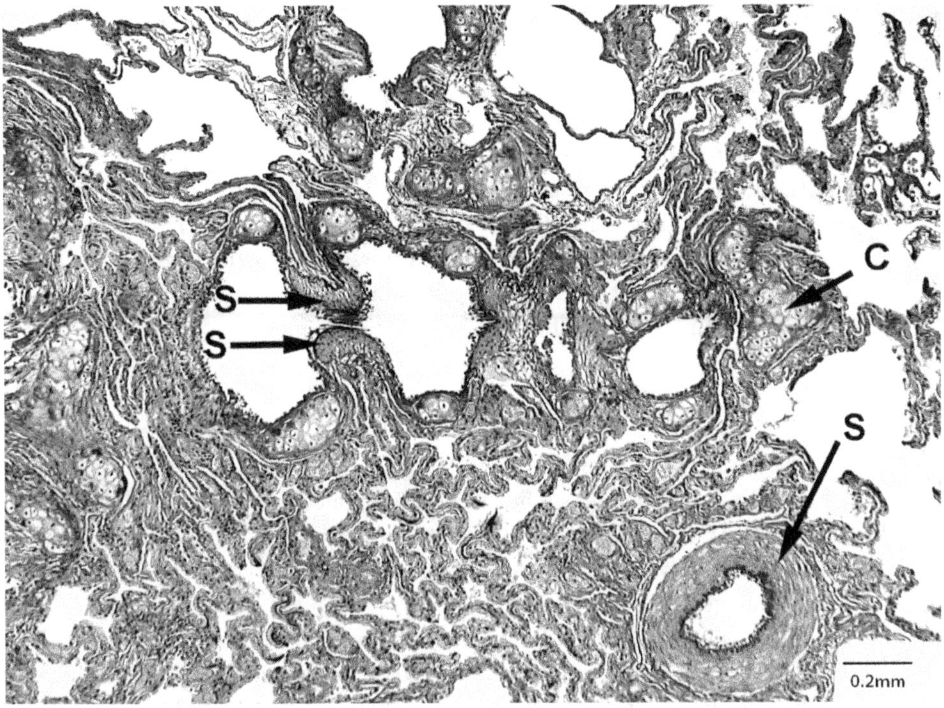

Figure 3.3 Lung micrograph of alveoli and distal bronchioles of the beluga whale, *Delphinapterus leucas*. Sagital sections and cross-sections of bronchiolar sphincter muscles are marked by arrows and the letter, S. One cross-section of cartilage is indicated by C. Photograph with permission and courtesy of M. Piscitelli.

The extensive cartilaginous reinforcement of the airways in cetaceans and sea lions should also allow higher flow rates and faster gas exchange during the brief surfacings of these animals. In cetaceans, it has been found that (1) the trachea is relatively short, (2) there is little change in the proximal vs. distal cross-sectional areas of the primary bronchi, (3) secondary bronchi are more numerous, and (4) cross-sectional areas of quaternary bronchi are high, 50–80% of that of the primary bronchi (Drabek and Kooyman, 1986).

Biomechanical compression studies have revealed that tracheas from bottlenose dolphins are relatively rigid, supporting the concept of alveolar collapse and the forcing of air into the trachea at depth (Bagnoli *et al.*, 2011). The trachea from a deep-diving pygmy sperm whale (*Kogia breviceps*) was found to be much more compressible, suggesting that eventual compression of the tracheobronchial tree further affords passage of respiratory air to the larynx and into nasal passages/sacs, where the air is necessary for formation of echolocation clicks (Davenport *et al.*, 2013).

A unique feature in the lungs of most dolphins and some odontocete whales is the presence of bronchiolar myoelastic sphincters (Belanger, 1940, Crespo and Lauria De Cidre, 2005, Goudappel and Slijper, 1958, Kooyman, 1973, Kooyman and Andersen, 1968, Kooyman and Sinnett, 1979, Ninomiya *et al.*, 2005, Piscitelli *et al.*, 2013, Wislocki, 1942). The function of these sphincters (Fig. 3.3) is unknown, although they have been postulated to (1) regulate gas distribution during diving, (2) play a role in alveolar re-expansion during ascent from depth, or (3) contribute to the rapid flow rates during exhalation.

Of historical interest, alveolar collapse as a mechanism for the prevention of caisson disease (decompression sickness) in whales was actually proposed by Damant in 1934 (Damant, 1934) in response to an article by Krogh (Krogh, 1934). In his article on whale physiology, Krogh had concluded it was unclear how whales avoided caisson disease, but speculated that the retia mirabilia had some role. Damant suggested that increased ambient pressure would compress the alveoli and block gas exchange and nitrogen uptake at depth.

3.1.2 Respiratory mechanics in marine mammals

The changes in respiratory mechanics suggested by the anatomical and histological findings in the above studies were investigated in the 1960s–1970s. Almost all of this research, which focused on flow rates and emptying of the lung, was conducted at or in association with Scholander's Physiological Research Lab at Scripps Institution of Oceanography.

In studies on excised lungs, it was found that dog lungs contained 27% of total lung capacity (TLC) at their relaxation volume (pleural pressure = 0), whereas sea lion lungs had a smaller relative relaxation volume (18% of TLC) (Denison *et al.*, 1971). Furthermore, when pleural pressure on the dog lungs was increased to +5 cm H_2O, only another 1% of TLC was expulsed before emptying stopped. In contrast, the sea lion lungs continued to empty with application of pleural pressures as high as +30 cm H_2O (the limit in the study). At that point, the mean gas volume of the lungs was 6% of TLC. A relaxation volume <17% of TLC in harbor porpoise lungs (*Phocoena phocoena*) (Kooyman and Sinnett, 1979), and low volumes in fin whales (*Balaenoptera physalus*) and sei whales (*B. borealis*) (Leith *et al.*, 1972) have also been reported. More recently, a mean 7% minimum air volume of excised lungs has been reported in a study of three phocid seal species and five odontocete species (Fahlman *et al.*, 2011). All these studies support the concept that reinforcement of the distal airways in diving mammals allows for the movement of air from the alveoli into the bronchi during lung compression, thus promoting collapse of the alveoli and cessation of gas exchange at depth.

Maximum expiratory flow rates, as high as 162 l s^{-1} in bottlenose dolphins, and 202 l s^{-1} in young gray whales (*Eschrichtius robustus*) (Kooyman and Cornell, 1981, Kooyman *et al.*, 1975), are necessary in cetaceans since exhalation and inhalation occur in less than 1 sec (Kooyman and Cornell, 1981, Kooyman *et al.*, 1975, Olsen *et al.*, 1969a, 1969b). Such flows allow for a tidal volume as high as 88% of TLC in the pilot whale (*Globicephala melena*) (Olsen *et al.*, 1969a, 1969b). In comparison, the average human tidal volume is less than 10% of TLC (Camporesi and Bosco, 2003). Such a high volume turnover is probably similar in other whales. In terms of vital capacity (VC, the volume of air from maximum inspiration to maximum expiration), maximum expiratory flow rates in cetaceans and sea lions are in the range

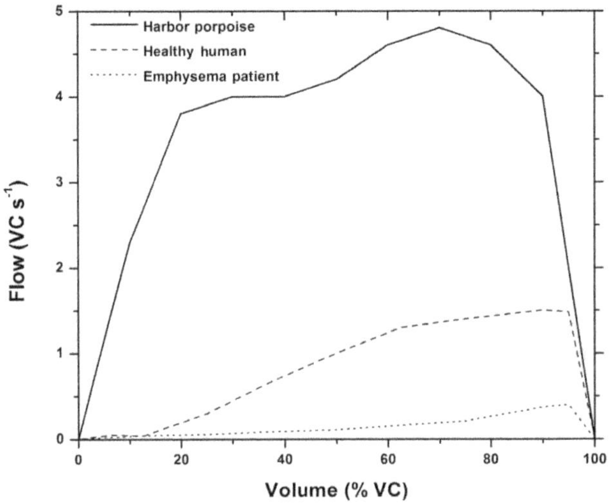

Figure 3.4 Flow volume loops (expiratory flow rate vs. lung volume) demonstrate the magnitude of high flow rates in the harbor porpoise (*Phocoena phocoena*) in comparison to a healthy human and to a patient with emphysema (with even lower flow rates). Importantly, cartilaginous reinforcement of the entire tracheo-bronchiolar tree in the porpoise allows for maintenance of maximum flow rates even at lower lung volumes (Kooyman and Sinnett, 1979). This contrasts markedly with the human emphysema patient and even with the normal, healthy human (Hyatt *et al.*, 1958). Maintenance of such high flow rates during a breath allows rapid respirations with tidal volumes equal to total lung volume in cetaceans. (VC = vital capacity, the volume of air exhaled from maximum inspiration to maximum exhalation). Adapted from above publications.

of 5–8 VC s^{-1} (Kerem *et al.*, 1975, Kooyman and Cornell, 1981, Kooyman and Sinnett, 1979). Such exhalation is not active as elastic recoil of the lung has been found to be the sole driving force during expiration of the pilot whale (Olsen *et al.*, 1969a, 1969b). These high flow rates minimize the time for exhalation/inhalation and thus allow animals to porpoise through the water. With such rapid breaths, the animals can spend most of their travel time below the surface, where drag is less (Williams *et al.*, 1992c). In contrast, maximum human expiratory flow rates are about 2 VC s^{-1} (Camporesi and Bosco, 2003).

Flow volume curves of the bottlenose dolphin and of excised harbor porpoise lungs are remarkable for the maintenance of high flows at low lung volumes (Kooyman and Cornell, 1981, Kooyman and Sinnett, 1979). This contrasts with human flow volume curves, in which expiratory flow decreases as volume decreases (Fig. 3.4). This difference, which contributes to the short exhalation time, is considered secondary to the cartilaginous reinforcement of distal airways in the dolphin.

3.1.3 Lung volumes of marine mammals

Lung volumes of diving mammals are in the general range of terrestrial mammals (Kooyman, 1973, Fahlman *et al.*, 2011, Piscitelli *et al.*, 2013) (Table 3.1). Notable exceptions are the small lungs of the deep-diving whales and the large lungs of the

Table 3.1 Lung volumes in diving mammals.

Species	TLC ml kg^{-1}	Reference	DLV ml kg^{-1}	Reference
Harbor porpoise *Phocoena phocoena*	80–130^{a}	A		
Bottlenose dolphin *Tursiops truncatus*	50–91^{a,f}	B	40–50^{d}	C
Minke whale *Balaenoptera acutorostrata*	72^{a}	D		
Sei whale *Balaenoptera borealis*	61–126^{a}	E		
Fin whale *Balaenoptera physalus*	61–126^{a}	E		
Pilot whale *Globicephala melena*	10^{b}	F		
Kogiid whales *Kogia* sp.	20–49^{a}	G		
N. bottlenosed whale *Hyperoodon ampullatus*	28^{a}	H		
Sperm whale *Physeter macrocephalus*			28^{d}	I
Manatee *Trichechus manutus*	65^{g}	J		
Northern fur seal *Callorhinus ursinus*	145^{a}	K		
Steller sea lion *Eumetopias jubata*	110^{a}	K		
California sea lion *Zalophus californianus*			48^{e}	L
Walrus *Odonbenus rosmarus*	116^{a}	K		
Harbor seal *Phoca vitulina*	91^{a}	K	23–39^{e}	M
Hooded seal *Cystophora cristata*	80^{a}			N
Ribbon seal *Histriophoca fasciata*	86^{a}	K		
Weddell seal *Leptonychotes weddellii*	48^{c}	O	22^{e} – 27^{c}	P
Elephant seal *Mirounga angustirostris*			20^{e}	Q
Sea otter *Enhydra lutris*	345^{a}	K	207^{e}	R

References: A: Kooyman and Sinnett, 1979; B: Kooyman and Cornell, 1981, Ridgway *et al.*, 1969; C: Skrovan *et al.*, 1999; D: Folkow and Blix, 1992; E: Leith *et al.*, 1972; F: Olsen *et al.*, 1969b; G: Piscitelli *et al.*, 2010; H: Scholander, 1940; I: Miller *et al.*, 2004b; J: Scholander and Irving, 1941; K: Lenfant *et al.*, 1970; L: Kooyman and Sinnett, 1982; M: Kooyman and Sinnett, 1982, Kooyman *et al.*, 1973b; N: Burns *et al.*, 2007; O: Kooyman *et al.*, 1971b; P: Kooyman *et al.*, 1973a, 1973b; Q: Kooyman *et al.*, 1973b; R: Ponganis *et al.*, 2003a.

Techniques: ^{a} inflation of excised lungs, ^{b} helium dilution, ^{c} nitrogen washout, ^{d} buoyancy – swim velocity calculations, ^{e} compression during simulated dives, ^{f} tidal volume measurement, ^{g} inspiratory capacity. Abbreviations: TLC: total lung capacity, DLV: diving lung volume.

shallow-diving sea otter. Inflation of excised lungs of bottlenosed whales (*Hyperoodon ampullatus*) and pygmy and dwarf sperm whales (*Kogia breviceps*, *K. sima*) revealed lung volumes of 28 and 21 ml kg^{-1}, respectively (Piscitelli *et al.*, 2010, Scholander, 1940). Allometric analyses of lung mass in cetaceans revealed that kogiids, physeterids, ziphiids, and mysticetes all had relative lung masses similar to terrestrial mammals, while delphinids, phocoenids, and monodonts had relatively larger lungs masses (Piscitelli *et al.*, 2010, 2013). Inflation of excised lungs of the sea otter yielded a volume of 345 ml kg^{-1} (Lenfant *et al.*, 1970). The high lung volume in the otter presumably contributes to its buoyancy at the surface, where it feeds, grooms, and cares for its young. Such buoyancy in the otter also elevates more of the body out of the water while the animal is at the surface; this should reduce body heat loss due to conduction in water.

In cetaceans and manatees, tidal volumes are large, 80–90% of TLC (Reynolds III and Odell, 1991, Ridgway, 1986), and in Weddell seals post-dive volumes are 75% of vital capacity (Kooyman *et al.*, 1971b). The large post-dive tidal volumes and increased ventilatory rates of Weddell seals allow for a maximum post-dive minute ventilation that is ten times the minimum value at rest, and a post-dive O_2 uptake rate that is eight times that at rest (Kooyman *et al.*, 1971b, 1973a). Harbor seals (*Phoca vitulina*) and California sea lions (*Zalophus californianus*) swimming vigorously in a water flume and taking a breath at each surfacing are able to attain maximum O_2 uptake rates that are 8–10 times that of the animals at rest (Ponganis *et al.*, 1990, 1991). These maximum O_2 uptakes in seals and sea lions are less than those of highly aerobic animals such as dogs and horses, but are equivalent to those of less specialized species such as goats and calves (Taylor *et al.*, 1987).

Post-dive respiratory rates in northern elephant seals (*Mirounga angustirostris*) average 22 breaths min^{-1} (Andrews *et al.*, 2000). Among odontocete cetaceans, post-dive breathing rates are 5–10, 6, and 5 breaths min^{-1}, respectively, in bottlenose dolphins (*Tursiops truncatus*), Blaineville's beaked whales (*Mesoplodon densirostris*), and sperm whales (*Physeter macrocephalus*) (Baird *et al.*, 2006b, Drouout *et al.*, 2004, Williams et al., 1999a). In baleen whales, post-dive respiratory rates in fin whales (*Balaenoptera physalus*) and humpback whales (*Megaptera novaeangliae*) are about 4–5 breaths min^{-1} (Goldbogen *et al.*, 2008, Lafortuna *et al.*, 2003). Blue whales (*Balaenoptera musculus*) average ten breaths during post-dive surface intervals of about 3-min duration (Goldbogen *et al.*, 2011).

Diving lung volumes, defined as the lung volume at the start of a dive, are important determinants of the size of the respiratory O_2 store during a dive. Cetaceans and otariid pinnipeds appear to dive on inspiration, while phocid pinnipeds usually dive on expiration (Kooyman, 1989, Ridgway, 1986). Consequently, the diving lung volumes of cetaceans are probably near TLC. This assumption is supported by the similarity of the calculated diving lung volume of the sperm whale, a deep-diving odontocete, to the measured TLC of another deep diver, the bottlenosed whale (Miller *et al.*, 2004b, Scholander, 1940). Determinations of diving lung volumes in pinnipeds during free dives and simulated dives have yielded values that are 40–50% of TLC (Kooyman and Sinnett, 1982, Kooyman *et al.*, 1971b, 1973b). More recently, blood oxygen profiles in deep-diving sea lions have provided evidence that lung collapse occurs at deeper depths in dives of greater maximum depth (McDonald and Ponganis, 2012). In other words, the sea lion appears to inhale deeper prior to deeper dives. Such increased lung volumes potentially increase the magnitude of the lung oxygen store, which has been most commonly calculated with values of 50% TLC in these species.

3.2 Seabird respiratory anatomy and function

3.2.1 Airway, air sac, and lung anatomy in diving birds

Although the data available for lung anatomy and function might be considered limited for marine mammals, there is even less description of respiratory anatomy and function in diving birds. Histological and detailed anatomical descriptions of upper airway structure in diving birds do not exist. In general, the avian trachea bifurcates into the primary bronchi, which extend into the lungs and also exit into the posterior air sacs. Within the lung, secondary bronchi branch off from the primary bronchi, eventually leading to the parabronchi, from which extend the air capillaries, which intertwine with pulmonary blood capillaries and are the site of gas exchange in the bird lung (Duncker, 1972, Duncker, 1974, Powell, 2000, Scheid, 1979, West *et al.*, 1977, Woodward and Maina, 2008). Secondary bronchi extend from the parabronchi back to the anterior air sacs. The air sacs of birds are poorly vascularized and not considered sites of gas exchange. There are nine air sacs: two cervical sacs, a single clavicular sac, two anterior thoracic sacs, two posterior thoracic sacs, and two abdominal sacs, all of which are connected to various secondary bronchi and some parabronchi. In general, air sac volumes are about ten times greater than lung air volumes (Duncker, 1972, Scheid *et al.*, 1974). Fig. 3.5 demonstrates the lung and air sac system of a penguin in a three-dimensional reconstruction from computerized tomographic (CT) scans.

Airflow through the various bronchi and air sacs is complex, but during inspiration and expiration air flow is unidirectional (caudal to cranial) in the parabronchi (Powell, 2000, Scheid, 1979, Scheid and Piiper, 1987). During inspiration, air flows both into the parabronchi and posterior air sacs, with air exiting the parabronchi into the anterior air sacs (Scheid, 1979). During exhalation, air flows predominantly from the posterior air sacs into the parabronchi, exiting into the anterior air sacs, and then out through secondary bronchi into the trachea (Scheid, 1979).

Inspiration and expirations are accomplished through activation of specific thoracic wall and abdominal muscles (Powell, 2000). Lung volume changes only minimally during the respiratory cycle (Jones *et al.*, 1985). During flight, wing upstrokes during inspiration and downstrokes during exhalation augment airflow through expansion and compression effects on the thoracoabdominal cavity (Boggs, 1997, Butler and Bishop, 2000). Differential air-sac pressures due to wing movements in swimming penguins also suggest that wing movements may induce air-sac pressure oscillations with secondary enhanced diffusion and/or airflow from the air sacs through the lung during the breath hold in diving penguins (Boggs *et al.*, 2001).

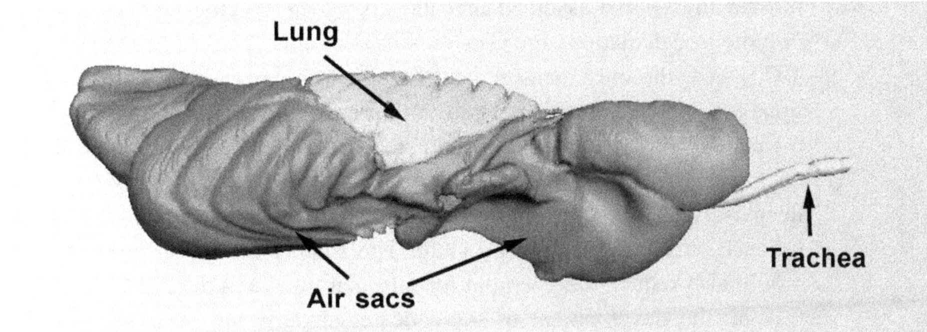

Figure 3.5 Three-dimensional computerized tomographic scan reconstruction of the air sacs and lungs of an emperor penguin (*Aptenodytes forsteri*). Adapted from Ponganis et al., 2015.

3.2.2 Lung/air-sac volumes of diving birds

Measurements of the combined gas volumes of air sacs and lung in birds are rare, and have usually been conducted by tracer gas dilution or inflation/cast injection to a given pressure. Indeed, the allometric equation relating air-sac/lung air volumes to body mass in all birds is based on samples collected in different studies from only five species (Lasiewski and Calder, 1971). This equation is often used to calculate diving air volume in seabirds because marine birds such as murres are reported to dive on inspiration (Croll *et al.*, 1992a). In the marine birds, total air-sac/lung volume should be equivalent to the diving air volume.

In aquatic birds, the most detailed examinations of air-sac and lung volumes have been in ducks. Ducks, however, usually exhale prior to diving (Butler and Woakes, 1979) Therefore, end-expiratory lung volumes have been used to estimate their diving air volumes. In the unrestrained lesser scaup (*Aythya affinis*), tidal volumes were 65–75 ml kg^{-1}, and the entire end-expiratory gas volume of the air sacs and lungs was 335 ml kg^{-1} (Stephenson, 1994). This total respiratory air volume was greater than that in restrained ducks in the same study (139 ml kg^{-1}). In this study, plumage air was 177 ml kg^{-1}; thus, in consideration of the contribution of respiratory and plumage air to body buoyancy, plumage air composed about 35% the total in an unrestrained duck. In the pekin duck (*Anas platyrhynchos*) under restrained conditions, respiratory air volume ranged from 108 to 220 ml kg^{-1} (Hudson and Jones, 1986). In unrestrained tufted ducks (*Aythya fuligula*), end-expiratory respiratory air volume was 165–232 ml kg^{-1} (Stephenson *et al.*, 1989b).

In penguins, diving air volumes (160–165 ml kg^{-1}) were first reported in gentoo and Adélie penguins (*Pygoscelis papua*, *P. adeliae*) from simulated dives in a pressure chamber (Kooyman *et al.*, 1973c). This value, determined from the volume of water pumped into the chamber from once it was filled until the chamber pressure began to rise, included both respiratory air and plumage air (the latter estimated to be about 10% of the total based on volume displacement measurements of birds washed with detergent). Gastrointestinal gas was assumed to be negligible. Because of these findings, most authors have assumed that the diving air volume represents the respiratory air volume in calculations for O_2 stores in penguins. These measurements, determined in 1973, were the only measurements available for penguin respiratory O_2 store calculations for 26 years, until diving air volumes of king penguins (*Aptenodytes patagonicus*, 69 ml kg^{-1}) were determined with the same technique and even the same pressure chamber (Ponganis *et al.*, 1999a). Because of the smaller diving air volume of the deeper-diving king penguin, it was suggested that deeper-diving penguins had less reliance on the respiratory O_2 store. The smaller diving air volume in the deep diver would also reduce the potential for nitrogen uptake at depth.

With the development of sophisticated underwater backpack recorders, it became possible to estimate diving air volume in free-diving penguins (Sato *et al.*, 2002). Based on buoyancy–swim speed calculations, and recorded swim speeds, body angles, and depth profiles during the final gliding ascents to the surface by penguins, Sato and colleagues estimated the air volume that would most closely predict the observed swim speed during that period. This air volume was taken to represent the start-of-dive respiratory volume, assuming that plumage air was minimal and no air was exhaled during the dive. In both Adélie and king penguins, diving air volume increased with maximum depth of dive, from 100 to 200 ml kg^{-1}, and from 50 to 125 ml kg^{-1}, respectively, in each species. Not only were these values variable with depth, but the highest values were approximately 20–60% greater than those measured in each species in the pressure chamber dives. Larger values in free-diving animals would be consistent with Stephenson's prior findings in restrained versus unrestrained lesser scaups.

More recently, similar techniques applied to emperor penguins (*A. forsteri*) at sea revealed a mean maximal air volume for deep dives of 117 ml kg^{-1}, while the mean value for the shallowest dives was 64 ml kg^{-1} (Sato *et al.*, 2011). There is at least one caveat, however, in using this technique to estimate start-of-dive air volumes and respiratory O_2 stores. These calculations are based on measurements collected during the final ascent to the surface. If the bird exhales prior to that final ascent, these volumes will underestimate the start-of-dive air volume. The data in emperor penguins were quite variable for any depth, and the calculated volumes actually decreased for the deepest dives. Either the penguins may exhale early, or they are, indeed, beginning deep dives with smaller air volumes. It should be noted that initial stroke rates on descent increased with dive depth, consistent with greater buoyancy and larger air volumes for deeper dives (Williams *et al.*, 2012).

As discussed above, estimations of respiratory O_2 stores in most diving birds are currently derived from respiratory air volumes predicted by allometric equations (Lasiewski and Calder, 1971). In ducks, estimations can be based on the results from the inert gas dilution methods reviewed above, and, in penguins, on results from swim speed–buoyancy calculations (Sato *et al.*, 2002, 2011, Scheid *et al.*, 1974, Stephenson, 1995, Stephenson *et al.*, 1989b). It is notable, however, that the end-expiratory air volume measured in the unrestrained diving duck is about twice the allometrically predicted value, and the highest diving air volumes calculated for free-diving Adélie and emperor penguins are about 30–50% greater than predicted values. These variations in values raise questions as to the exact size of the air sacs and lungs in diving birds.

Most recently, lung and air sac volumes in Adélie, king and emperor penguins were determined from 3D reconstructions from CT scans (Ponganis *et al.*, 2015). The volumes of the lungs scaled allometrically according to published equations (Lasiewski and Calder, 1971). However, maximal air-sac volumes during positive pressure breath holds were 2.2–3 times greater than allometrically predicted values. Maximal total air volumes of the lungs and air sacs were calculated to be 311, 368, and 374 ml kg^{-1} in Adélie, king, and emperor penguins, respectively. Whether the birds can spontaneously inspire to that volume is uncertain because it is well known that the volume of the abdominal air sacs during spontaneous ventilation is much less than their maximal capacity (Scheid, 1979). Although uncertainty still exists as to the exact volume of the air sacs prior to a dive, certainly all the values estimated in free-diving penguins are less than the maximal value measured by CT scan in each species (see Chapter 12 for further review). If start-of-dive air-sac volumes are, indeed, greater than calculated end-of-dive values, the O_2 store will be greater as will initial buoyancy that must be overcome as the bird descends.

In regard to ventilation, the tracheal volumes of birds are about 4.5 times greater than those of similarly sized mammals (Hinds and Calder, 1971). As a consequence, at rest, birds have a larger tidal volume than similarly sized mammals (Bouverot, 1978). During exercise, respiratory rate increases and tidal volume may increase or remain unchanged, dependent on the species (Butler, 1991). In most species studied, both minute ventilation and oxygen uptake increase similarly during flapping flight to values as high as ten times resting rates (Butler, 1991, Butler and Bishop, 2000). Ventilation in emperor penguins swimming in a flume was sufficient to allow a maximal O_2 uptake about eight times the measured resting rate (Kooyman and Ponganis, 1994).

There are few data available for post-dive breathing rates in birds. Respiratory rate after the longest reported dive (27.6 min) of an emperor penguin decreased from 22 to 16 breaths min^{-1} over the first 5 min of the post-dive interval; the initial rate was similar to that reported after shorter-duration dives, but respiratory rate remained elevated longer after the long dive (Kooyman *et al.*, 1971a, Sato *et al.*, 2011). Respiratory rates of emperor penguins at rest have been reported as low as 3 breaths min^{-1} (Kooyman *et al.*, 1971a, Meir *et al.*, 2008).

4 Oxygen storage and transport

Oxygen transport and increased O_2 storage within the body are fundamental components of the breath-hold capacities of marine mammals and birds. The body O_2 stores are located in the respiratory system, blood, and muscle. The magnitude and distribution of these O_2 stores vary among species, and are primarily dependent on diving lung volume, blood volume, hemoglobin (Hb) concentration, muscle mass, and myoglobin (Mb) concentration. As will be reviewed, Hb and Mb are the primary O_2-binding proteins in the body.

The precise functions of other O_2-binding proteins, such as neuroglobin and cytoglobin, have not been defined yet (Burmester and Hankeln, 2009, Burmester *et al.*, 2000, Weber and Fago, 2004). Potential roles include oxygen transport and storage, and scavenging of reactive oxygen species and reactive nitrogen species (Burmester and Hankeln, 2009, Reeder, 2010, Weber and Fago, 2004). The neuroglobin amino acid sequences of seals and whales differed by only two to three amino acids from those of terrestrial controls, suggesting that differences in neuroglobin function (i.e., O_2 affinity) between terrestrial and marine mammals were unlikely (Schneuer *et al.*, 2012). The potential role of neuroglobin in the hypoxemic tolerance of marine mammals is discussed in Chapter 11.

The functional size of an O_2 store is also affected by how much O_2 can be extracted from those stores, i.e., the initial and final O_2 concentrations during a dive. This is exemplified in a classic forced submersion study of pekin ducks (*Anas platyrhynchos*) taken to their maximum breath-hold capacity (Hudson and Jones, 1986). At that end point, although blood O_2 was nearly depleted, 25% of the original respiratory O_2 store still remained. As will be reviewed later in the chapter, that large percentage of respiratory O_2 was unusable because of the O_2-binding properties of duck hemoglobin. At that O_2 fraction in the lung, the resulting P_{O2} was low enough that duck Hb was essentially devoid of O_2, resulting in "imminent cardiovascular collapse." Thus, all the O_2 could not be extracted from the lung. The final O_2 concentration and actual size of the respiratory O_2 store in the duck were dependent ultimately on the biochemical properties and O_2 affinity of Hb.

This chapter on O_2 storage and transport is divided into four sections. First, the structure and function of Hb and Mb in diving mammals and birds will be reviewed. Then, the classic assumptions used in the calculation of body O_2 stores will be examined. Third, the many measurements of the determinants of O_2 stores in different species will be compiled (i.e., diving air volume, blood volume, Hb concentration, muscle mass, Mb concentration). In addition, because of the high hematocrits and Hb concentrations in the blood of many divers, blood rheology and coagulation will be reviewed in this section. Lastly, the magnitudes and distribution of O_2 stores in different species will be reviewed in relation to their known diving behaviors.

4.1 Hemoglobin structure and function

Hemoglobin (Hb) in diving birds and mammals is similar in structure to that of other mammals and birds in that it is composed of four polypeptide chains, each with an iron-containing heme group that can bind an O_2 molecule. When all four heme sites are bound with O_2, Hb is fully saturated. At 100% saturation, there are 1.34 ml O_2 per gram of Hb. The concentration of Hb in blood is the primary determinant of blood O_2 content because the solubility of O_2 in blood at body temperature is quite low (0.00124 mM mm Hg^{-1} or 0.003 ml O_2 dl^{-1} mm Hg^{-1}) (Powell, 2000, West, 1972).

There are two forms of Hb which cannot bind to O_2. Methemoglobin, which contains iron in the ferric state (Fe^{+++}) as opposed to the usual ferrous state (Fe^{++}), is unable to bind O_2, but is usually found in only minimal concentrations (Reeder, 2010). Carboxyhemoglobin is also incapable of binding O_2, but again is usually only present in very low concentrations in most animals. It is formed when carbon monoxide, which has 240 times greater affinity for Hb than O_2, is present in the blood (West, 1972).

4.2 O_2–hemoglobin dissociation curves

The O_2–Hb dissociation curve (Fig. 4.1) describes the reversible binding of O_2 to Hb as a function of Hb saturation and the partial pressure of O_2 (P_{O2}). Its sigmoidal shape is due to allosteric interactions and cooperativity between the four polypeptide subunits of the Hb molecule (Powell, 2000, West, 1972). P_{50}, the P_{O2} at which Hb is 50% saturated, is used as an index of Hb's affinity for O_2, and its reference value is usually taken at normal body temperature and blood pH (pH 7.4 in mammals and 7.5 in birds). The O_2–Hb dissociation curve (and the P_{50}) can be shifted to the left or right by changes in three primary factors that regulate the O_2 affinity of Hb.

The three important factors that affect Hb's affinity for O_2 are pH, temperature, and the concentration of organic phosphates inside the red blood cell (Powell, 2000, West, 1972). Acidosis (decreased pH) and elevated temperatures both decrease the affinity of Hb, shifting the curve to the right and increasing the P_{50}. The effect of pH on Hb affinity for O_2 is known as the Bohr effect. Both acidosis and elevated temperature occur in exercising muscle and should facilitate unloading of O_2 from Hb. Alkalosis (increased pH) and lower temperature both increase O_2 affinity, shifting the curve to the left and decreasing the P_{50}. These factors in the lung (an increased pH due to low CO_2 and a normal as opposed to higher temperature in muscle) are considered to facilitate the loading of Hb with O_2. In mammals, 2,3-diphosphoglycerate (2,3-DPG) binds to Hb

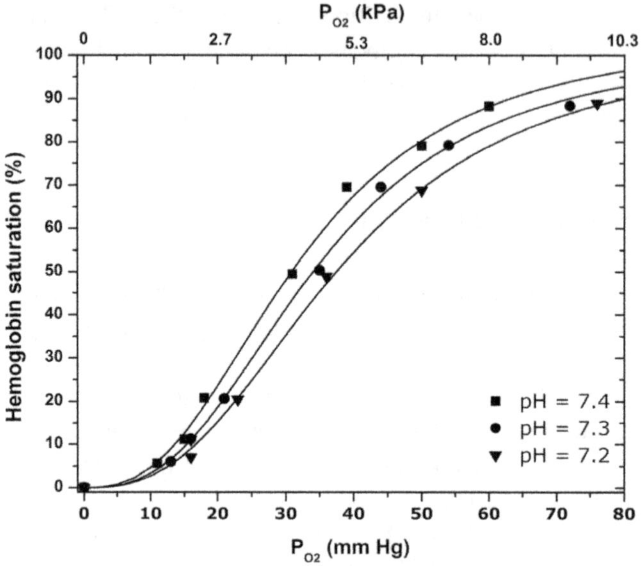

Figure 4.1 O_2–hemoglobin dissociation curves of the northern elephant seal, *Mirounga angustirostris*, at pHs, 7.4, 7.3, and 7.2 shifted to the right with acidosis, demonstrating an increase in the P_{50} value and a decrease in O_2 affinity. The mean P_{50} of elephant seal hemoglobin at pH 7.4 was 30.5 mm Hg (4.06 kPA). Adapted from Meir et al. 2009.

and also affects its O_2 affinity. In high-altitude-adapted animals, for instance, 2,3-DPG is elevated and decreases O_2 affinity, shifting the curve to the right, increasing the P_{50} and improving O_2 unloading to the tissues (West, 1972). In bird red blood cells, myoinositol 1,3,4,5,6-pentophosphate (IPP) binds to Hb and can decrease its affinity for O_2, shifting its curve to the right and increasing the P_{50} (Powell, 2000). Although high-altitude-adapted birds such as the bar-headed goose (*Anser indicus*) have Hbs with greater O_2 affinity (shift to the left, lower P_{50}) than low-altitude birds, IPP concentrations are similar (Petschow *et al.*, 1977). Instead, changes in the amino acid sequence of the Hb polypeptide chains alter the binding site and affinity for IPP (Powell, 2000, Weber, 2007, Weber and Fago, 2004).

In light of reports of elevated carbon monoxide and carboxyhemoglobin in deep-diving seals (Pugh, 1959, Tift *et al.*, 2014), it should also be noted that although carboxyHb cannot transport O_2, these compounds can increase the O_2 affinity of the remaining Hb, and shift the dissociation curve to the left (Hlastala *et al.*, 1976, Roughton and Darling, 1944). Roughton and Darling suggested that the left shift could be especially beneficial in increasing O_2 at low P_{O2} values and that this could account for the 1898 observation by J. S. Haldane and L. Smith of improved hypoxic tolerance of mice on exposure to low carbon monoxide concentrations. Low concentrations of carbon monoxide have also been found to enhance oxygen delivery in more recent experimental models of brain blood flow (Koehler *et al.*, 1984). In addition, low concentrations of carbon monoxide may also contribute to prevention of reperfusion injury in divers, as well as at least partially contribute to elevated tissue mitochondrial volume densities in pinnpeds through effects of carbon monoxide on mitochondrial biogenesis (Lancel *et al.*, 2009, Piantadosi, 2008, Piantadosi *et al.*, 2008, Rhodes *et al.*, 2009). See Chapters 7, 9, 11 and 13 for further review.

4.2.1 O_2–hemoglobin dissociation curves: marine mammals

With this as background, it is now possible to review Hb function in diving mammals and birds. The concentration of Hb in blood is its most notable aspect in diving mammals. Many marine mammals have been found to have exceptionally high Hb concentrations in comparison to terrestrial mammals. However, the O_2 affinity of Hb in these animals was not that different from many of their terrestrial counterparts (Lenfant, 1969, Lenfant *et al.*, 1970, Meir *et al.*, 2009, Qvist *et al.*, 1981). The P_{50} values were in the range of 26–30 mm Hg. The one exception was the manatee (*Trichechus manutus*) with a P_{50} of 16 mm Hg (Farmer *et al.*, 1979, White *et al.*, 1976) The magnitude of the Bohr effect (Δ log P_{50} / Δ log pH) in marine mammals was also similar to that found in terrestrial mammals (Lenfant, 1969, Lenfant *et al.*, 1970, Meir *et al.*, 2009, Qvist *et al.*, 1981). Hill's coefficient (n), an index of cooperativity between the Hb subunits, was again within the normal range (Lenfant *et al.*, 1970).

Investigations of fetal blood in Weddell seals (*Leptonychotes weddellii*) revealed that fetal blood had greater O_2 affinity and a lower Bohr effect than maternal blood, which is consistent with its lower concentration of 2,3-DPG (Qvist *et al.*, 1981). The difference in P_{50} between maternal and fetal blood was 5–6 mm Hg (Lenfant *et al.*, 1969a, Qvist *et al.*, 1981). This difference in P_{50} was attributed to the lower 2,3-DPG concentrations in the fetus. Although amino acid sequencing was not performed, there did not appear to be any significant differences in the molecular structures or binding properties of isolated Hb between the fetus and the adult seal. Thus, there was no evidence in seals for a distinct fetal versus adult Hb.

4.2.2 O_2–hemoglobin dissociation curves: seabirds

Examination of Hb in diving birds has revealed that their Hb concentrations are in the upper range of avian values (see Table 4.3). However, the most remarkable finding is the increased O_2 affinity of penguin Hbs in comparison to that of flighted birds (Fig. 4.2). This shift in O_2 affinity is probably due to specific amino acid substitutions in the polypeptide chains of Hb (Tamburrini *et al.*, 1994, 1999). The P_{50} values of Adélie, gentoo, chinstrap, and little blue penguins (*Pygoscelis adeliae*, *P. papua*, *P. Antarctica*, *Eudyptula minor*) are 30–35 mm Hg, and that of the emperor penguin (*Aptenodytes forsteri*) is 28 mm Hg (Lenfant *et al.*, 1969b, Meir and Ponganis, 2009, Milsom *et al.*, 1973). In comparison, the P_{50} values of most flighted birds, including ducks and murres, are in the range of 40–55 mm Hg (Black and Tenney, 1980, Johansen *et al.*, 1987, Lenfant *et al.*, 1969b, Milsom *et al.*, 1973, Petschow *et al.*, 1977). In addition to penguins, a left-shifted dissociation curve, lower P_{50}, and higher affinity of Hb for O_2 occur in high-altitude birds (Black and Tenney, 1980, Petschow *et al.*, 1977).

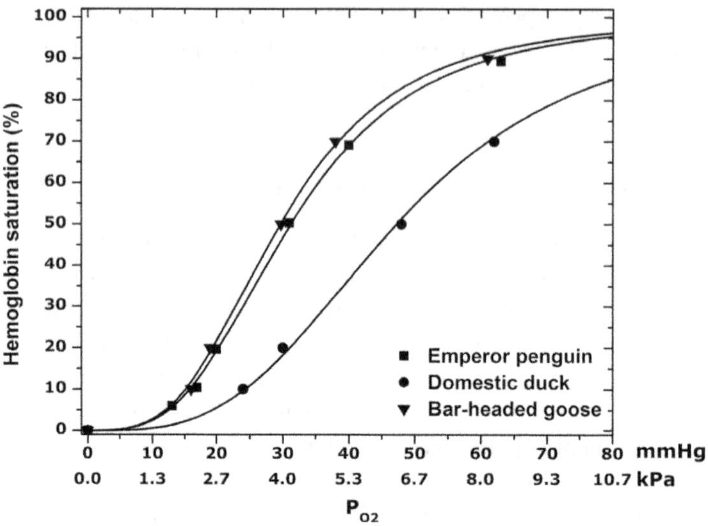

Figure 4.2 O_2–hemoglobin dissociation curves of the emperor penguin (*Aptenodytes forsteri*) and bar-headed goose (*Anser indicus*) demonstrated a marked shift to the left, decreased P_{50}, and increased oxygen affinity in comparison to that of the domestic duck (*Anas platyrhynchos*). There was about a 14 mm Hg (1.87 kPa) difference between the P_{50} of the duck and those of the penguin and high-altitude goose. Adapted from Meir and Ponganis, 2009, and Black and Tenney, 1980.

As regard other aspects of Hb function, Hill coefficients and Bohr effects in penguins are similar to those of other birds (Lenfant *et al.*, 1969b, Meir and Ponganis, 2009).

The higher O_2 affinity of penguin Hb has several important implications in relation to the depletion of O_2 stores. In contrast to the pekin duck at the brink of "imminent cardiovascular collapse," but with 25% of its respiratory O_2 store unused, the emperor penguin can fully utilize its respiratory O_2 store (Hudson and Jones, 1986, Stockard *et al.*, 2005). One adaptation which facilitates depletion of air-sac O_2 in penguins is the higher O_2 affinity of Hb (Meir and Ponganis, 2009, Milsom *et al.*, 1973). At a P_{O2} of 20 mm Hg, for example, the Hb of the pekin duck is devoid of O_2, whereas that of the emperor penguin is still 27% saturated. Thus, a higher affinity Hb allows the emperor penguin to more fully deplete its O_2 store as well as to have a higher blood O_2 content at low P_{O2}.

It should also be noted that it is probably necessary to take into account the P_{50} of Hb during calculation of the respiratory O_2 store of a given avian species. In penguins, the entire respiratory store is available, whereas in flighted birds only 75% is available (Croll *et al.*, 1992a, Stephenson *et al.*, 1989b). However, P_{50} values for most flighted, diving birds are unavailable.

4.3 Myoglobin structure and function

Myoglobin consists of a single polypeptide chain and heme group. Consequently, one Mb molecule binds one molecule of oxygen (Antonini, 1965, Wittenberg and Wittenberg, 1989). Because the molecular weight of Mb (17,000–18,000 daltons) is approximately

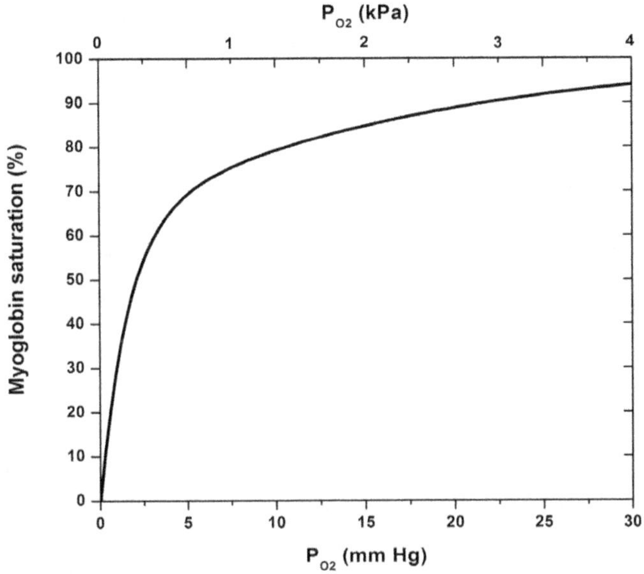

Figure 4.3 O_2–myoglobin dissociation curve of the horse (*Equus ferus caballus*) demonstrated a sigmoidal shape with P_{50} near 2.4 mm Hg (0.32 kPa). The P_{50} of myoglobin was about ten times less than that of hemoglobin, demonstrating the much greater affinity of myoglobin for O_2. Adapted from Schenkman et al., 1997.

one-fourth that of Hb, the volume of O_2 bound per gram of Mb at 100% saturation is usually considered by most authors to be the same as Hb, i.e., 1.34 ml O_2 g Mb^{-1}. In actuality, the value for either Hb or Mb may vary dependent on the exact molecular weight of Hb or Mb in a given species. As an example, values of 1.20–1.24 ml O_2 g Mb^{-1} have been used in some estimations of O_2 stores in birds (Croll *et al.*, 1992a, Stephenson *et al.*, 1989b).

The P_{50} of Mb at 37 to 40 °C is <4 mm Hg (0.53 kPa) in both terrestrial and most diving mammals and penguins (Antonini, 1965, Helbo and Fago, 2012, Nichols and Weber, 1989, Schenkman *et al.*, 1997, Suzuki and Imai, 1998, Weber *et al.*, 1974, Wright and Davis, 2015). In melon-headed and minke whales (*Peponocephala electra*, *Balaenoptera acutorostrata*), the P_{50} values are slightly higher at 4.9 and 4.6 mm Hg, respectively (Helbo and Fago, 2012, Wright and Davis, 2015). In general, the affinity of Mb for O_2 is at least ten-fold greater than that of Hb in most species. Once bound to Mb, O_2 in muscle is lost from the blood O_2 store and no longer available for other tissues during a dive.

The shape of the O_2–Mb dissociation curve is hyperbolic (Fig. 4.3) (Antonini, 1965, Schenkman *et al.*, 1997). Mb's affinity for O_2 is increased by lower temperatures (its P_{50} decreases). Although it was originally reported that the P_{50} of Mb did not shift with changes in pH, a small shift to the right (decreased affinity for O_2) has been detected with acidosis. The P_{50} of horse Mb at 37 °C changes from 2.39 mm Hg at pH 7.0 to

2.46 mm Hg at pH 6.5 (Schenkman *et al.*, 1997). However, unlike Hb, the affinity of Mb is not affected by changes in any known co-factors (Antonini, 1965).

In addition to its role as an O_2 store, Mb has potential roles in (a) facilitation of diffusion of O_2 to mitochondria in muscle; (b) nitric oxide production via nitrite reductase activity; and (c) the scavenging of reactive O_2 species. Readers are referred to reviews and Chapter 13 for further discussion (Flogel *et al.*, 2010, Garry *et al.*, 2003, Gros *et al.*, 2010, Helbo and Fago, 2012, Helbo *et al.*, 2013, Lin *et al.*, 2007, Ordway and Garry, 2004, Weber, 2007, Weber and Fago, 2004, Wittenberg, 1970, Wittenberg and Wittenberg, 1989, 2003).

4.4 Calculation of O_2 stores

The calculation of O_2 stores is based on a variety of anatomical and physiological parameters, many of which have not been measured in a given species. Consequently, assumptions are common, and revisions are necessary as new data become available. The earliest estimations of O_2 storage in seals and penguins were probably those of Irving and Scholander in the 1930s (Irving, 1934, 1939, Irving *et al.*, 1935b, Scholander, 1940). Primary assumptions and potential sources of error include: (1) the diving air volume (air volume in the respiratory system at the start of a dive); (2) blood volume and Hb concentration in animals with large blood volumes; (3) muscle mass; (4) the frequent assumption that Mb concentration is the same in all muscles; and (5) the net extraction of O_2 from either the respiratory system or blood during a dive. These limitations will be demonstrated in the paragraphs below as the calculation of O_2 stores is reviewed.

4.4.1 Respiratory O_2 stores

The respiratory O_2 store is determined by the diving air volume and the net extraction of O_2 from that air volume during a dive. Chapter 3 reviews diving air volumes and provides a list of measured values in marine mammals (see Table 3.1). In general, cetaceans are considered to dive at full lung capacity, and pinnipeds at 50% total lung volume (Gentry and Kooyman, 1986, Kooyman, 1989, Ponganis, 2011). However, more recent evidence suggests that sea lions inspire deeper and have larger diving air volumes for deeper dives (McDonald and Ponganis, 2012). Manatees are considered to dive on inspiration with full lung volumes, and sea otters with 60% total lung capacity (Ponganis *et al.*, 2003a).

Most ducks have been considered to dive on exhalation, and diving air volumes used in O_2 store calculations have been end-expiratory values (Keijer and Butler, 1982, Stephenson, 1995, Stephenson et al., 1989b). As reviewed in Chapter 3, penguins dive on inspiration and, in free dives, diving air volume (calculated at the end of a dive) appears to increase with depth (Sato *et al.*, 2002, 2011, Wilson *et al.*, 2003). The diving air volume at the start of a penguin's dive is still unknown. In other diving birds, it is usually assumed that the allometrically predicted air-sac/lung volume represents the diving air volume (Croll *et al.*, 1992a).

The net extraction of oxygen from the diving air volume is based on the difference between the initial O_2 fraction and the lowest possible end-of-dive value. In diving mammals, this difference has been assumed to be 15% (Kooyman, 1989). This value is reasonable, especially in seals which dive on exhalation and probably have start-of-dive O_2 fractions below 20%. In addition, this assumption is consistent with expiratory and end-tidal O_2 fractions that were in the range of 2–4% after dives and forced submersions (Kooyman *et al.*, 1973a, Ponganis *et al.*, 1993a, Ridgway *et al.*, 1969, Scholander, 1940).

In ducks, respiratory O_2 fractions have been determined in the air sacs. Given that the O_2 fraction is known to vary between anterior and posterior air sacs, O_2 fractions in representative air sacs have been determined, and then averaged, based on the relative volumes of the anterior and posterior air sacs. Such calculations in ducks have provided a mean O_2 fraction of 17.6% prior to the start of diving (Keijer and Butler, 1982, Stephenson *et al.*, 1989b). In forcibly submerged ducks at maximum breath-hold duration, the mean end-of-submersion O_2 fraction was 4.1% (Hudson and Jones, 1986). Thus, the maximum decrease in the O_2 fraction is 13.5%, which represents about 75% of the available O_2 in the respiratory system. No data exist for diving seabirds, so this 13.5% change in the O_2 fraction has been used in the calculation of their respiratory O_2 stores (Croll *et al.*, 1992a).

In penguins, a change of 15% in air-sac O_2 fraction has been used in O_2 store calculations, based upon final air-sac O_2 fractions of 2–4% during simulated dives in a pressure chamber (Kooyman, 1989, Kooyman and Ponganis, 1990, Kooyman *et al.*, 1973b). Air-sac O_2 fractions have now also been determined in diving emperor penguins, and the maximum change between initial and end-of-dive O_2 fractions was 19%, with end-of-dive O_2 fractions near 0 after some dives (Ponganis *et al.*, 2010a, Stockard *et al.*, 2005). The near-complete depletion of respiratory O_2 in emperor penguins is at least partially attributable to the previously discussed increased O_2 affinity of penguin versus duck Hb. Although other penguin species also have Hbs with increased O_2 affinity, it is not known whether they tolerate as low a P_{O2} and whether they are capable of such complete O_2 extraction from the respiratory store.

In summary, the respiratory O_2 store can be calculated as 13.5%, 15%, and 19% of the diving air volume in flighted diving birds, marine mammals, and penguins, respectively. Apart from emperor and possibly king penguins, a 15% value may also be more appropriate for the other penguin species as the limits of their hypoxemic tolerance are not known. Most estimates of respiratory O_2 stores in marine mammals and all diving birds prior to 2010 utilized either the 13.5% or 15% values. Due to the rarity of its measurement, diving air volume is probably the greater source of potential error in this calculation. For many species, this air volume still remains a dilemma just as it was for the Nobel Prize-winning physiologist August Krogh in his 1934 estimation of the lung O_2 store of the blue whale (Krogh, 1934).

4.4.2 Blood O_2 stores

Blood O_2 stores have been calculated on the basis of blood volume, Hb concentration, the net desaturation of Hb during a dive, and the assumption that one-third of the blood volume is arterial and two-thirds venous (Kooyman, 1989, Lenfant *et al.*, 1970). Although such a calculation appears straightforward, there is potential for error in the measurement of the individual parameters

4.4.2.1 Blood O_2 stores: blood volume and Hb measurement techniques

The determination of blood volume is classically conducted by measurement of (1) red cell volume (usually by dilution of radio-labeled red cells, i.e., chromium 51); and (2) plasma volume (usually by dilution of a marker or dye such as radio-labeled albumin or Evan's blue dye). Due to the difficulty of using radio-labeled compounds in remote field locations, the majority of blood volume determinations in diving mammals and birds have been conducted with use of a formula, in which blood volume is calculated by dividing the plasma volume by the term (1-hematocrit). Plasma volume is usually measured by the Evan's blue dye technique (El-Sayed *et al.*, 1995), and the hematocrit (the packed cell volume) is determined by centrifugation.

In animals with large blood volumes and large spleens, the accurate determination of the red cell volume is potentially difficult due to prolonged mixing times for dilution of labeled red blood cells (Persson *et al.*, 1973) and to fluctuations in hematocrit secondary to the level of splenic contraction (Cross *et al.*, 1988, Turner and Hodgetts, 1959). In addition, splenic dilatation and sequestration of red blood cells during anesthesia of animals with large spleens will also contribute to prolonged mixing times, as well as to lower hematocrits. In the Weddell seal (*Leptonychotes weddellii*), for example, the differencc between the hematocrit under anesthesia and during diving led to a 50% reduction in the estimated blood volume (Ponganis *et al.*, 1993a). Lastly, use of the dye dilution–hematocrit technique is dependent on the assumption that the measured hematocrit is representative of a uniform hematocrit throughout the entire circulation.

Blood volume is usually assumed to be one-third arterial and two-thirds venous in blood O_2 store calculations, (Lenfant *et al.*, 1970). This assumption is reasonable and probably not a major source of error in blood O_2 store calculations. There are few data available on the distribution of blood volume between the arterial and venous systems. Estimates for the venous fraction of blood volume in mammals have been as high as 70–80% (Alexander, 1963, Wiedeman, 1963). In animals such as seals, with large spleens and/or large central venous capacitance, the fractions of blood volume considered arterial and venous may vary dependent on the physiological state of the animal.

Hemoglobin is typically determined with commercially available spectrophotometric techniques, with the most common procedure measuring a cyanomethemoglobin derivative. Again, for purposes of O_2 store calculations, the highest circulating Hb concentration should be used, but this may be difficult to obtain as it is rarely possible to obtain blood samples during or immediately after dives (when the spleen is presumably fully contracted). Many researchers will try to obtain samples immediately after capture because presumed activation of the sympathetic nervous system should constrict the spleen and elevate both hematocrit and Hb concentration during the capture.

4.4.2.2 Blood O_2 stores: initial and final Hb saturation assumptions

In oxygen store calculations, it is typically assumed that arterial blood is 95% saturated initially, and that it can decline to about 20% saturation. The latter figure is based on the hypoxemic tolerance studies of Elsner and co-workers (Elsner *et al.*, 1970b, Kerem and Elsner, 1973). The 95% value is reasonable, given that there may be some degree of pulmonary shunting or ventilation–perfusion mismatch (Kooyman and Sinnett, 1982). In addition, at least in seals, some carboxy-Hb may be present due to a six-fold elevation in carbon monoxide content in blood of seals as compared to humans (Pugh, 1959). In adult elephant seals (*Mirounga angustirostris*), carboxy-Hb can be as high as 10% (Tift *et al.*, 2014). This elevation in carbon monoxide has been attributed to the high concentrations of hemoglobin in seals and the carbon monoxide generated during the metabolic breakdown of hemoglobin to bilirubin. Although this appears counterproductive in terms of oxygen storage, it may be that low levels of carbon monoxide or carboxy-Hb contribute to regulation of vascular tone and prevention of re-perfusion injury (Herrera *et al.*, 2008, Kajimura *et al.*, 2010, Motterlini and Otterbein, 2010, Nakaoa *et al.*, 2005).

The initial venous Hb saturation level during a dive is typically considered to be the Hb saturation equivalent to 5 ml O_2 dl^{-1} less than the initial arterial O_2 content. It has been assumed that venous blood oxygen can be completely depleted based on forced submersion findings in both ducks and seals (Hudson and Jones, 1986, Kerem and Elsner, 1973). It should be noted that the initial venous Hb saturation (at 5 ml O_2 dl^{-1} less than the initial arterial O_2 content) is dependent on the Hb concentration. At 15 g dl^{-1} Hb, the initial venous Hb saturation would be 74%, while at an Hb concentration of 26 g dl^{-1}, the corresponding value would be 85%. Thus, for a typical arterio-venous O_2 content difference of 5 ml O_2 dl^{-1}, the expected venous Hb saturation as well as P_{O2} will vary dependent on the Hb content of the blood. This is important because, as will be seen in Chapter 5, venous Hb saturations and P_{O2} can be elevated above the expected resting values. This can occur both prior to and during dives of emperor penguins and sea lions, and during dives of elephant seals (McDonald and Ponganis, 2013, Ponganis *et al.*, 2011). Such an increase in venous oxygen content is called arterialization of venous blood. If this occurs prior to diving, as in emperor penguins, the venous blood O_2 store is increased by 5 ml O_2 dl^{-1} (Ponganis *et al.*, 2010a).

4.4.2.3 Blood O_2 stores: blood volumes, Hb concentrations

Hemoglobin concentrations and blood volumes of marine mammals may be increased as much as 50–70% above the typical human Hb concentration and blood volume of 15 g dl^{-1} and 70 ml kg^{-1}, respectively (Tables 4.1, 4.2). The greatest increases in both Hb and blood volume are in the longest-duration divers and in highly active species (Ponganis, 2011, Ridgway and Johnston, 1966). This is exemplified in the phocid seals, in which blood volumes may be 2–3 times that of the standard human value. In contrast, Hb concentrations and blood volumes of avian divers are only at the upper range of bird values (Table 4.3). The highest reported Hb concentrations and blood volumes of avian divers are not in the penguins, but in small, flighted diving birds such as Cassin's auklet (*Ptychoramphus aleuticus*) and the ancient murrelet (*Synthliboramphus antiquus*)

Table 4.1 Hemoglobin (Hb) concentrations and blood volumes (BV) in cetaceans and manatees.

Species	Hb ($g\ dl^{-1}$)	BV ($ml\ kg^{-1}$)	Reference
Bottlenose dolphin *Tursiops truncatus*	14	71	A
Pacific white-sided dolphin *Lagenorhynchus obliquidens*	17	108	A
Dall porpoise *Phocoenoides dalli*	20	143	A
Commerson's dolphin *Cephalorhynchus commersoni*	18		B
Harbor porpoise *Phocoena phocoena*	19		C
Beluga whale *Delphinapterus leucas*	21	128	D
Sperm whale *Physeter macrocephalus*	22	200	E, F
Killer whale *Orcinus orca*	18	90	G, H, I, J
Pilot whale *Globicephala scammoni*	16		G, K
Gray whale *Eschrictius robustus*	13	81	G
Bowhead whale *Balaena mysticetus*	20		L
Sei whale *Balaenoptera borealis*	16		M
Manatee *Trichechus manutus*	15	80	N

References: A: Ridgway and Johnston, 1966; B: Hedrick and Duffield, 1991; C: Reed *et al.*, 2000; D: Ridgway *et al.*, 1984; E: Ridgway, 1986; F: Sleet *et al.*, 1981; G: Gilmartin *et al.*, 1974; H: Dhindsa *et al.*, 1974; I: Lenfant *et al.*, 1968; J: Ridgway, 1972; K: Medway and Moldovan, 1966; L: Castellini *et al.*, 2006; M: Lenfant, 1969; N: Blessing, 1972b, Gallivan and Best, 1980, White *et al.*, 1976.

(Elliott *et al.*, 2010). The latter values are similar to those of the kittiwake (*Rissa tridactyla*), a non-diver (Table 4.3).

As discussed above, increased blood O_2 storage is partly achieved through elevated Hb concentrations. High blood Hb concentrations can be achieved through increased Hb content within the red blood cell as well as by an increased hematocrit (Hct, packed red blood cell volume). Both of these factors can affect blood flow properties (rheology) and the efficient transport of O_2. In this last section on the blood O_2 store, consideration will be given primarily to blood flow, blood viscosity, their effects on arterial impedance (cardiac workload), and the potential for thrombosis (clot formation). Readers are referred to an excellent review of blood rheology in marine mammals for further details (Castellini *et al.*, 2010).

Table 4.2 Hemoglobin (Hb) concentrations and blood volumes (BV) in pinnipeds and sea otters.

Species	Hb (g dl^{-1})	BV (ml kg^{-1})	Reference
California sea lion *Zalophus californianus*	18	120	A
Steller sea lion *Eumetopias jubata*	17	120	B, C
Northern fur seal *Callorhinus ursinus*	17	109	B
Antarctic fur seal *Arctocephalus gazella*	18	132	D
Australian sea lion *Neophoca cinerea*	19	178	D, E
New Zealand sea lion *Phocarctos hookeri*		150	F
Galapagos sea lion *Zalophus wollebaeki*	23	186	G
Walrus *Odobenus rosmarus*	16	106	B
Harbor seal *Phoca vitulina*	21	132	B, H
Leopard seal *Hydrurga leptonyx*	15	131	I
Gray seal *Halichoerus grypus*	20	213	J
Ribbon seal *Histriophoca fasciata*	24	132	B
Harp seal *Phoca groenlandica*	23	168	K
Hooded seal *Cystophora cristata*	23	106	K
Ringed seal *Phoca hispida*	25	158	L
Baikal seal *Phoca sibirica*	27	177	M
Weddell seal *Leptonychotes weddellii*	26	210	N
Northern elephant seal *Mirounga angustirostris*	25	216	O
Sea otter *Ehnydra lutris*	19	174	P

References: A: Weise and Costa, 2007; B: Lenfant *et al.*, 1970; C: Mellish *et al.*, 2007, Richmond *et al.*, 2006; D: Costa *et al.*, 2001; E: Fowler *et al.*, 2007; F: Costa *et al.*, 1998; G: Villegas-Amtmann and Costa, 2010; H: Burns *et al.*, 2005; I: Kuhn *et al.*, 2006; J: Noren *et al.*, 2005; K:Burns *et al.*, 2007; L: Lydersen *et al.*, 1992, St. Aubin *et al.*, 1978; M: Petrov and Shoshenko, 1987, Ponganis *et al.*, 1997b; N: Ponganis *et al.*, 1993a; O: Simpson *et al.*, 1970, Thorson and Le Boeuf, 1994; P: Thometz, 2014, Thometz *et al.* 2015.

Table 4.3 Hemoglobin (Hb) concentration and blood volume (BV) in aquatic and marine birds.

Species	Hb (g dl^{-1})	BV (ml kg^{-1})	Reference
Tufted duck *Aythya fuligula*	15–18	107–141	A
Mallard duck (pekin duck) *Anas platyrhynchos*	17	91–113	A, B
Black-legged kittiwake *Rissa tridactyla*	16	123	C
Coot *Fulica americana*	17	95	B
Red-throated loon *Gavia stellata*	21	132	B
White pelican *Pelecanus onocrotalus*	14		D
Rhinoceros auklet *Cerorhinca monocerata*	17	127	E
Cassin's auklet *Ptychoramphus aleuticus*	21		C
Ancient murrelet *Synthliboramphus antiquus*	20		C
Thick-billed murre *Uria lomvia*	18	123	F
South Georgia diving petrel *Pelecanoides georgicus*	20		G
Cormorant *Leucocarbo fuscescens*	15		H
Japanese cormorant *Phalacocrorax capilllatus*		139	E
Little penguin *Eudyptula minor*	18		I
Rockhopper penguin *Eudyptes crestatus*	16		K
Magellanic penguin *Spheniscus magellanicus*	14		K
Humboldt penguin *Spheniscus humboldti*	15		L
Adélie penguin *Pygoscelis adeliae*	16	93	M
Gentoo penguin *Pygoscelis papua*	16		M
Chinstrap penguin *Pygoscelis antarctica*	20		M
King penguin *Aptenodytes patagonicus*	18	83	N

Table 4.3 *(cont.)*

Species	Hb (g dl^{-1})	BV (ml kg^{-1})	Reference
Emperor penguin *Aptenodytes forsteri*	18	100	O

References: A: Keijer and Butler, 1982; B: Bond and Gilbert, 1958; C: Elliott *et al.*, 2010; D: Puerta *et al.*, 1991; E: Yamamoto *et al.*, 2011; E: Yamamoto *et al.*, 2011 F: Croll *et al.*, 1992a; G: Kooyman, 1989; H: Melrose and Nicol, 1992; I: Mill and Baldwin, 1983; J: Hawkey *et al.*, 1989; K: Villouta *et al.*, 1997; L: Lenfant *et al.*, 1969b; M: Milsom *et al.*, 1973; N: Ponganis *et al.*, 1999a; O: Ponganis *et al.*, 1997a.
The kittiwake is a non-diver, the mallard duck and coot are dabblers, and the pelican is a plunge diver.

4.4.2.4 Blood O_2 stores: rheology

As highlighted in the Castellini *et al.* review (2010), blood viscosity can be affected by temperature, shear rate, Hct level, and red blood cell characteristics (shape, size, deformability, aggregability, membrane composition, and cytoplasmic viscosity). A variety of adaptations in marine mammals and diving birds appear to optimize their blood rheology. For instance, although decreased temperature increases viscosity, the increase in blood viscosity in bowhead whales (*Balaena mysticetus*) was less than in human blood in samples exposed to low temperatures that would be found in the flukes of the whale (Elsner *et al.*, 2004b). Similarly, blood viscosities of little blue (*Eudyptula minor*) and Adélie penguins were less than that of the chicken at all temperatures tested (Clarke and Nicol, 1993).

Shear rate is essentially analogous to blood flow rate, and viscosity decreases curvilinearly as shear rate increases (Castellini *et al.*, 2010). Thus, viscosity will be greater under low-flow conditions (lower heart rates) and lower during periods of high flow. Furthermore, as Hct increases, viscosity will also increase, with greater increases in viscosity at lower shear rates (Castellini *et al.*, 2010). These principles were confirmed in several species of marine mammals and seabirds in early studies (Guard and Murrish, 1975, Hedrick and Duffield, 1991, Hedrick *et al.*, 1986, Wickham *et al.*, 1989, 1990a, 1990b). Notably, blood viscosity at low shear rates and high Hct was lower in seals than in pigs (Wickham *et al.*, 1989), leading to the suggestion that relatively lower viscosities in seals under these conditions lowered cardiac workload during a dive, especially when heart rate began to increase near or at the end of a dive (see Chapter 5 for cardiac responses during dives). This suggestion is based on the fact that blood viscosity contributes to systemic vascular resistance (Klabunde, 2011, Murray *et al.*, 1969), and hence to the workload of the heart (see Chapter 5). However, to make matters more complex, later studies revealed that blood viscosities at various shear rates, including low shear rates, varied in different seal species (Elsner and Meiselman, 1995, Meiselman *et al.*, 1992). For example, the ringed seal had low viscosities, especially at low shear rates, while the Weddell seal had high viscosities. This led to the suggestion that such high viscosities in some species would lead to splenic storage of red blood cells during non-diving periods (i.e., at a time with no need for a large circulating blood O_2 store) (Elsner and Meiselman, 1995). (See Chapter 5 for Hct variation and splenic contraction in diving seals.)

Differences in blood viscosities may also be affected by many other factors, including red cell size, shape, and deformability, as well as the concentration of Hb within the cell, which affects the cell's cytoplasmic viscosity (and, in turn, its deformability) (Castellini *et al.*, 2010). Differences in these variables, as well as differences in plasma proteins, may affect viscosity (Wickham *et al.*, 1989, 1990b); increased red cell aggregation in Weddell seals is probably associated with the increased fibrinogen concentrations in their blood (Meiselman *et al.*, 1992).

Increased viscosity, high Hct, periods of low blood flow, and elevated fibrinogen levels in Weddell seals all raise the question of increased risk of thrombosis in diving animals. In human patients, endothelial injury from hypoxemia and ischemia (decreased organ perfusion) is associated with changes in coagulation status and the potential for thromboembolic events (Weidman *et al.*, 2014). There have been relatively few studies of coagulation in diving mammals and birds. In northern elephant seals (*Mirounga angustirostris*), fibrinogen, antithrombin III, and platelets were reported as similar to other domestic species; the activated clotting time (ACT), prothrombin time (PT), and activated partial thromboplastin time (PTT) were relatively short in comparison to other species (Gulland *et al.*, 1996).

Factor XII of the coagulation cascade is absent in bottlenose dolphins (*Tursiops truncatus*), killer whales (*Orcinus orca*), and sei whales (*Balaenoptera borealis*); both their PT and activated PTT are prolonged (Robinson *et al.*, 1969, Saito *et al.*, 1976, Tibbs *et al.*, 2005). It has been suggested that lack of Factor XII and absence of an active extrinsic coagulation pathway may serve to decrease risk of thrombosis in stagnant, pooled blood during the dive response (Saito *et al.*, 1976, Tibbs *et al.*, 2005). In contrast, increased fibrinolytic activity does not appear to play a protective role against thrombus formation in stagnant blood in seals. Clot lysis time increased during forced submersions of gray seals (*Halichoerus grypus*), which was consistent with reduced rather than increased fibrinolytic activity (Lohman *et al.*, 1998).

In addition to potential effects of coagulation factors, decreased temperatures and decompression are known to activate platelets (Field and Tablin, 2012, Field *et al.*, 2001, Patterson *et al.*, 1993). Decreased responses of platelets to various agonists have been documented in killer whale and elephant seal platelets (Field and Tablin, 2012, Field *et al.*, 2001, Patterson *et al.*, 1993). Although elephant seal platelets are activated by exposure to 4 °C temperature, decompression-induced activation does not occur (Field and Tablin, 2012). Differences in the lipid compositions of platelet membranes of the marine mammals versus terrestrial mammals were suggested as possible mechanisms underlying the decreased sensitivities of their platelets to various agonists (Field and Tablin, 2012, Patterson *et al.*, 1998).

4.4.3 Muscle O_2 stores

Muscle O_2 stores are estimated on the basis of muscle mass, Mb concentration, and 1.34 ml O_2 g^{-1} Mb. Although this calculation appears straightforward, there are potential sources of error.

4.4.3.1 Muscle O_2 stores: muscle mass and Mb distribution

Muscle mass for a given species is not always known, and an assumption of 30% of body mass has often been used. Although a 30% value appears reasonable for many animal species (see Tables 4.4, 4.5), the balaenopterid whales are an exception. Muscle mass in these large, lunge-feeding whales comprises 45–62% of body mass. Muscle mass is 25–29% of body mass in flighted avian divers, but 33–38% in penguins. Most measurements of muscle mass have been made by anatomical dissection. Muscle mass can also be determined through 3D reconstructions of CT or magnetic resonance scans (Ponganis *et al.*, 1997a).

Most authors have also assumed a uniform distribution of Mb concentration throughout all muscles of the body. This is not necessarily true. In emperor penguins, for example, there is a three-fold difference in Mb concentration between the leg muscles and the primary locomotory muscles, the pectoralis and supracoracoideus muscles (Ponganis *et al.*, 1997a). In Baikal seals (*Phoca sibirica*), not only are there differences in Mb contents of individual muscles, but there are also differences between superficial and deep portions of muscles as well as changes in Mb concentration with the season of the year (Neshumova *et al.*, 1983, Neshumova and Cherepanova, 1984, Petrov and Shoshenko, 1987). Mb concentrations in the epaxial muscles of dolphins and seals also vary in the cranio-caudal direction, with the highest concentrations in the caudal regions (Polasek and Davis, 2001, Polasek *et al.*, 2006). However, at least in the Baikal seal, the overall average Mb concentration for all muscles was only about 10% less than the highest value measured deep in the primary locomotory muscle (Neshumova and Cherepanova, 1984). This is at least partially due to the fact that the primary locomotory muscles are the most massive muscles in the body. Therefore, in most cases, reasonable estimates of muscle O_2 stores are possible with measurement of Mb concentration in a single muscle.

4.4.3.2 Myoglobin analyses

Most authors in the field of diving physiology have determined Mb concentrations with a spectrophotometric method (Reynafarje, 1963). In this technique, which involves measurement and subtraction of tissue absorbances at two wavelengths, the effect of Hb on the measurements is eliminated due to the assumed identical extinction coefficients of Hb at those wavelengths. However, these assumptions may not always be correct. The Hb extinction coefficients at those wavelengths can vary in different species, and, in addition, the Mb extinction coefficients used in the calculation are also not constant among species (Masuda *et al.*, 2008).

Masuda *et al.* found that use of the Reynafarje technique overestimated rat heart cytoplasmic Mb concentrations by 0.10 mM, which was 37% greater than the 0.26 mM value determined by ^1H NMR spectroscopy. The authors attributed this overestimation to differences in extinction coefficients as well as to baseline errors in spectral subtraction, and suggested use of a baseline corrected spectra of a standard Hb and Mb solution to correct such errors. Whether these findings invalidate the many Mb determinations with the Reynafarje technique in diving animals is unclear. Marine mammals have much higher Mb concentrations (i.e., 4.5 g 100 g^{-1} muscle is equivalent to 3.8 mM in

Table 4.4 Measured muscle mass as percentage of body mass in marine mammals.

Species	Muscle (%)	Reference
Bottlenose dolphin *Tursiops truncatus*	36	A
Sperm whale *Physeter macrocephalus*	34	B
Sei whale *Balaenoptera borealis*	58	B
Fin whale *Balaenoptera physalus*	45	B
Bryde's whale *Balaenoptera brydei*	46	B
Minke whale *Balaenoptera acutorostrata*	62	B
Pacific right whale *Balaena glacialis siboldi*	31	B
Humpback whale *Megaptera novaeangliae*	30	B
Manatee *Trichechus manutus*	40	C
California sea lion *Zalophus californianus*	37	D
Harp seal *Phoca groenlandica*	26	E
Hooded seal *Cystophora cristata*	28	E
Ringed seal *Phoca hispida*	30	F
Baikal seal *Phoca sibirica*	30	G
Weddell seal *Leptonychotes weddellii*	35	H
Southern elephant seal *Mirounga leonina*	28	I
Crabeater seal *Lobodon carcinophagus*	44	J
Sea otter *Enhydra lutris*	33	K

References: A: Goforth, 1986; B: Lockyer, 1976; C: Scholander and Irving, 1941; D: Ponganis *et al.*, 1997c; E: Burns *et al.*, 2007, 2010; F: Lydersen *et al.*, 1992; G: Neshumova and Cherepanova, 1984; H: Fujise *et al.*, 1985; I: Bryden, 1972; J: Bryden and Erickson, 1976; K: Thometz, 2014, Thometz *et al.* 2015.

In most calculations of muscle oxygen stores in species for which muscle mass has not been determined, a value of 30% of body mass is usually assumed.

Table 4.5 Muscle mass (as a percentage of body mass) in aquatic and marine birds.

Species	Muscle (%)	Reference
Tufted duck *Aythya fuligula*	25	A
Black-legged kittiwake *Rissa tridactyla*	21	B
Cassin's auklet *Ptychoramphus aleuticus*	26	B
Ancient murrelet *Synthliboramphus antiquus*	25	B
Rhinoceros auklet *Cerorhinca monocerata*	27	C
Thick-billed murre *Uria lomvia*	29	B, D
King penguin *Aptenodytes pagagonicus*	33 – pre-breeding 37 – pre-molt	E
Emperor penguin *Aptenodytes forsteri*	38	F

References: A: Keijer and Butler, 1982; B: Elliott *et al.*, 2010; C: Yamamoto *et al.*, 2011; D: Croll *et al.*, 1992a; E: Cherel *et al.*, 1993; F: Ponganis *et al.*, 1997a.

The kittiwake represents a non-diver.

cytoplasm), and Masuda *et al.* estimated that use of the Reynafarje technique would only lead to a 3.6% error in sperm whale (*Physeter macrocephalus*) myoglobin concentration. Such an error would have minimal effect on O_2 store calculations given the high concentrations found in sperm whale muscle. Clearly, however, a combined spectrophotometric and ^{1}H NMR study in marine mammal muscles would be valuable in the evaluation of this technique and the confirmation of past determinations of Mb concentrations in marine mammals and diving birds.

Thus, it is important to remember potential limitations of available data in the calculation of muscle O_2 stores. These include available muscle mass data, variation in Mb concentrations in different muscles, and even potential limitations in the analytic technique used to determine Mb concentration. In addition, in muscle samples that have been preserved frozen for long time periods (i.e., months or longer), it is important to verify that dehydration has not occurred. This is most easily accomplished by freeze drying fresh and stored samples and comparing water contents.

4.4.3.3 Myoglobin concentrations

In marine mammals, Mb concentrations vary almost 100-fold, from 0.1 g 100 g^{-1} muscle in manatees to 9.5 g 100 g^{-1} muscle in hooded seals (see Table 4.6, 4.7). It is in the deep, long-duration divers that Mb concentrations are highest. These animals

Table 4.6 Myoglobin (Mb) concentrations in cetaceans and the manatee.

Species	Mb (g 100 g^{-1})	Reference
Bottlenose dolphin *Tursiops truncatus*	2.7–3.2	A, G
Pacific white-sided dolphin *Lagenorhynchus obliquidens*	3.5	A
Killer whale *Orcinus orca*	3.1	B
Northern right whale dolphin *Lissodelphis borealis*	1.8	A
Indus River dolphin *Platanista indi*	2.6	C
Common dolphin *Delphinus delphis, D. capensis*	3.6	A
Striped dolphin *Stenella coeruleoabla*	5.8	A
Spinner dolphin *Stenella longirostris*	5.5	D
Spinner dolphin *Stenella attenuata*	2.5	E
Fraser's dolphin *Lagenodelphis hosei*	7.1	D
Harbor porpoise *Phocoena phocoena*	4.0	A, F
Beluga whale *Delphinapterus leucas*	3.4	A
Short-finned pilot whale *Globicephala macrorhynchus*	6.8	G
Pygmy sperm whale *Kogia breviceps*	4.3–5.9	A, H
Cuvier's beaked whale *Ziphius cavirostris*	4.3	A
Beaked whale (four species) *Mesoplodon* sp.	7.3	H
Narwhal *Monodon monoceros*	7.9	I
Northern bottlenose whale *Hyperoodon ampullatus*	6.3	J
Sperm whale *Physeter macrocephalus*	5.4	J
Sei whale *Balaenoptera borealis*	0.9	K
Fin whale *Balaenoptera physalus*	1.1–2.4	A, K, L
Blue whale *Balaenoptera musculus*	0.8	M
Bowhead whale *Balaena mysticetus*	3.5	A

Table 4.6 *(cont.)*

Species	Mb (g 100 g^{-1})	Reference
Manatee *Trichechus manutus*	0.1	F

References: A: Noren and Williams, 2000; B: Noren *et al.*, 2012b; C: Blessing, 1972a; D: Dolar *et al.*, 1999; E: Castellini and Somero, 1981; F: Blessing, 1972b; G: Velten *et al.*, 2013; H Kielhorn *et al.*, 2013; I: Williams *et al.*, 2011b; J: Scholander, 1940; K: Tawara, 1950; L: Hochachka and Foreman, 1993; M: Lawrie, 1953. Data in this table are from primary locomotory muscles in adult animals. Values are in g 100 g^{-1} muscle wet weight.

include phocid seals (hooded, harp, ribbon, Weddell, and elephant seals), otariids (Galapagos (*Zalophus wollebaeki*) and California sea lions (*Z. californianus*)) and cetaceans such as the sperm whale, bottlenose whales, and the narwhal (*Monodon monoceros*). Among species with limited documentation of diving behavior, high Mb concentrations suggest deep-diving capacity in striped (*Stenella coeruleoabla*), spinner (*Stenella longirostris*), and Fraser's (*Lagenodelphis hosei*) dolphins. However, not all deep divers have the most extreme Mb concentrations. Cuvier's beaked whale (*Ziphius cavirostris*), the mammalian dive record holder, is an example; Mb content is high, but not the highest among marine mammals (Table 4.6). And, conversely, in the Baikal seal, elevated Mb concentrations may support yet undocumented, long-duration dives during the Siberian winter when the lake freezes over.

Elevated Mb concentrations are also a hallmark of the penguins, with the highest values in the deep-diving king and emperor penguins (Table 4.8). Although Mb concentrations are higher in the primary swimming muscles versus the legs of penguins, the Mb concentration is still considerably elevated in the leg muscles. In flighted diving birds, Mb concentrations are elevated only two- to three-fold, with higher concentrations in the primary underwater locomotory muscles. The Japanese cormorant, however, is an exception in regard to Mb distribution.

4.5 Magnitude and distribution of total body O_2 stores

Both the magnitude and distribution of total body O_2 stores of divers versus non-divers vary between mammals and birds (Tables 4.9–4.11). In comparison to humans ("a non-diver"), mass specific body O_2 stores are elevated 1.5- to almost five-fold in all marine mammals except the manatee. In general, diving mammals with greater diving capacities have significantly larger O_2 stores.

In contrast, total body O_2 is not exceptionally elevated in diving birds. Even in the premier avian diver, the emperor penguin, the total mass-specific O_2 store is only elevated about one-third above that of the non-diving black-legged kittiwake (*Rissa tridactyla*). The distribution of O_2 stores between mammals and birds also differs, and the most striking difference in is the respiratory compartment. In all diving mammals

Table 4.7 Myoglobin (Mb) concentrations in pinnpeds and sea otters.

Species	Mb (g 100 g^{-1})	Reference
California sea lion *Zalophus californianus*	5.4	A
Steller sea lion *Eumetopias jubata*	2.7	B
Northern fur seal *Callorhinus ursinus*	3.5	C
Antarctic fur seal *Arctocephalus gazella*	2.4	D
Australian sea lion *Neophoca cinerea*	2.7	E
Galapagos sea lion *Zalophus wollebaeki*	5.3	F
Walrus *Odobenus rosmarus*	3.0	C
Harbor seal *Phoca vitulina*	5.5	C
Gray seal *Haliochoerus grypus*	5.4	D
Leopard seal *Hydrurga leptonyx*	5.1	G
Harp seal *Phoca groenlandica*	8.6	H
Hooded seal *Cystophora cristata*	9.5	H
Ribbon seal *Histriophoca fasciata*	8.1	C
Ringed seal *Phoca hispida*	4.1	I
Baikal seal *Phoca sibirica*	6.0	J
Weddell seal *Leptonychotes weddellii*	5.4	K
Northern elephant seal *Mirounga angustirostris*	7.8	L
Sea otter *Enhydra lutris*	3.3	M

References: A: Weise and Costa, 2007; B: Richmond *et al.*, 2006; C: Lenfant *et al.*, 1970; D: Reed *et al.*, 1994a; E: Fowler *et al.*, 2007; F: Villegas-Amtmann and Costa, 2010; G: Kuhn *et al.*, 2006; H: Burns *et al.*, 2007, 2010; I: Lydersen *et al.*, 1992; J: Neshumova and Cherepanova, 1984; K: Ponganis *et al.*, 1993a; L: Hassrick *et al.*, 2010, Thorson and Le Boeuf, 1994; M: Thometz, 2014, Thometz *et al.*, 2015.

Data in this table are from primary locomotory muscles in adult animals. Values are in g 100 g^{-1} muscle wet weight.

Table 4.8 Myoglobin (Mb) concentrations in pectoral and leg muscles of aquatic and marine birds.

Species	Mb Concentration (g 100 g^{-1} muscle)		Reference
	Pectoral	Leg	
Tufted duck *Aythya fuligula*	0.6–0.7	1.0	A
Black-headed gull *Larus ridibundus*	0.6		B
Black-legged kittiwake *Rissa tridactyla*	0.4		C
Manx shearwater *Puffinus puffinus*	0.6		B
Rhinoceros auklet *Cerorhinca monocerata*	1.8	1.2	D
Cassin's auklet *Ptychoramphus aleuticus*	1.1	0.8	C
Ancient murrelet *Synthliboramphus antiquus*	1.0	0.9	C
Atlantic puffin *Fratercula arctica*	1.3	0.8	E
Common murre *Uria aalge*	1.4	0.6	E
Thick-billed murre *Uria lomvia*	1.9		F
Japanese cormorant *Phalacocrorax capilllatus*	1.6	1.1	D
Little penguin *Eudyptula minor*	2.8		G
Rockhopper penguin *Eudyptes crestatus*	3.7		B
Royal penguin *Eudyptes schlegeli*	4.3		B
Adélie penguin *Pygoscelis adeliae*	3.0		H
Gentoo penguin *Pygoscelis papua*	4.4		H
King penguin *Aptenodytes patagonicus*	4.3		B
Emperor penguin *Aptenodytes forsteri*	6.4	2.0	I

References: A: Keijer and Butler, 1982; Stephenson *et al.*, 1989b; B: Baldwin *et al.*, 1984; C: Elliott *et al.*, 2010; D: Yamamoto *et al.*, 2011; E: Davis and Guderley, 1987; F: Croll *et al.*, 1992a; G: Mill and Baldwin, 1983; H: Weber *et al.*, 1974; I: Ponganis *et al.*, 1997a.

The gull and kittiwake represent non-divers.

Table 4.9 Mass specific total body oxygen stores and percentage distribution among lung, blood, and muscle O_2 stores in humans, some cetaceans, manatees, and sea otters.

Species	Total O_2 store ($ml\ O_2\ kg^{-1}$)	Lung (%)	Blood (%)	Muscle (%)	Reference
Human	24	42	44	14	A
Bottlenose dolphin *Tursiops truncatus*	34	27	33	40	B
Pacific white-sided dolphin *Lagenorhynchus obliquidens*	40	23	49	28	B
Killer whale *Orcinus orca*	36	23	36	41	
Beluga whale *Delphinapterus lucus*	51	17	51	32	C
Narwhal *Monodon monoceros*	75	12	38	50	D
Sperm whale *Physeter macrocephalus*	81	5	64	30	E
Manatee *Trichecus manutus*	21	33	60	7	F, G
Sea otter *Enhydra lutris*	69	45	33	21	H

A: Ponganis *et al.*, 2011; B: Noren *et al.*, 2012b; C: Shaffer *et al.*, 1997; D: Williams *et al.*, 2011b; E: Miller *et al.*, 2004b, Sleet *et al.*, 1981; F: Gallivan *et al.*, 1986; G: Lenfant *et al.*, 1970; H: Thometz, 2014, Thometz *et al.* 2015).

Calculations and values as per cited references.

except the sea otter (*Enhydra lutris*), the lung O_2 stores comprise a smaller percentage of the total O_2 store than in humans. Diving birds, on the other hand, especially flighted diving birds, have about 50% of the total body O_2 store in the respiratory system. This is similar to that of non-divers such as the black-legged kittiwake and the dabbling mallard duck. In the deep-diving king and emperor penguins, about one-third of the total body O_2 store is still located in the airs sacs and lungs. In part, this difference in the distribution of O_2 stores between mammals and birds is due to the relative magnitudes of their respective respiratory air volumes (Lasiewski and Calder, 1971). Importantly, these differences in the magnitudes and distribution of O_2 stores between mammals and birds have implications for the cardiovascular responses and pulmonary gas exchange required during dives for the effective utilization of those O_2 stores (see Chapter 5).

Despite these differences in O_2 store distribution between diving mammals and birds, there are also similar trends among them. In penguins and all marine mammals except the manatee, muscle is a significant component of the total O_2 store in comparison to their non-diving counterparts. A large muscle O_2 store, and especially a high muscle O_2 content concentrated in the primary underwater locomotory muscles, will again have

Table 4.10 Mass specific total body oxygen stores and percentage distribution among lung, blood, and muscle O_2 stores in pinnipeds.

Species	Total O_2 store (ml O_2 kg^{-1})	Lung (%)	Blood (%)	Muscle (%)	Reference
California sea lion *Zalophus californianus*	55	13	39	48	A
Steller sea lion *Eumetopias jubata*	40	20	45	35	B
Australian sea lion *Neophoca cinerea*	56	10	70	20	C
New Zealand sea lion *Phocarctos hookeri*	48	12	63	25	D, E
Galapagos sea lion *Zalophus wollebaeki*	74	8	67	25	F
Northern fur seal *Callorhinus ursinus*	42	24	43	33	E, G
Antarctic fur seal *Arctocephalus gazella*	44	25	55	20	H
Walrus *Odobenus rosmarus*	38	24	50	26	G, I
Harbor seal *Phoca vitulina*	62	7	57	35	G, I, J
Crabeater seal *Lobodon carcinophagus*	43	12	67	21	E
Leopard seal *Hydrurga leptonyx*	51	7	46	47	K
Gray seal *Halichoerus grypus*	61	6	68	26	L
Ribbon seal *Histriophoca fasciata*	72	8	49	43	E, G
Harp seal *Phoca groenlandica*	73	7	49	44	E
Hooded seal *Cystophora cristata*	90	7	51	42	E
Ringed seal *Phoca hispida*	52	12	44	44	E, M
Baikal seal *Phoca sibirica*	79	5	66	29	N
Weddell seal *Leptonychotes weddellii*	89	4	66	30	O
Northern elephant seal *Mirounga angustirostris*	94	3	71	26	P

A: Weise and Costa, 2007; B: Richmond *et al.*, 2006; C: Fowler *et al.*, 2007; D: Costa *et al.*, 1998; E: Burns *et al.*, 2007; F: Villegas-Amtmann and Costa, 2010; G: Lenfant *et al.*, 1970; H: Costa *et al.*, 2001; I: Ponganis, 2011; J: Burns *et al.*, 2005; K: Kuhn *et al.*, 2006; L: Noren *et al.*, 2005; M: Lydersen *et al.*, 1992; N: Neshumova and Cherepanova, 1984, Neshumova *et al.*, 1983, Petrov and Shoshenko, 1987, Ponganis *et al.*, 1997b; O: Ponganis *et al.*, 1993a; P: Bryden, 1972, Simpson *et al.*, 1970, Thorson and Le Boeuf, 1994.

Calculations and values as per cited references.

Table 4.11 Mass specific total body oxygen stores and percentage distribution among respiratory, blood, and muscle O_2 compartments in aquatic and marine birds.

Species	Body mass	Total O_2 store	Respiratory	Blood	Muscle	Reference
	(g)	ml O_2 kg^{-1}	(%)	(%)	(%)	
Black-legged kittiwake *Rissa tridactyla*	359	50	56	42	2	A
Mallard duck (pekin duck) *Anas platyrhynchos*	1080	29	43	51	6	B
Tufted duck *Aythya fuligula*	674	42	48	48	4	B
Thick-billed murre *Uria lomvia*	1029	45	47	44	6	C
Ancient murrelet *Synthliboramphus antiquus*	197	60	50	44	6	A
Cassin's auklet *Ptychoramphus aleuticus*	171	61	50	44	6	A
Rhinoceros auklet *Cerorhinca monocerata*	560	55	49	40	11	D
Adélie penguin *Pygoscelis adeliae*	3800	63	48	26	26	E, F
King penguin *Aptenodytes patagonicus*	12 000	55	34	30	39	F
Emperor penguin *Aptenodytes forsteri*	25 000	68	33	31	36	F

A: Elliott *et al.*, 2010; B: Keijer and Butler, 1982; C: Croll *et al.*, 1992a; D: Yamamoto *et al.*, 2011; E: Chappell *et al.*, 1993; F: Ponganis *et al.*, 2011.

Calculations and values as per cited references. Typical body masses are included since many readers will probably not be familiar with the body sizes of these species.

implications for the nature of cardiovascular responses required for effective utilization of O_2 stores during dives (Chapter 5).

In marine mammals, there is decreased dependence on the respiratory O_2 store in deeper divers. This is especially evident in deep-diving mammals such as the sperm whale, elephant seal, hooded seal, and Weddell seal. Less reliance on the respiratory O_2 store decreases the need for pulmonary gas exchange at depth and should decrease the risks of excess blood N_2 uptake at depth, including the risk of decompression sickness. The same argument, although to a lesser extent, could be applied to the smaller mass-specific end-of-dive air volumes in king (*Aptenodytes patagonicus*) and emperor penguins (*A. forsteri*). However, start-of-dive air volumes in penguins are not documented, and recent air-sac volume measurements suggest that air-sac volumes may be much greater than expected in penguins (Ponganis *et al.*, 2015). If penguins do inhale to such large volumes, the total body O_2 store and respiratory O_2 fraction would increase greatly. Further assessments of start-of-dive air volumes in penguins are needed.

As a percentage of the total body O_2 store in diving mammals, the blood and muscle O_2 stores range from 54% in the sea otter to 97% in the northern elephant seal (see Tables 4.9, 4.10). The relative distribution of O_2 between these two stores is affected by more than just the Hb and Mb concentrations. This is exemplified in three deep-diving phocid seals with total body O_2 stores near 90 ml O_2 kg^{-1} and more than 93% of that O_2 in blood and muscle. Hb concentrations and blood volumes are especially elevated in the elephant seal and Weddell seals (Table 4.2), while Mb content of hooded seal muscle is extremely elevated (Table 4.7). Such differences may be significant from a physiological perspective. Provided the metabolic cost to muscle during dives is similar among the three species, the need for delivery of O_2 from blood to muscle to maintain aerobic metabolism during long dives would appear to be less in the hooded seal than in the Weddell seal and elephant seal. This, in turn, may be reflected in the nature of cardiovascular responses during the dives of these animals. Following the same argument but in the opposite direction, the very small muscle O_2 store and low Mb concentration in manatees suggest that maintenance of muscle blood flow and blood-to-muscle O_2 transfer is necessary in order to maintain aerobic muscle metabolism during their dives.

At least two features of the distribution of O_2 stores in diving birds are particularly notable. The first is the large respiratory O_2 store, especially in flighted diving birds. The large respiratory O_2 store is most likely a consequence of avian respiratory anatomy and ventilation requirements during flight. The structure of the air-sac–lung system, its volume, and a high respiratory rate all contribute to ventilation during flight (Butler, 1991, Lasiewski and Calder, 1971, Maina and King, 1987). Because of this large respiratory store and also the small muscle O_2 store, both pulmonary gas exchange and blood-to-muscle O_2 transfer might be expected to occur during diving in flighted birds in order to maintain aerobic metabolism. This may account for the minimal elevations in Mb concentrations in flighted diving birds.

Second, it should also be noted that the relatively reduced size of the respiratory O_2 store in king and emperor penguins is not due to a decrease in their respiratory air volumes. The estimated volume of the respiratory system (diving air volume) in diving king and emperor penguins is similar to that predicted by avian allometric equations (Sato *et al.*, 2002, 2011). In these birds, the relative magnitude of the respiratory O_2 stores is reduced primarily because of an increased muscle O_2 store. The latter is secondary to both a much higher Mb concentration and a larger muscle mass in the penguins as compared to flighted diving birds. It should also be noted that the size of the respiratory O_2 store may also vary with depth of dive as estimated diving air volume increases with maximum depth of dive in penguins (Sato *et al.*, 2002, 2006, 2011). An increase in air volume with depth has also been suggested in deep-diving shags (Cook *et al.*, 2008, 2010).

4.6 Advantage of size in the rate of O_2 store utilization

The maximum four- to five-fold difference in mass-specific total body O_2 stores between humans and the best-diving mammals cannot account for the differences in their respective breath-hold capacities. The same argument applies to the minimal increases in mass-specific O_2 storage in diving birds. Scholander emphasized this in his 1940 monograph. At least two additional factors contribute to that breath-hold capacity. The first is the advantage of increased body size (Noren and Williams, 2000, Noren *et al.*, 2012b). The second is the regulation of the depletion rate of those O_2 stores through changes in heart rate and organ perfusion. The latter is the subject of the next chapter. Increased body size, as noted by Krogh in 1934, confers an advantage in divers through the well-known relationship of metabolic rate to body mass (Kleiber, 1975, Krogh, 1934). Furthermore, not only is the resting or basal mass-specific metabolic rate lower in larger animals, but the metabolic cost of locomotion is also less in larger animals (Heglund *et al.*, 1982a, 1982b , Taylor *et al.*, 1982). While the former is at least partially a function of surface area–volume relationships, the latter is considered dependent on the velocity of shortening of active muscle units and the rate at which those units are activated (both higher in smaller animals). Thus, a given-sized O_2 store will last longer in the larger animal both at rest and during locomotion. Conversely, high metabolic rates secondary to small body size may also account for the high mass specific O_2 stores of the smallest marine mammal, the sea otter.

5 Cardiovascular dive response

The physiological hallmark of diving in marine mammals and birds is a decrease in heart rate relative to the pre-dive or surface heart rate (Irving *et al.*, 1941b, Scholander, 1963). Regulation of heart rate, cardiac output, and the degree of peripheral vasoconstriction during dives is essential to the management and utilization of body O_2 stores because (a) the magnitude and distribution of cardiac output to peripheral tissues contributes to rates of tissue O_2 delivery and tissue O_2 consumption; and (b) cardiac output contributes directly to the rate of blood O_2 uptake from the lungs (Hogan *et al.*, 1993, Kvietys and Granger, 1982, Lutz *et al.*, 1975, Taylor *et al.*, 1987, Valtin, 1973). Although these responses were investigated as early as 1870 by Bert (see Irving *et al.*, 1941b for review) and also examined under conditions of asphyxia (Irving, 1934, 1938, 1939, Irving *et al.*, 1935a), the slowing of heart rate to below resting levels (bradycardia) and the constriction of peripheral blood vessels (vasoconstriction) of seals and penguins during forced submersions were first thoroughly documented in Scholander's 1940 monograph.

During Scholander's forced submersion experiments, heart rates of seals were as low as 10 beats min^{-1} (bpm). Vasoconstriction and the circulatory isolation of muscle were convincingly demonstrated by (a) the depletion of muscle O_2 with a concomitant increase in muscle lactate concentration during the submersion, and (b) the subsequent wash-out of lactate into blood during the post-submersion period (Fig. 5.1). This dive reflex (severe bradycardia in combination with peripheral vasoconstriction) isolated peripheral organs and tissues from the circulation, decreased the rate of blood O_2 depletion (Fig. 5.2), and conserved that blood O_2 for the heart and brain, thus prolonging the duration of the breath hold (Irving *et al.*, 1941b, Scholander, 1940). It was not until the 1960s that Elsner found that the reductions in heart rates of seals and sea lions during trained submersions were not as severe as during forced submersions (Elsner, 1965, Elsner *et al.*, 1964a). Since that time, diving physiologists have continued to investigate the nature, plasticity, and consequences of dive responses in multiple species under different conditions.

Chapter 5 is the longest and probably most detailed section of this book because the dive response represents the core of diving physiology. As such, it is essential that students and researchers understand the physiological mechanisms and implications of the dive response. The chapter will lay the ground work for understanding the role of heart rate and vasoconstriction in (a) the conservation and preservation of O_2 for the heart and brain; (b) the regulation of metabolic rate and the depletion of O_2 stores,

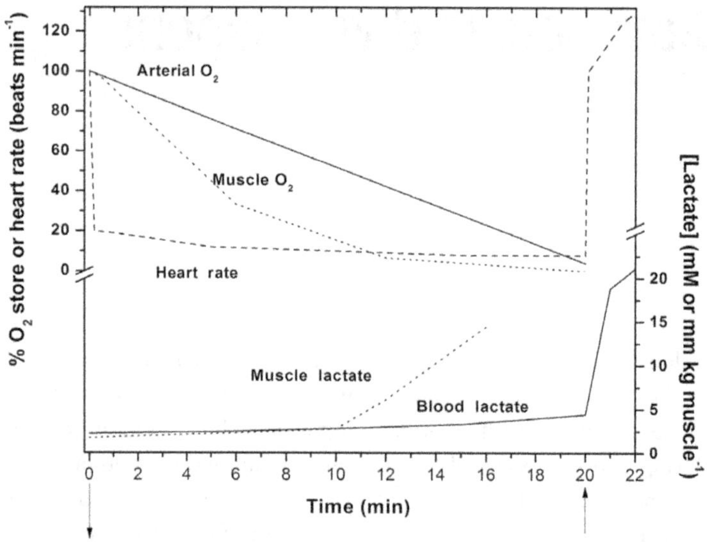

Figure 5.1 Heart rate and blood/muscle oxygen and lactate profiles during forced submersions of seals. Note the depletion of muscle oxygen and isolated intramuscular accumulation of lactate during the severe bradycardia of the submersion with subsequent wash-out of lactate into the blood during the increase in heart rate after the submersion. These profiles provided the evidence of widespread vasoconstriction during the bradycardia of forced submersion. Arrows indicate start and end of submersion. Adapted from Irving et al., (1941b), Scholander (1940, 1963), Scholander et al. (1942a).

especially the muscle O_2 reservoir (Chapters 9 and 10); (c) determination of the duration of aerobic metabolism and aerobic dive limits (Chapter 10); (d) thermoregulation (Chapter 8); and (e) nitrogen uptake/distribution and the risk of decompression sickness (Chapter 12).

In order to review the extensive literature on this subject, mammalian and avian diving cardiovascular physiology are each considered separately in this chapter. The last section of the chapter will outline the neuroregulatory control mechanisms for the dive response in both mammals and birds. Other topics in the cardiovascular physiology of these animals, including unique anatomical adaptations and the optimization of myocardial oxygen supply and demand, will be reviewed in Chapter 6.

In this chapter, there are 25 subsections for diving mammals and 12 subsections for diving birds. Each subsection covers a distinct topic and is meant to provide the reader with the research findings in that specific area. Topics include heart rate, blood flow, and organ perfusion patterns during (a) forced submersions, (b) simulated dives, (c) trained submersions, (d) spontaneous breath holds and sleep apnea (breath holding during sleep), (e) surface swimming, and (f) free dives. The variability and intensity of heart rate and peripheral vasoconstriction will be examined in different species and under different circumstances. The physiological and anatomical mechanisms underlying the peripheral vascular response (vasoconstriction) in these animals will also be reviewed. Particular attention will be focused on how heart rate and peripheral blood flow are

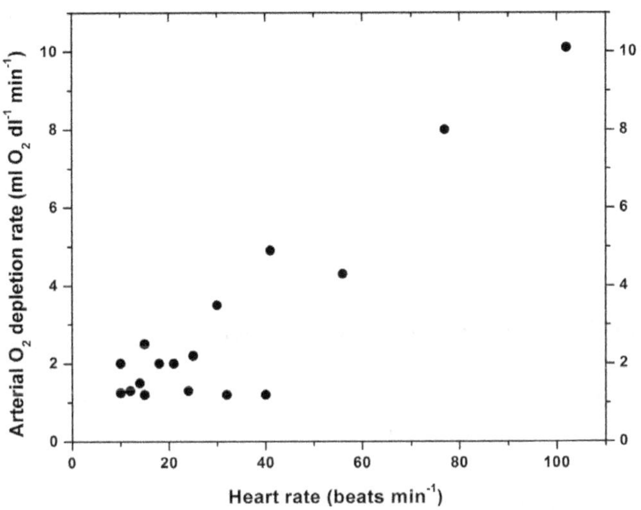

Figure 5.2 Heart rate and the rate of blood oxygen depletion during forced submersion. This graph demonstrates the dependence of the blood oxygen depletion rate on heart rate and represents the end result of the "dive response." Adapted from Irving et al. (1941b).

regulated during different underwater activities of these animals and how the "dive response" might vary as a function of exercise and other parameters during the dive.

5.1 Cardiovascular physiology in marine mammals

5.1.1 Blood flow distribution in terrestrial mammals at rest and during exercise

As a reference for comparison to blood flow distribution during diving, the distribution of cardiac output to peripheral tissues for humans at rest (Williams and Leggett, 1989) can be summarized as follows: brain 12%, heart 4%, kidneys 19%, liver 7% hepatic artery and 15% portal vein (via gastrointestinal tract), gastrointestinal tract 15%, muscle 17%, skin/skeleton 10%, fat 5%, and other 11%. Values vary in different studies as well as different species (Adachi *et al.*, 1976, Behrman and Lees, 1971, Hohimer *et al.*, 1983, Musch *et al.*, 1987a, 1987b, Upton, 2008), but this summary provides a convenient starting point in consideration of the redistribution of blood flow during both diving and exercise.

As will be seen in this chapter, a key question raised by more recent cardiovascular studies of diving animals is the degree of peripheral blood flow reduction and tissue ischemia (low blood flow) during moderate as opposed to severe reductions in heart rate. In particular, regulation of muscle blood flow in an actively swimming, diving animal is especially important in regard to the management of blood and muscle O_2 stores. Do heart rate and muscle blood flow correlate with the stroke rate of propulsive muscle so as to supply blood O_2 to muscle during diving? Is muscle blood flow tailored to muscle O_2 consumption to maximize the duration of aerobic metabolism and allow simultaneous depletion of the blood and muscle O_2 stores (Davis and Kanatous, 1999)? Could there be low-grade or intermittent muscle perfusion independent of muscle workload? Or, is there no muscle blood flow during active diving, and is muscle solely dependent on its myoglobin-bound O_2 store for aerobic metabolism?

An important point to note at this time is that the classic dive response during forced submersion differs radically from the classic exercise response. In contrast to the description at the start of this chapter for the dive response, elevated heart rates during exercise of birds and terrestrial mammals increase cardiac output, and deliver the major portion of that blood flow to exercising muscle (Taylor *et al.*, 1987). The magnitudes of blood flow to non-exercising muscle, splanchnic organs (gastrointestinal tract, liver), and the kidneys are either unchanged or decreased dependent on the species and severity of exercise (Armstrong and Laughlin, 1984, Armstrong *et al.*, 1987, 1992, Butler, 1991, Butler *et al.*, 1988, Hohimer *et al.*, 1983, Laughlin *et al.*, 2010, Musch *et al.*, 1987a, Perko *et al.*, 1998, Sanders *et al.*, 1976). Significant species differences do exist. For example, in contrast to most mammals, splanchnic and renal blood flow during even the heaviest workloads are maintained in Alaska sled dogs, animals which have been selected and trained for long-duration, continuous aerobic exercise (Van Citters and Franklin, 1969). Similarly, there may well be differences in diving heart rate and blood flow distribution patterns in diving animals with different activity patterns (i.e., "divers" versus "surfacers" – see Chapter 1 for this classification).

The physiological processes that underlie the exercise response are also relevant to the mechanisms controlling the dive response. During exercise, activation of the sympathetic nervous system with subsequent vasoconstriction of arterial blood vessels is thought to limit and/or decrease flows to visceral organs and non-exercising muscle, despite the increase in heart rate. In contrast, the increase in blood flow to working muscle occurs despite an increased peripheral sympathetic response. The increased blood flow to working muscle is considered to be mediated by several possible factors, including (a) myogenic relaxation of smooth muscle in blood vessels compressed by contracting muscles, (b) release of endothelial vasodilators in the vasculature of active muscles (nitric oxide), (c) release of "metabolic vasodilators" (adenosine and adenine nucleotides, prostaglandins, kinins, potassium), and (d) a muscular pump mechanism due to contraction of active muscle (expulsion of venous blood and decreased venous pressures, and increased kinetic energy in blood in the vascular bed imparted from contracting muscle) (Laughlin *et al.*, 2010, Saltin, 2007). In addition, hypoxia augments vasodilation even further (Casey *et al.*, 2010, Casey and Joyner, 2011, Park *et al.*, 1992, Skinner and Marshall, 1996). Whether and why blood flow to exercising muscle may or may not occur during dives of diving mammals (and birds) will be the subject of review later in this chapter.

5.1.2 Forced submersions of seals: bradycardia and vasoconstriction

The initial investigations into diving physiology utilized forced submersion, a technique in which the animal is restrained and submerged involuntarily for a period of time that is unknown to the subject. This approach in Scholander's experiments yielded a maximum dive response as described previously in this chapter. The extreme slowing of heart rate to far below resting levels (bradycardia) and the intense constriction of peripheral arteries (vasoconstriction) slowed the rate of blood O_2 depletion and redistributed blood flow to essentially the heart and brain, two of the most O_2-dependent organs in the body. The intensity of vasoconstriction during the submersion was demonstrated by a lack of bleeding in muscle biopsies, by the inability to draw blood samples through peripheral arteries, and, most significantly, by the concentration profiles of muscle O_2 content/lactate concentration during and after the submersion. As demonstrated in Fig. 5.1, during the submersion muscle O_2 declined more rapidly than blood O_2, resulting in eventual muscle lactate accumulation while at the same time blood O_2 was still preserved and there was only minor elevation in blood lactate concentration (Scholander, 1940, Scholander *et al.*, 1942a). After the submersion, when heart rate increased and blood flow returned to muscle, the wash-out of lactate into the blood confirmed that muscle had been isolated from the circulation during the submersion. Use of a hot wire anemometer demonstrated that muscle blood flow paralleled changes in heart rate during submersions and even during respiratory sinus arrhythmias (spontaneous oscillations in heart rate during the breathing cycle) (Grinnell *et al.*, 1942). Similar muscle blood flow patterns were also later confirmed with tissue laser Doppler flowmetry in submerged harbor seals (*Phoca vitulina*) (Jobsis *et al.*, 2001).

In addition to muscle ischemia during forced submersions, widespread vasoconstriction to other organs was consistent with (a) the maintenance of blood pressure in central arteries, (b) observations of decreased mesenteric blood flow, (c) decreased urine production, and (d) decreased renal clearances of urea, creatine, and p-aminohippurate (Irving *et al.*, 1935a, 1942b, Murdaugh *et al.*, 1961, Schmidt-Nielsen *et al.*, 1959, Scholander, 1940). Because blood pressure is the product of cardiac output (flow) and peripheral vascular resistance (arterial constriction), a constant blood pressure during forced submersions, when heart rate was reduced at least six-fold below resting levels, would suggest that peripheral vasoconstriction had increased resistance by a similar factor.

5.1.3 Forced submersions: angiography and Doppler flow probe measurements

In the 1960s, both angiography and the use of perivascular flow probes demonstrated the dramatic reductions in blood flow in the aorta and peripheral arteries during forced submersions of seals (Bron *et al.*, 1966, Elsner, 1969, Elsner and Gooden, 1983, Elsner *et al.*, 1966a, 1970a, 1985, Van Citters et al., 1965). Decreases in both heart rate and peripheral blood flow were usually nearly instantaneous with the start of submersions in smaller seals such as harbor seals (*Phoca vitulina*). In elephant seals (*Mirounga angustirostris*), now known to be the premier pinniped divers, the decrease in heart rate could be more variable (Elsner *et al.*, 1966a, Van Citters *et al.*, 1965). Perhaps, this may be related to their greater dive capacity or to their dive behavior (i.e., a surfacer versus a diver, see Chapter 1). For example, in one elephant seal heart rate eventually reached 4 bpm late in a 40-min submersion, but in the first minute it only slowly declined about 50% from eupneic levels to 37 bpm (Van Citters *et al.*, 1965). By 5 min, heart rate was still about 22 bpm, and it only reached 10 bpm at 20–25 minutes.

5.1.4 Forced submersions: microsphere studies

The actual decreases in tissue perfusion during the bradycardia and vasoconstriction of forced submersions were documented with radio-labeled microsphere studies in the late 1970s (Blix *et al.*, 1983, Zapol *et al.*, 1979). During these severe bradycardias, 80–100% reductions in tissue blood flow were verified in all tissues except the brain, lung, and adrenal glands. Overall, brain blood flow was preserved during these submersions. Sequential studies revealed that transient reductions in cerebral cortical and cerebellar blood flow occurred early in the submersion, but that, later in the submersion period, flow in these regions and the hypothalamus were increased above pre-submersion levels. Adrenal blood flow was preserved during forced submersions at 40–60% of pre-submersion levels. Maintenance of adrenal perfusion was consistent with the observed elevations in plasma cortisol and catecholamines during the submersion period (Cherepanova *et al.*, 1993, Hance *et al.*, 1982, Liggins *et al.*, 1979, 1993).

5.1.5 Forced submersions: arterio-venous shunts

Lastly, increased microsphere deposition in the lungs, especially early during the submersion, suggested to both teams of investigators that blood circulated through arterio-venous (A-V) shunts early during submersion (Blix *et al.*, 1983, Zapol *et al.*, 1979). The spleen might act as such a conduit, but A-V anastomoses also exist in the skin of seals (Bryden and Molyneux, 1978, Molyneux and Bryden, 1975, 1978). Notably, maintenance of blood flow in the peripheral toe arteries of the hind flipper had been occasionally observed in forcibly submerged seals (Irving *et al.*, 1942b). However, A-V anastomoses have extensive sympathetic nerve innervation and, at least in the tail fluke of the bowhead whale (*Balaena mysticetus*), these anastomoses are especially sensitive to norepinephrine levels (Elsner *et al.*, 2004a). So, it would be expected that these potential shunts would usually constrict during a forced submersion.

On the other hand, flow may continue through A-V shunts in the flippers of seals even during forced submersions (Blix *et al.*, 2010). The foreflipper, for example, has large superficial veins not associated with the deeper countercurrent heat exchanger (Blix *et al.*, 2010). Flipper temperatures were also maintained during forced submersions despite declines in skin temperatures of the back (Hammel *et al.*, 1977). On this basis, Blix and co-workers have hypothesized that such shunting through A-V anastomoses in the flippers could contribute to heat loss and body temperature declines observed during forced submersions (Blix *et al.*, 2010, Hammel *et al.*, 1977, Hol *et al.*, 1975, Odden *et al.*, 1999, Scholander *et al.*, 1942b).

Interestingly, the early opening of an A-V shunt during a submersion could also account for the observations by Van Citters *et al.* (1965) of a large reduction in iliac artery blood flow in the presence of a constant heart rate and blood pressure at the start of forced submersions of elephant seals. For example, iliac artery blood flow could decrease by 50% at the onset of a submersion even before significant changes occurred in heart rate (Van Citters *et al.*, 1965). In one instance, iliac artery flow decreased by 1200 ml min^{-1}, a 75% reduction from eupneic levels, despite no changes in heart rate or blood pressure. Such changes were consistent with an intense vasoconstriction response in the iliac artery. However, remember that blood pressure is the product of cardiac output and peripheral vascular resistance, and that cardiac output is the product of heart rate and stroke volume. Arterial vasoconstriction and a reduction in iliac blood flow of 1200 ml min^{-1} in the presence of a constant blood pressure and heart rate would argue that either stroke volume decreased at the start of the submersion and/or that vasoconstriction to other tissues decreased. Otherwise, blood pressure should increase. In this regard, it is notable that during breath holds of human divers, skin sympathetic nerve activity decreased while that in muscle increased, consistent with vasodilation to skin and vasoconstriction to muscle (Fagius and Sundolf, 1986). Thus, those reductions in iliac artery blood flow in the presence of a constant heart rate and blood pressure are consistent with the use of A-V shunts during forced submersion. These findings also suggest that the magnitude of vasoconstriction to a particular tissue and the subsequent reduction in that tissue's blood flow may not always be quantitatively proportional to the change in the heart rate profile. Such a lack of correlation between heart rate and a given tissue's blood flow during a dive has potential limitations on assumptions used in the numerical modeling of tissue blood flow distribution and oxygen consumption during diving.

5.1.6 Forced submersion: the physiology and anatomy of peripheral vasoconstriction

In regard to the widespread vasoconstriction that occurred during the forced submersion of seals, it is notable that local vasodilation and muscle blood flow did not occur in ischemic, hypoxic muscle that had elevated tissue lactate concentrations during the latter segments of forced submersions. The release of local vasodilators in hypoxic or exercising muscle in most mammals usually represents a mechanism to overcome increased sympathetic vasoconstriction at the arteriolar level during exercise and to increase blood flow selectively to working muscle (Casey and Joyner, 2011, Joyner and Wilkins, 2007, Laughlin and Armstrong, 1987, Laughlin *et al.*, 2010, Saltin, 2007).

The maintenance of intense vasoconstriction in the presence of oxygen depletion and lactate accumulation in muscle during forced submersions is considered secondary to the dense distribution of sympathetic nerve fibers penetrating into the walls of major supply arteries that are far from the capillary beds of skeletal muscle and other major organs in seals (White *et al.*, 1973). It is the differential distribution of arterial sympathetic nerve innervation that most probably contributes to the pattern of redistribution of blood flow during the submersion. Dense sympathetic innervation occurs in the arteries supplying spinal/hindlimb muscles, splanchnic organs, and the kidneys of seals. Activation of sympathetic nerve fibers in the proximal segments of these arteries would constrict these vessels independent of the accumulation of local vasodilators in distal tissue, i.e., the extramuscular throttle of Gooden and Elsner (Cherepanova *et al.*, 1993, Gooden and Elsner, 1985). In addition, such constriction in the proximal arteries may also be enhanced because of the elevated circulating catecholamine levels during the submersion and the presence of adrenergic receptors in those arterial segments. In contrast to the arteries supplying the kidneys, gastrointestinal tract, and hind flipper muscles, such innervation patterns of sympathetic fibers were not observed in the aorta, carotid arteries, coronary arteries, or pulmonary arteries. The lack of such fibers would allow for the maintenance of blood flow to the brain and heart despite intense activation of the sympathetic nervous system.

Differences in sympathetic innervation of major supply arteries to various muscle groups may also account for the differential muscle blood flow patterns observed in Baikal seals (*Phoca sibirica*) during forced submersions (Cherepanova *et al.*, 1993, Matyukhin *et al.*, 1988, Neshumova and Cherepanova, 1984). During these submersions, the decline in muscle blood flow was most intense to muscles with the highest myoglobin concentrations, namely the muscles of the back and hind limb. The decline in blood flow to muscles of the pelvis and forelimb was intermediate, while there was no change in blood flow to muscles of the neck. Further investigation of the physiological basis of this differential muscle blood flow pattern during forced submersion has not been conducted. Perhaps there are differences in the densities of sympathetic nerve fibers to arteries supplying different muscle groups. It is also noteworthy that muscle "electrical" activity during swimming was greatest in the muscles with the highest myoglobin concentrations (Neshumova *et al.*, 1986). Thus, the most active swimming muscles, those with the highest myoglobin concentrations, have the most intense arterial vasoconstrictor responses during forced submersions. Although hidden away in the Russian literature, these observations have important implications not only for understanding regulatory mechanisms, but also for the better understanding and modeling of oxygen store utilization during diving (Chapters 9 and 10).

During dives or forced submersions, sympathetic vasoconstriction presumably overrides effects of any potential vasodilators, such as nitric oxide or carbon monoxide, that may be circulating in the blood. However, as discussed in Chapters 11 and 13, such compounds may contribute to maintenance of blood flow in non-constricted or partially constricted blood vessels, such as those to the brain, lung, heart, and, in pregnant animals, the placenta.

5.1.7 Forced submersion: the physiology and anatomy of pulsatile myocardial blood flow

In contrast to the complete ischemia of skeletal muscle, it is notable that pulsatile flow patterns have been observed in seal myocardial blood flow during forced submersions (Elsner *et al.*, 1985). Such pulsations in coronary flow may be due to the lack of sympathetic innervation in the proximal coronary artery and the competing effects of distal coronary arteriolar constriction and transient, localized, hypoxia-linked release of vasodilators at the arteriolar level. In addition, distal coronary arteriolar constriction in the seal appears to be mediated by cholinergic fibers and acetylcholine release, and not by the sympathetic nervous system (Elsner and de la Lande, 1998). Such cholinergic activity mediated by the parasympathetic system via the vagus nerve provides a mechanism to match myocardial oxygen supply (blood flow) and demand (cardiac work) during the vagally induced bradycardia of diving (Elsner and de la Lande, 1998).

5.1.8 Forced submersion of other mammals

In addition to the studies above on seals, this "dive reflex" has also been confirmed in forced submersion studies of other diving mammals, including manatees, fur seals, muskrats, beavers, mink, nutria, and even neonatal and fetal seals (Elsner *et al.*, 1970a, Ferrante, 1970, Ferrante and Opdyke, 1969, Folkow *et al.*, 1971, Hammond *et al.*, 1969, Irving *et al.*, 1963, Liggins *et al.*, 1980, McKean, 1982, Neshumova and Cherepanova, 1984, Scholander, 1940, Scholander and Irving, 1941, West and Van Vliet, 1986). As already reviewed in Chapter 2, this pattern of bradycardia and vasoconstriction is a general response to asphyxia; it occurs to some degree not only in humans and non-diving mammals such as armadillos and sloths, but also in a wide range of other vertebrates, including reptiles, amphibians, and even fish out of water (Garey, 1962, Irving *et al.*, 1942a, Johansen, 1959, Murdaugh and Jackson, 1962, Scholander, 1963, Scholander and Irving, 1941, Shelton and Jones, 1965, White and Ross, 1966, Wilber, 1960).

5.1.9 Simulated dives of pinnipeds

Simulated dives differ from forced submersions in that the animal is also exposed to increased ambient pressure. Such studies in pinnipeds have been conducted in pressure chambers designed by G. L. Kooyman in the 1960s (Kooyman and Sinnett, 1982, Kooyman *et al.*, 1976a, 1970, 1973b, Sinnett *et al.*, 1978). The goals of these experiments have primarily focused on the respiratory system and blood gas analyses. Some heart rate profiles were available, however, and the data demonstrated that the heart rate response during simulated dives was similar to that during forced submersions. Heart rates during simulated dives of Weddell seals, northern elephant seals, harbor seals, and sea lions were reported in the range of 5–20 bpm, similar to those in forced submersions of pinnipeds (Kooyman and Sinnett, 1982, Sinnett *et al.*, 1978).

5.1.10 Trained submersions of pinnipeds: moderate bradycardia

Trained submersions differ from forced submersions in that the animal is trained or conditioned to hold its breath for fixed durations. Under such conditions, the intensity of the bradycardia is usually much less, i.e., the decline in heart rate is not as great. This was demonstrated in sea lions and seals trained to immerse their heads in water (Elsner, 1965, Elsner *et al.*, 1964a), and also in sea lions trained to hold their breath in air (Ridgway *et al.*, 1975a). As will be reviewed below and later in the chapter, trained submersions as well as spontaneous breath holds are invaluable in that they provide the opportunity to examine these physiological parameters with sophisticated techniques in a non-moving and often dry animal. Elsner and co-workers, for example, were able to demonstrate that stroke volume, the volume of blood ejected from the heart per heartbeat, was constant between apnea (breath-hold period) and eupnea (breathing period) in the sea lion (Elsner *et al.*, 1964a). In 1977, Dormer and colleagues confirmed that cerebral blood flow actually increased during the bradycardia of trained head immersions of sea lions (Dormer *et al.*, 1977). As will be reviewed in Chapter 6, trained submersions of young elephant seals in a magnetic resonance imaging (MRI) scanner allowed assessments of cardiac output, stroke volume, aortic flow, and contraction/relaxation of the spleen (Thornton *et al.*, 1997, 2001, 2005).

5.1.11 Trained submersion: muscle blood flow during moderate bradycardias

The observation of higher heart rates during a trained breath hold raised the question as to the degree of vasoconstriction under such conditions, and, importantly, the isolation of muscle from the circulation and from the blood O_2 store during the breath hold. This question became even more relevant when heart rates during free dives and spontaneous breath holds of harbor seals were found to be quite variable and again often not as low as during forced submersions (Jones *et al.*, 1973).

In actuality, evidence for less intense vasoconstriction to muscle during higher heart rates was noted as early as the 1940s in a study on muscle blood flow during forced submersions (Grinnell *et al.*, 1942). Notably, the authors reported that the seal exhibited an anticipatory increase in both heart rate and muscle blood flow when the experimenter's hand was raised at the end of the submersion. During the course of the study, the seal appeared to have become conditioned to the raising of the experimenter's hand prior to elevation of its head above water at the end of the submersion. It was demonstrated that this increase in heart rate and muscle blood flow even occurred when the hand was raised in the middle of a submersion; however, both heart rate and muscle blood flow declined quickly when the breath hold was not ended.

Further investigation of the relation of heart rate, muscle blood flow, and muscle oxygenation was conducted in the 1990s during naïve and trained three-minute submersions of harbor seals (Jobsis *et al.*, 2001). In comparison of trained vs. naïve submersions (Fig. 5.3), submersion heart rate almost doubled to 35 bpm, muscle blood flow as measured by laser Doppler flowmetry increased three-fold, the deoxygenation rate of muscle, as determined by near-infrared spectroscopy, slowed by about 25%, and the rate of decline in extradural vein P_{O2} increased. Although the seals were still bradycardic during the trained submersions, the relative increase in muscle blood flow presumably allowed for blood-to-muscle O_2 transfer, accounting for both the decrease in the rate of muscle deoxygenation and the increase in the rate of venous P_{O2} decline during the submersion. In contrast to a forced submersion, muscle was not completely isolated from the circulation in the trained submersions of this study. However, despite supplementation from the blood O_2 store, the muscle O_2 store still declined, albeit at a slower rate. The frequency and magnitude of such blood O_2 supplementation of muscle metabolism during free diving remain key questions in understanding the management of O_2 stores during dives.

As in Grinnell's study almost 60 years earlier, the conscious, exquisite control of heart rate and peripheral vasoconstriction was also demonstrated in the Jobsis study when the last trained submersion of the study was not ended at the usual three minutes (Fig. 5.4). Without any cues from the investigators, the seal immediately lowered its

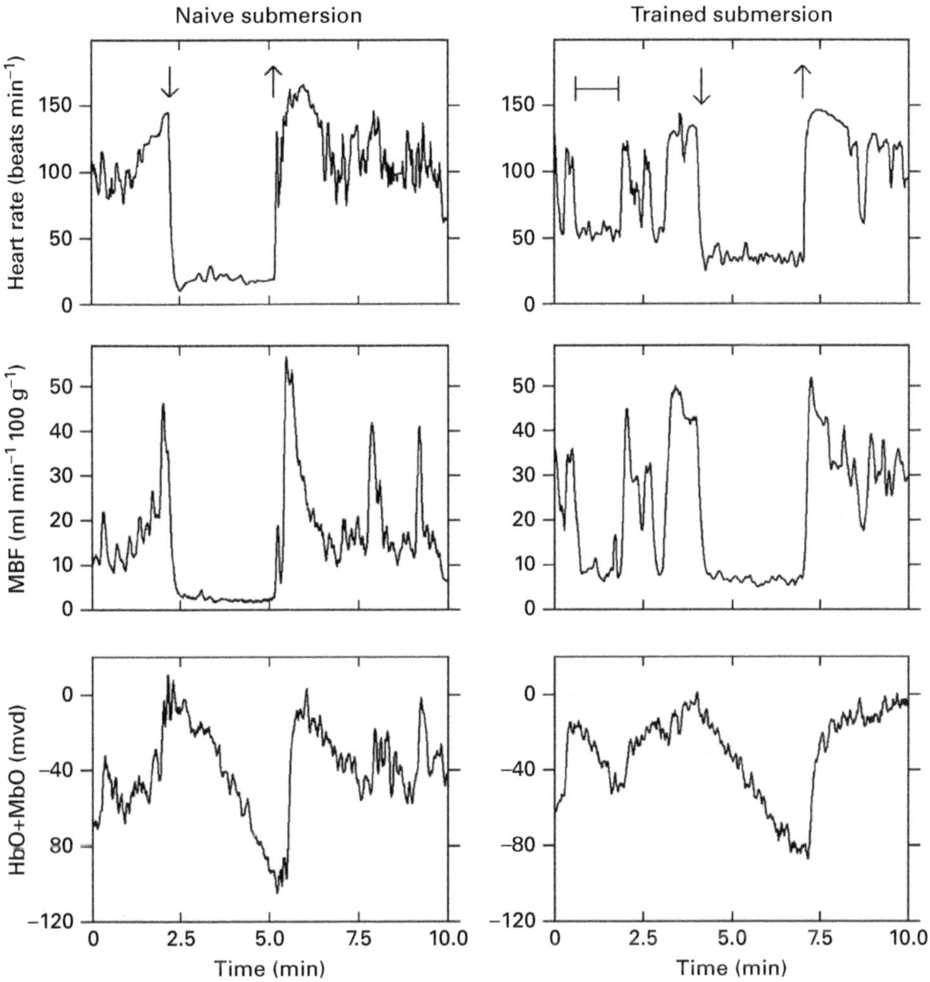

Figure 5.3 Heart rate, muscle blood flow, and muscle oxygenation during naïve and trained three-minute submersions of a harbor seal (*Phoca vitulina*). Heart rate and muscle blood flow are greater and muscle desaturation rate less during the trained submersions. Arrows indicate start and end of submersion. Bracket indicates spontaneous breath hold. Muscle oxygenation (near-infrared spectroscopy) is the saturated myoglobin and hemoglobin ($MbO_2 + HbO_2$) signal in vander units. Reproduced with permission from the *Journal of Experimental Biology* (Jobsis et al., 2001).

heart rate and muscle blood flow to naïve levels at the three-minute mark, and maintained those low levels until the end of the submersion at five minutes.

5.1.12 Spontaneous breath holds and sleep apnea

Marine mammals breathe intermittently, and it had long been known that less intense declines in heart rate occurred during short, spontaneous breath holds in seals, dolphins,

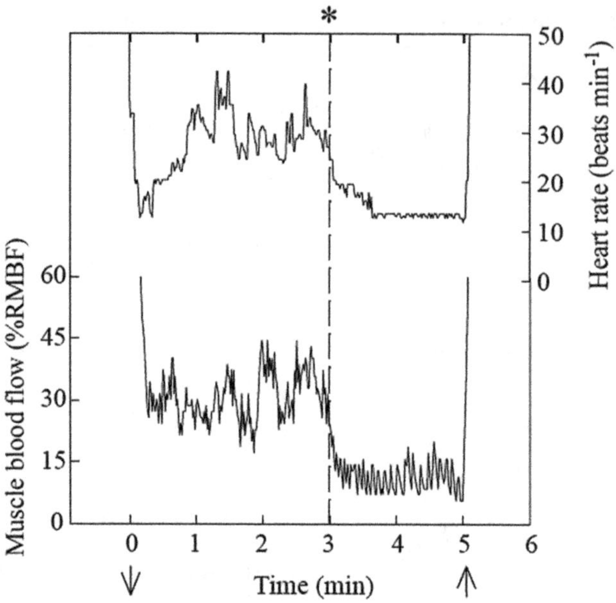

Figure 5.4 Regulation of heart rate and muscle blood flow in a young harbor seal (*Phoca vitulina*) during a trained submersion when submersion was not ended at the usual three minutes (*). Immediate declines in heart rate and muscle blood flow occurred at three minutes, and were maintained at forced submersion levels until the end of the submersion. Reproduced with permission from the *Journal of Experimental Biology* (Jobsis et al., 2001).

and manatees (Grinnell *et al.*, 1942, Irving *et al.*, 1941a, Scholander and Irving, 1941). Muscle blood flow also decreased slightly as heart rate declined during the respiratory arrhythmia of the seal (Grinnell *et al.*, 1942). Such heart rate patterns during the breathing cycle are classified as respiratory sinus arrhythmias, and are associated with higher heart rates during inspiration and slower heart rates during exhalation or during the breath hold. Sinus arrhythmias were reviewed in diving mammals and birds as early as 1939 (Irving, 1939). Similar respiratory variations in heart rate also occurred in cormorants, penguins, porpoises, and whales (Enstipp *et al.*, 1999, Halsey *et al.*, 2008, Meir *et al.*, 2008, Ponganis and Kooyman, 1999, Reed *et al.*, 2000).

The sleep apneas of phocid seals (long, spontaneous breath holds during sleep) are probably the best-known and most-studied spontaneous, non-diving breath holds in marine mammals (Andrews *et al.*, 1997, Bartholomew, 1954, Blackwell and Le Boeuf, 1993, Castellini, 1994, Castellini *et al.*, 1986, 1987, 1994a, 1994b, Falabella *et al.*, 1999, Kenny, 1979, Kooyman and Campbell, 1972, Ponganis *et al.*, 2006a, 2006b, 2008, Ridgway *et al.*, 1975b, Williams and Bryden, 1993). In sleeping sub-adult and adult northern elephant seals, mean apneic durations ranged from 7.4 to 9.7, minutes with an observed maximum duration of 23.1 min (Blackwell and Le Boeuf, 1993). Breathing only occurred during slow-wave sleep (Castellini *et al.*, 1994a).

In contrast to humans (Dempsey *et al.*, 2010), this was not a pathological condition. Such long, spontaneous breath holds in a sleeping, non-mobile, and approachable animal provided the opportunity to examine apneic cardiovascular responses in great detail.

In young elephant seals, heart rate during sleep apnea, although variable, was near 40–50 bpm, and was essentially an extension of the baseline (minimum) heart rate of the seal's sinus arrhythmia during respiration (Castellini *et al.*, 1994a; see Fig. 5.5a). Overall, mean heart rate, cardiac output, and muscle blood flow during sleep apnea were about 50% of the mean eupneic values; stroke volume was not significantly different between apnea and eupnea (Ponganis *et al.*, 2006b, 2008,). Notably, apneic cardiac outputs, although less than eupneic values, were typical of similarly sized terrestrial mammals at rest. With maintenance of a "normal" cardiac output and some muscle blood flow during sleep apnea, most organs were probably well-perfused and muscle was not isolated from the blood O_2 store during the breath hold. This maintenance of tissue perfusion is in stark contrast to the situation during forced submersions, and is more similar to that just reviewed for trained submersions. On the basis of heart rate and blood flow patterns, one would predict that blood O_2 depletion should be faster and muscle O_2 depletion slower during sleep apnea than during forced submersion. And, indeed, that is what was found (Ponganis *et al.*, 2008, Stockard *et al.*, 2007) (see Chapter 11).

Captive, wild Baikal seals in an aquarium also exhibited long, spontaneous breath holds, even when they were not asleep (Ponganis *et al.*, 1997b). When these shy animals were aware of human presence, they remained submerged at rest for up to 25 min. Apneic heart rates were 30–50 bpm during short breath holds of less than 10 min, but heart rate progressively decreased during longer breath holds – to 5–10 bpm for the last five minutes of a 25-minute breath hold. A wash-out of lactate into the blood after submersions greater than 15 min was also consistent with vasoconstriction and tissue O_2 store depletion. Although access to Baikal seals is difficult, this range of breath-hold durations and cardiovascular adjustments make this smallest of phocid seals a potentially valuable model for future studies.

5.1.13 Surface swimming: heart rate, cardiac output, and O_2 consumption

Cardiovascular responses and swim patterns during surface swimming in a swim flume have been most studied in pinnipeds. As workload increased in phocid seals, percentage time submerged decreased, but surface heart rate and submerged heart rate remained constant and distinct (Fedak *et al.*, 1988, Williams *et al.*, 1991). For example, in harbor seals, submerged and surface heart rates averaged 50 and 137 bpm, respectively, regardless of workload (Williams *et al.*, 1991). The high surface heart rate and an increased stroke volume at the surface resulted in a four-fold greater cardiac output at the surface (Ponganis *et al.*, 1990). However, similar to sleep apnea in seals, the submerged (apneic) cardiac output and stroke volume remained typical of seals and other mammals at rest.

Maintenance of significant cardiac output in shallow-swimming phocid seals was also supported by specific activity decay curves of bolus injections of labeled substrates (Castellini *et al.*, 1985). The decay curves were similar to those in seals at rest. The similarity of submerged heart rate during surface swimming to that during sleep apnea of phocid seals again would imply that there is some muscle blood flow during surface swimming. However, unlike exercise in terrestrial mammals (Musch *et al.*, 1987a), submerged heart rate, and presumably muscle blood flow, did not increase with workload (Williams *et al.*, 1991). Under such conditions, muscle blood flow during submergence could only increase if there were a selective redistribution of blood flow away from central organs and toward working muscle. On a conceptual basis, enhanced muscle blood flow might occur if there were only partial sympathetic vasoconstriction of proximal artery segments in combination with dilation of distal intramuscular arterial segments secondary to localized vasodilator release in working muscle.

In flume-swimming sea lions (*Zalophus californianus*), the swim pattern and heart rate responses differed from those in harbor seals in that percentage time submerged did not decrease with workload (Williams *et al.*, 1991). In addition, both surface and submerged heart rates increased with workload. There was a constant, but small, difference between surface and submerged heart rates. Another difference was that submerged heart rates in sea lions were near 100 bpm, much higher than in harbor seals. Cardiac output, which at rest was equivalent to that of harbor seals, increased with exercise workload in the sea lion (Ponganis *et al.*, 1991).

The increase in submerged heart rate of the sea lion in relation to workload contrasted with the constant submerged heart rate of the harbor seal. The sea lion's response was more typical of exercise in terrestrial mammals. This raised the question of whether muscle blood flow and oxygen delivery during the submergence differed between the two pinniped species. Theoretically, the higher myoglobin concentrations in the longer-diving harbor seal might make it less reliant than the young sea lion on muscle blood flow for O_2 supply during submergence. The differences in heart rate also emphasized potential differences in the magnitude of pulmonary gas exchange during submergence between the two species. Higher cardiac outputs and more gas exchange during submergence may be necessary in the sea lion due to its short surface periods at high workloads.

Regardless of the differences in cardiovascular responses in flume-swimming harbor seals and sea lions, they both exhibited about an 8–10-fold metabolic scope, which was normal to high for most mammals but less than those of "elite" athletes such as dogs and horses (Ponganis *et al.*, 1990, 1991, Taylor *et al.*, 1987, Williams *et al.*, 1991). Maximum O_2 consumption in the flume was in the range of 35–40 ml O_2 kg^{-1} min^{-1}.

Exercise research on cetaceans has focused on the bottlenose dolphin (*Tursiops truncatus*). Maximum O_2 consumption in exercising dolphins (pushing against a load cell) was about 30 ml O_2 kg^{-1} min^{-1} (Williams *et al.*, 1993). At low workloads and swim speeds, heart rate oscillated between a short post-inspiratory tachycardia and an apneic bradycardia near 50 bpm (Williams *et al.*, 1992c, 1993). This difference disappeared as heart rate increased at higher workloads and swim speeds. This cardiovascular response paralleled that in the sea lion. Notably, both surface and submerged heart rates during wave-riding were less than those during surface swimming at similar speed in the dolphin. This finding provided evidence for the lower energetic costs of wave-riding.

5.1.14 Free dives: early studies of marine mammals

Despite the significance of heart rate and cardiovascular responses to the regulation of O_2 consumption and dive duration, investigation of these physiological parameters in free-diving animals has been limited. Measurements of heart rate and other indices of cardiovascular function in free dives of aquatic mammals began with the use of a standard electrocardiogram (ECG) recorder and application of long ECG leads to allow acquisition of the ECG while the animals swam in relatively close confines nearby. Demonstration of a diving bradycardia with heart rates as low as 40 bpm were actually reported from partial ECG records as early as 1941 during shallow dives of dolphins in tanks and at sea (Irving *et al.*, 1941a). Elsner later documented that heart rate decreased from pre-dive rates of 100 bpm to submerged heart rates of 20–40 bpm in a dolphin diving in a tank (Elsner *et al.*, 1966b).

Reports of heart rate in unrestrained, diving seals were based on this same technique and were limited to short-duration, shallow dives of Weddell seals (*Leptonychotes weddellii*) and harbor seals and to the initial descent of long dives in Weddell seals (Jones *et al.*, 1973, Kooyman and Campbell, 1972). Both of these studies found that diving heart rates were variable, often about 50% the eupneic heart rate, and much greater than the 10 bpm rates reached in early forced submersion studies. The higher heart rates in unrestrained shallow dives of seals again raised the question as to the nature of the peripheral vascular response and the degree of restriction of blood flow to tissues under such conditions. In contrast to short dives, Kooyman and Campbell also reported that initial heart rates of longer dives were lower than those during shorter dives, consistent with more intense vasoconstriction and greater conservation of blood O_2 during the longer dives.

5.1.15 Weddell seals: heart rate and splanchnic/renal blood flows

In a later study designed by Kooyman's team to assess the degree of vasoconstriction and magnitude of organ perfusion in free-diving Weddell seals (Davis *et al.*, 1983), the clearances of inulin and indocyanine green (ICG) were determined as indices of renal glomerular filtration rate (GFR) and hepatic blood flow, respectively. Both inulin and ICG clearances were maintained during shorter-duration dives, suggesting that GFR and hepatic perfusion did not decrease during those dives in which Kooyman and Campbell had observed higher initial heart rates. In addition, the plasma became lipemic, indicative of maintenance of gastrointestinal blood flow. In longer dives, inulin clearance (GFR) decreased, again consistent with the observed lower heart rates and presumed greater vasoconstriction during such dives. However, ICG clearance did not decline in long dives, which suggested that hepatic perfusion was maintained even during long dives. Davis and co-workers pointed out that there was one caveat to this conclusion – namely, it was unknown if hepatic extraction of ICG was constant under conditions of decreased liver perfusion. If the extraction efficiency increased during hypoperfusion, then the ICG clearance technique would overestimate hepatic blood flow under such conditions. Thus, the issue of hepatic blood flow during longer dives was unresolved.

These studies of heart rate and perfusion were conducted on the sea ice of McMurdo Sound, Antarctica, with use of an isolated dive hole as pioneered by Kooyman in his behavioral and respiratory physiology investigations of free-diving Weddell seals (Kooyman, 1968, 1981). In this approach, a seal was allowed to dive in a man-made ice hole to which it had to return because there were no other cracks or holes in the ice which it could reach to escape. This technique, which allowed data collection from dives with a wide range of depths and durations, provided the basis to further investigate cardiovascular responses during diving in the 1980s. Such data collection became feasible with the development of microprocessor-based heart rate recorders and blood samplers (Hill, 1986). In a series of studies in the 1980s through 1990s, Zapol and associates assessed heart rate profiles and perfusion patterns of the kidneys, liver and muscle in diving Weddell seals (Zapol, 1996). These findings will be reviewed in detail below as these projects, together with work by the Kooyman teams, provide the most detailed assessment available of cardiovascular responses in any free-diving marine mammal.

Heart rates during dives of 10–15-min duration of free-diving Weddell seals were 35–45 bpm, and those of dives longer than 20 minutes were 29–36 bpm (Hill *et al.*, 1987). Heart rates during post-dive periods were 66–98 bpm, and at rest heart rates ranged from 60 to 78 bpm. Diving heart rates were variable and clearly below resting heart rates (Fig. 5.5b); in addition, heart rate decreased with dive duration. However, even during long dives, heart rates were twice the 15 bpm value of Weddell seals during forced submersions (Zapol *et al.*, 1979). Notably, the decline in fetal heart rate during free dives of pregnant seals was also not as severe and abrupt as during forced submersions (Hill *et al.*, 1987, Liggins *et al.*, 1980).

In order to assess the degree of vasoconstriction and perfusion/function of the liver and kidneys during these moderate bradycardias in diving Weddell seals, renal and hepatic clearance rates were assessed with use of radio-labeled tracers and sequential blood sampling in two studies. In a companion study to Hill *et al.*'s heart rate investigation, serial blood samples were obtained with use of a backpack microprocessor-controlled blood sampler (Guppy *et al.*, 1986, Hill, 1986). In long exploratory dives, prolonged equilibration times and markedly decreased clearance rates of p-amino-hippuric acid and inulin were consistent with decreased renal perfusion and GFR, respectively. Although measurements were only obtained for one dive, the results agreed with the prior findings of Davis *et al.* and with the heart rate observations of Hill *et al.* Clearance of cholate, considered an index of hepatic blood flow (Shrestha *et al.*, 1997), was successfully measured in one short-duration dive. Cholate clearance during the dive was slower than during rest periods at the surface, suggesting decreased hepatic function even during short dives. This contrasted with Davis *et al.*'s work, in which ICG clearance was maintained during short dives. However, Guppy *et al.* also reported that the equilibration time for cholate during the short dive was short in comparison to the longer equilibration times for tracers in longer dives, and that this was consistent with more flow and less vasoconstriction of the liver during short-duration dives. Thus, precise evaluation of liver blood flow and function during dives was again difficult to achieve.

Nonetheless, the combined work of both the Davis and Guppy papers remain remarkable achievements, and even 25–30 years later remain the only studies of renal and hepatic perfusion in free-diving animals. These papers demonstrated that splanchnic (hepatic and gastrointestinal) and renal perfusion were maintained in short-duration, aerobic dives of Weddell seals, and that significantly decreased perfusion was primarily associated with the lower heart rates found in longer dives (Davis, 2014). In addition, maintenance of aerobic metabolism even at lower heart rates may be facilitated by the high O_2 capacity of Weddell seal blood, which potentially allows for adequate O_2 extraction at lower tissue perfusion rates (Davis and Kanatous, 1999).

5.1.16 Weddell seals: heart rate, myoglobin saturation and muscle blood flow

Indirect evaluation of muscle blood flow patterns in diving Weddell seals was conducted in the 1990s with monitoring of muscle myoglobin (Mb) saturation via application of a backpack near-infrared spectrophotometer (Guyton *et al.*, 1995). This again was a remarkable technical achievement of the Zapol group. Myoglobin saturation profiles during dives were quite variable, and, overall, myoglobin desaturated at rates of 5.1% and 2.5% min^{-1} during dives less than and greater than 17 minutes, respectively. The authors concluded that muscle blood flow decreased but continued to some degree in the latissimus dorsi muscle during the mild bradycardias of these diving seals because Mb saturation profiles revealed that (a) the myoglobin desaturation rate was much less than expected and myoglobin never completely desaturated even during long dives; (b) myoglobin desaturation rates leveled off during some dives; (c) on occasion, myoglobin partially resaturated during a dive; and (d) relative blood volume in muscle (as indexed by the absorption profile of 810 nm wavelength light) often increased during ascent from long dives.

Maintenance of some blood flow to muscle during the relatively mild bradycardias in diving Weddell seals is supported by Grinnell *et al.*'s earlier observations of the coupling of muscle blood flow and heart rate during forced submersions, and with Jobsis *et al.*'s later observations of increased muscle blood flow during the higher heart rates of trained versus naïve submersions (Grinnell *et al.*, 1942, Jobsis *et al.*, 2001). It is also consistent with the intermediate declines in muscle blood flow in the mid-back and forelimb regions of Baikal seals during forced submersions (Cherepanova *et al.*, 1993).

However, interpretation of these data and conclusions in regard to the primary locomotory muscle O_2 store is complex for several reasons. First, the actual O_2 consumption rate of muscle during a dive is not known, especially for the latissimus dorsi muscle on which the near-infrared probe was implanted. That muscle extends from the mid-back to the shoulder, and functions in movement of the foreflipper. Both on an anatomical basis (Howell, 1930), and as indicated in the electromyogram studies of Baikal seals (Neshumova *et al.*, 1986), the latissimus dorsi muscle is not the primary propulsive muscle of the phocid seal. The O_2 consumption of the latissimus dorsi during a dive might well be near that of muscle at rest, i.e., 2 ml O_2 kg^{-1} min^{-1} (Blei *et al.*, 1993); at that rate, its myoglobin O_2 store could last 30 min even in the absence of any blood flow. The well-developed longissimus dorsi–iliocostalis complex is considered to provide the primary propulsion movements of the hind flippers (Howell, 1930, Pierard, 1971); it was also the most active muscle in the Neshumova *et al.* electromyogram study. This muscle complex also has the highest myoglobin concentration, and, in Baikal seals it had the most intense vasoconstriction during forced submersions (Cherepanova *et al.*, 1993). The myoglobin desaturation profile and the muscle blood flow profile of this muscle during free dives will probably have the most implications for utilization and management of O_2 stores.

5.1.17 Weddell seals: heart rate and flipper stroke rate

Examination of the relationship of heart rate to flipper stroke rate of diving Weddell seals has provided another indirect assessment of the relation of muscle blood flow to workload in diving Weddell seals (Davis, 2014, Davis and Williams, 2012, Williams *et al.*, 2015). In short-duration dives of Weddell seals, heart rate was low but was found to increase linearly with stroke rate. The authors concluded that the cardiovascular response in these free-diving animals involved a combination of both a dive response and an exercise response. This combined response was proposed to support aerobic muscle metabolism by both depletion of the Mb-bound O_2 store and blood-to-muscle O_2 transfer during the dive. This hypothesis and maintenance of some degree of muscle blood flow during short-duration, aerobic dives were consistent with (a) the above Mb saturation profiles of Guyton *et al.*; (b) absence of locomotory muscle temperature elevation in free-diving seals (Ponganis *et al.*, 1993b); (c) maintenance of muscle blood flow during Jobsis' trained submersions of seals; (d) Grinnel *et al.*'s findings that, even during forced submersions, small increases in heart rate were accompanied by increases in muscle blood flow; and (e) effects of central command and muscle mechanoreceptor feedback during exercise (see Section 5.3.6).

A positive correlation of heart rate with stroke rate has also been reported in shallow dives of bottlenose dolphins (Davis and Williams, 2012, Noren *et al.*, 2012a, Williams *et al.*, 2015). The observed increase in heart rate with stroke rate in seals and dolphins implies a linkage of heart rate and presumably muscle blood flow to muscle workload. However, in contrast to a classic exercise response, the increase in heart rate was small. In addition, dependent on the level of peripheral vasoconstriction prior to such an increase in heart rate, the expected elevation in blood flow would presumably be to all tissues and not just exercising muscle. Rather than a selective, localized increase in blood flow to working muscle as in a classic exercise response, a decrease in peripheral sympathetic tone should increase flow to all tissues due to the distribution of sympathetic nerve fibers on large, proximal arteries supplying multiple tissues in seals (White *et al.*, 1973). In addition, blood flow to splanchnic organs may increase more than that to muscle for a given change in sympathetic nerve discharge. As demonstrated in nutria (*Myocastor coypus*), muscle blood flow was more sensitive than renal blood flow to sympathetic nerve stimulation (i.e., the kidney required a higher nerve stimulation rate for equivalent blood flow reduction) (Folkow *et al.*, 1971). Based on that observation, any small reduction in peripheral sympathetic discharge during a dive would presumably increase blood flow to the kidney more than to muscle. See Sections 5.35–5.3.7 for further discussion.

5.1.18 Free dives: heart rates of other marine and aquatic mammals

Researchers have also investigated heart rate responses during free dives of other marine mammals. In the 1980s ECG telemetry investigations of manatees demonstrated that heart rate usually remained near 40 bpm during spontaneous dives (see Fig. 5.5f), and only transiently reached 50 bpm during eupnea (Gallivan *et al.*, 1986). At least in the manatee, 40 bpm appeared to be the "normal" resting heart rate, leading these authors to suggest that the manatee did not have a diving bradycardia, but rather a respiratory tachycardia. This heart rate profile in the manatee reflects the description of the sinus arrhythmia previously discussed in elephant seals and other animals. And, again, as in the elephant seal during sleep apnea, these moderate heart rates may maintain blood flow to many organs and tissues during routine diving.

It was not until the 1980s–1990s that application of either microprocessor-based heart rate recorders, Holter monitors, or ECG telemetry units allowed documentation of heart rate profiles in pinnipeds diving to depth (Ponganis, 2007). These studies all confirmed that (a) the degree of bradycardia was variable, but, overall, dive heart rate decreased as dive duration increased; (b) diving heart rates were frequently 25–50% of the eupneic level; (c) at times, diving heart rates could approach the levels observed during forced submersions; and (d) heart rate usually increased during the ascent (the so-called ascent or anticipatory tachycardia) (Fig. 5.5). The species examined included Weddell seals, harbor seals, gray seals (*Haliocherus grypus*), ringed seals (*Phoca hispida*), elephant seals (*Mirounga angustirostris, M. leonina*), Antarctic fur seals (*Arctocephalus gazella*), Steller sea lions (*Eumetopias jubatus*), and California sea lions (Andrews *et al.*, 1997, Boyd *et al.*, 1999, Elsner *et al.*, 1989, Fedak, 1986, Fedak *et al.*, 1988, Hill *et al.*, 1987, Hindell and Lea, 1998, Hindle *et al.*, 2010, Ponganis *et al.*, 1997d, Thompson and Fedak, 1993). The Antarctic fur seal was a relative exception to low diving heart rates, with its lowest heart rates during dives near 80 bpm, about two-thirds the eupneic level, even during its long dives of three-minute duration (Boyd *et al.*, 1999); see Fig. 5.5e.

5.1.19 Gray seals and elephant seals: heart rates during free dives

The variability in heart rates during dives of pinnipeds were well-demonstrated in gray seals and elephant seals (Fig. 5.5c, a). Mean dive heart rates of gray seals were 40–50 bpm during 1–3 min dives, but decreased to 10 bpm for dives of 20–26 min duration (Thompson and Fedak, 1993). During one 14-minute dive, heart rate during 90% of the dive was less than 4 bpm! In this dive, peripheral vasoconstriction and conservation of blood O_2 presumably were maximized just as during forced submersions. In comparison, during 30 min dives of northern elephant seals, mean heart rates were still 25–30 bpm (Andrews *et al.*, 1997). Although heart rate also declined with dive duration in elephant seals, the relatively high heart rates in 20–30-minute dives of elephant seals probably allowed for much more tissue perfusion and O_2 delivery than

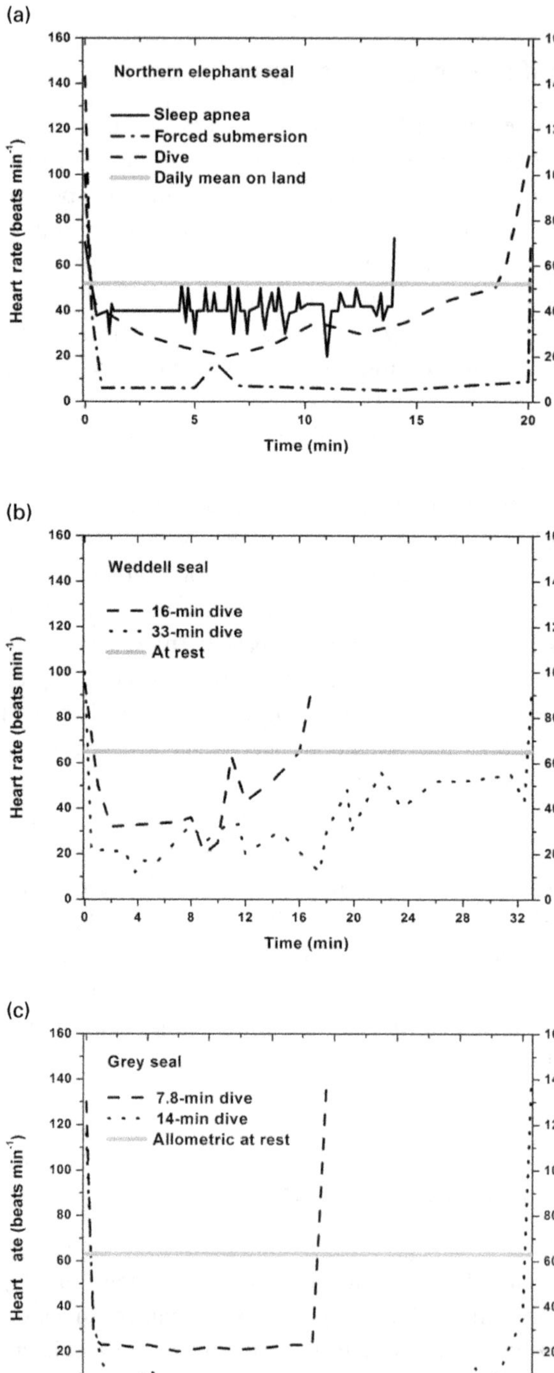

Figure 5.5 Heart rate responses during free dives in relation to heart rate at rest in five pinnipeds and the manatee. Graphs illustrate generalized pattern of heart rate profiles relative to reported heart rate at rest. **(a)** Elephant seal (*Mirounga angustirostris*), free dive, sleep apnea and forced submersion, adapted from Andrews *et al.* (1997) and Elsner *et al.* (1966a). **(b)** Weddell seal (*Leptonychotes weddellii*), adapted from Hill *et al.* (1987). **(c)** Gray seal (*Halichoerus grypus*),

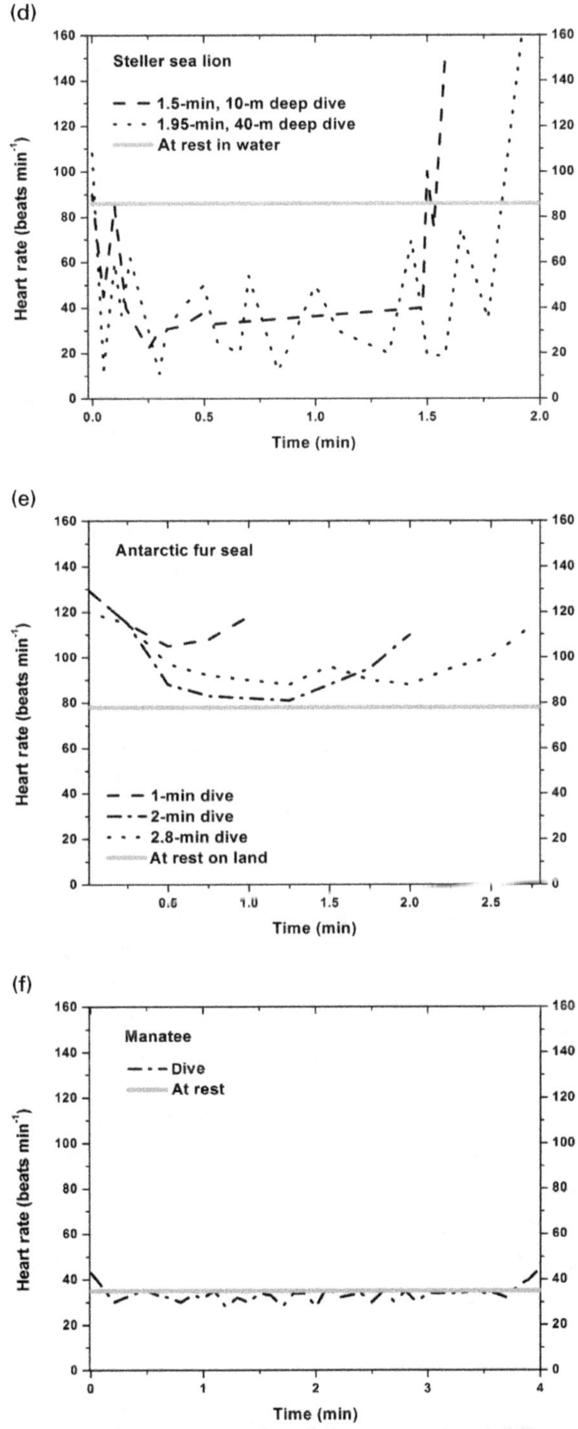

Figure 5.5 (*cont.*) adapted from Thompson and Fedak (1993); heart rate at rest from allometric equation in Stahl (1967). **(d)** Steller sea lion (*Eumetopias jubatus*), adapted from Hindle *et al.* (2010). **(e)** Antarctic fur seal (*Arctocephalus gazella*), adapted from Boyd *et al.* (1999). **(f)** Manatee (*Trichechus manutus*), adapted from Gallivan *et al.* (1986).

during that 14-min dive of the gray seal. However, as reviewed in Chapter 1, the elephant seal is capable of 40–60-minute and even 120-minute dives. During much of the time course of these extreme dives, one would expect that heart rate would be similar to the 4 bpm rate exhibited by the gray seal in the 14-min dive reported by Thompson and Fedak.

5.1.20 Elephant seals: venous oxygen profiles and blood flow implications

In addition to heart rate and blood flow measurements, blood P_{O2} and Hb saturation profiles can also provide insight into possible blood flow patterns during dives. Venous profiles are especially significant because blood O_2 extraction by perfused organs and muscle decreases venous P_{O2} and Hb saturation. In particular, increased or even constant P_{O2} values during a dive are not consistent with extraction of O_2 by working muscle. Yet, increased P_{O2}s and Hb saturations occur early during dives in both the hepatic sinus and extradural vein of elephant seals (Meir *et al.*, 2009). These profiles suggest that working muscle is ischemic during early descent, and, in fact, these observations are consistent with the rapid onset of iliac artery constriction reported by Van Citters *et al.* during forced submersions of elephant seals. Furthermore, in the initial portions of some dives, venous blood actually became arterialized (P_{O2} and saturations equivalent to arterial values). Such high values are not only consistent with muscle ischemia, but they also suggest A-V shunting. Otherwise, such high arterial values could not be achieved in venous blood.

At this point, it is worthwhile to remember that (a) A-V shunting was suggested in the forced submersion microsphere studies of Zapol and Blix; and (b) A-V shunting would be consistent with a constant blood pressure in the presence of a constant cardiac output and a 75% reduction in iliac artery blood flow at the start of a forced submersion as reported by the study by Van Citters *et al.* Thus, studies spanning almost 50 years and ranging from forced submersions to free dives all suggest the possible utilization of A-V shunts during dives.

In contrast to the maintenance or increase in venous O_2 during early descent, blood O_2 profiles decreased during the ascent phase of elephant seal dives (Meir *et al.*, 2009). Such patterns are consistent with increased tissue perfusion, and blood O_2 extraction by muscle and other organs during the increase in heart rate commonly observed during ascent. It has been proposed that such blood O_2 depletion serves to increase the lung to blood P_{O2} gradient and enhance O_2 uptake during the subsequent surface interval (Thompson and Fedak, 1993).

5.1.21 Steller sea lions and California sea lions: heart rate during free dives

In Steller sea lions, a moderate bradycardia occurred and the relationship of heart rate to stroke rate appeared variable and dependent on the nature of different dives. Such variation was reported in a paper which used ODBA (overall dynamic body

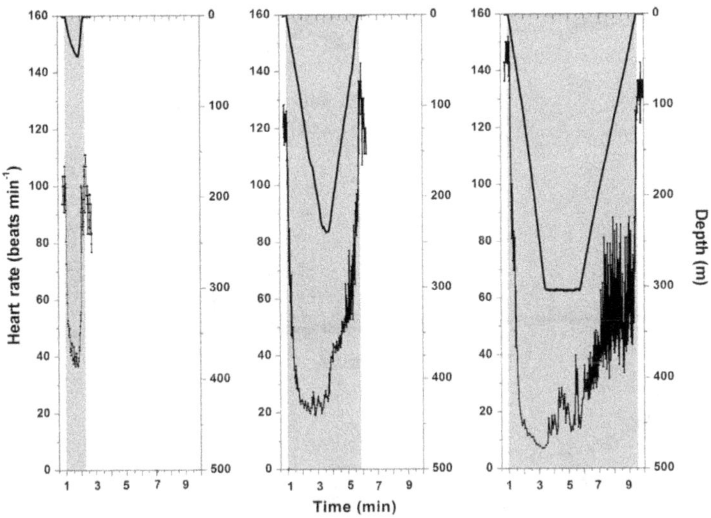

Figure 5.6 Beat-to-beat heart rate profiles of California sea lions (*Zalophus californianus*) during shallow, intermediate, and deep dives. Heart rate at rest on the beach was 52 bpm. Shaded area indicates dive interval, solid line is depth, and dotted line is heart rate. Adapted from McDonald and Ponganis (2014).

acceleration) as an index of stroke rate and muscular effort in dives of Steller sea lions (Hindle *et al.*, 2010). In shallow dives (10 m) of Steller sea lions, heart rate and ODBA correlated whereas, in deeper dives (40 m), there was no correlation between heart rate and ODBA.

Heart rate data in California sea lions during maternal foraging trips to sea revealed U-shaped profiles with higher initial heart rates, lower minimum heart rates, and lower overall dive heart rates as depth of dive and dive duration increased (McDonald and Ponganis, 2014). Dive heart rates were near or above resting heart rate in the typical short-duration (<3 min) dives of sea lions, but were less than resting rates in long-duration, deep (300–400 m) dives (Fig. 5.6). Heart rates were as low as 10 bpm during late descent and the bottom phase of the deep dives.

The U-shaped heart rate profiles during deep dives of sea lions should contribute to greater lung O_2 utilization at shallower depths during the higher heart rates of both early descent and late ascent. However, during the deeper and bottom phases of the dive, the extreme bradycardias should minimize lung N_2 and O_2 uptake even before "lung collapse" at depth. This would allow the sea lion to take advantage of its large respiratory O_2 store at shallower depths, but minimize N_2 absorption even prior to lung collapse during the deepest portions of the dive. Besides limiting pulmonary gas exchange at depth, the severe bradycardias during late descent and the bottom phase of deepest dives should also conserve blood O_2 stores and help maintain arterial oxygenation during periods of lung collapse. In contrast, higher heart rates during shallow dives would provide for greater utilization of the lung O_2 store as well as potential delivery of O_2 to muscle.

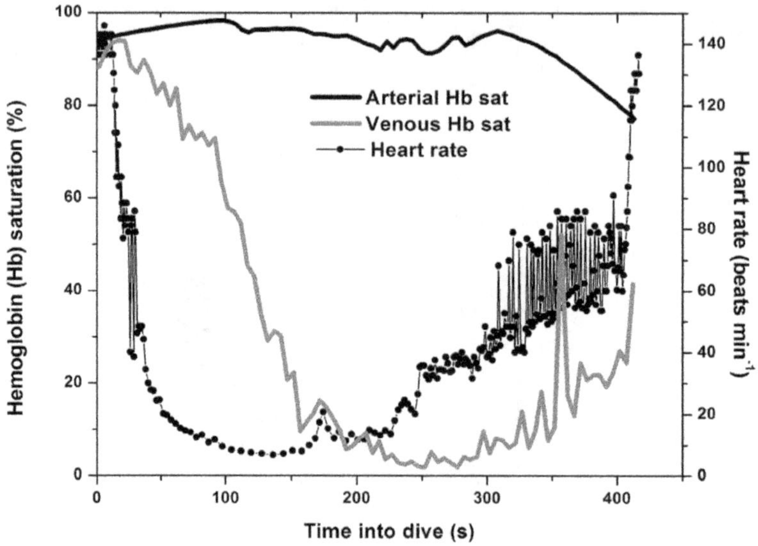

Figure 5.7 Arterial and venous hemoglobin (Hb) saturations relative to the heart rate profile in three 7-min dives by different sea lions (*Zalophus californianus*) to depths of 310–350 m. Although not simultaneous records from one animal, these profiles demonstrate characteristic hemoglobin saturation (S_{O2}) and heart rate profiles during such dives. Adapted from McDonald and Ponganis (2014).

5.1.22 California sea lions: venous oxygen profiles and blood flow implications

Increases in venous P_{O2} and Hb saturation during early descent of diving sea lions were not consistent with maintenance of muscle blood flow during that segment of the dive (McDonald and Ponganis, 2013). This is analogous to the observations and arguments made for elephant seal venous oxygen profiles in Section 5.1.20. During the mid-dive and latter segments of shallow dives of California sea lions, venous P_{O2} and Hb saturation in the posterior vena cava declined, suggesting that peripheral tissue perfusion and depletion of the lung and blood O_2 stores continued during the moderate bradycardias observed during these dives. However, the venous O_2 depletion patterns were highly variable, and the distribution pattern of blood flow to different tissues, muscle, or even A-V shunts is still unknown.

During the latter descent and early bottom phases of deep (>250 m) dives, venous O_2 declined rapidly to near-zero levels and then, during ascent, gradually increased (Fig. 5.7). In light of the maintenance of arterial Hb saturation and also a severe bradycardia during late descent and the early bottom phase of these deep dives, the rapid and almost complete decline in venous O_2 was interpreted to be secondary to extreme tissue ischemia and almost complete tissue extraction of O_2 from any perfused blood (i.e., severe vasoconstriction in those tissues draining into the posterior vena cava). In contrast, during ascent, the increase in venous O_2 content was postulated to be due to an increase in peripheral blood flow secondary to the observed increase in heart rate during ascent.

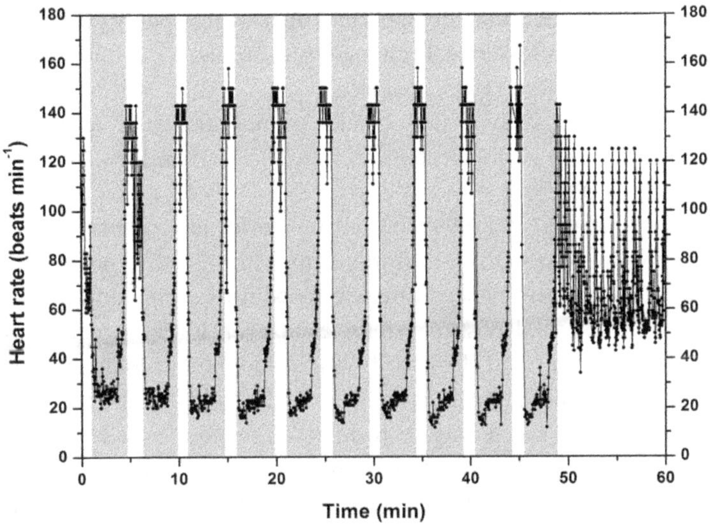

Figure 5.8 Heart rate profiles of a bottlenose dolphin (*Tursiops truncatus*) trained to station at depth and perform ten serial 100-m deep dives. Shaded areas represent dive time. During the recovery period after the session, heart rate returned into a typical sinus arrhythmia pattern. Adapted from Houser et al. (2010).

5.1.23 Dolphins: heart rates during free dives

In diving cetaceans, heart rate research has been focused on the bottlenose dolphin (Houser et al., 2010, Noren *et al.*, 2004, Williams et al., 1999). Beat-to-beat heart rate profiles of dolphins trained to perform serial dives to 100 m are illustrated in Fig. 5.8. The findings in the paper by Williams *et al.* documented the heart rate response of dolphins and, for the first time, provided insight into the relationship of diving heart rate to muscular work effort. In free-dives to 60-m and 210-m depth, heart rates declined from pre-dive average rates of 101–111 bpm to 20–30 bpm within 1 min of submergence; during the bottom phases of these dives, heart rates averaged 37 and 30 bpm for the 60-m and 210-m dives, respectively (Williams *et al.*, 1999). As in pinnipeds, the dolphins had an increase in heart rate during ascent. The authors reported that the bradycardias and heart rate patterns of these diving dolphins were similar to those reported previously for a sedentary dolphin stationing at 2-m depth (Elsner *et al.*, 1966b) despite the fact that their dolphins were actively swimming during the dives (fluke stroke rate was also monitored). They concluded that the heart rate response in diving dolphins was dominated by the dive response and not by an exercise response (in which heart rate would increase in relation to work effort (stroke rate)). Indeed, during the higher heart rates during the final ascent, stroke rates were actually decreased due to periods of prolonged gliding in this phase of the dive. As has already been mentioned, the issue of a linkage between heart rate (i.e., muscle blood flow) and muscular work in diving animals and the question of whether exercise modification of the dive reflex predominates in these animals have become central to the understanding and management of O_2 stores during diving. This paper was probably the first that examined heart rate in relation to stroke rate. Prior papers had relied on swim speed as an index of stroke effort (Guyton *et al.*, 1995, Hill *et al.*, 1987), but as will be seen in Chapter 7, swim speed is not necessarily a reliable index of muscular work because these animals can glide effortlessly at swim speeds similar to those when they are actively stroking (Williams *et al.*, 2000).

Although Williams *et al.* (1999) did not find evidence for an exercise response in their first study, this does not necessarily imply that there is no muscle blood flow during diving. Indeed, persistence of muscle blood flow during dives was suggested by documentation of elevated post-dive muscle nitrogen levels in deep-diving dolphins 20 years earlier (Ridgway and Howard, 1979). In addition, in contrast to those heart rate findings in deep open-ocean dives, it has now been found that there is some correlation of heart rate with stroke rate during short-duration, shallow dives of bottlenose dolphins and in trained, shallow dives of dolphins at sea (Davis and Williams, 2012, Noren *et al.*, 2012a, Williams *et al.*, 2015). Better understanding of the tentative linkage of heart rate and muscle workload awaits further study of the cardiovascular response of dolphins diving to different depths and durations.

5.1.24 Other mammals: heart rates during free dives

Free-diving heart rates have also been recorded in the hippopotamus and in several aquatic, small mammals including the mink, muskrat, and platypus (Elsner, 1966, Evans *et al.*, 1994, MacArthur and Karpan, 1989, Stephenson *et al.*, 1988). Similar to marine mammals, heart rates decrease significantly on submersion, and in some circumstances may be quite low. In the diving muskrat swimming against higher water current flows, submerged heart rates are greater than those at lower swim speeds, again suggesting that, at least in this species, increased muscle workload is accompanied by increased muscle blood flow (i.e., an exercise-modified response).

5.1.25 Marine mammal cardiovascular response during free dives: summary

In summary, heart rate responses in free-diving marine mammals can vary considerably, dependent not only on the species but also on the nature of a given dive. In general, except for perhaps the manatee, heart rates during most dives decrease to below "resting" levels. Species with higher Mb concentrations can exhibit more severe bradycardias, but such slow heart rates are not always utilized. The regulation of peripheral blood flow distribution appears primarily controlled by sympathetic tone via a central vascular "throttle" created by the innervation pattern of sympathetic nerve fibers to large, proximal arteries.

At least in some situations, especially in short-duration dives, heart rate and stroke rate appear to correlate, suggesting that cardiac output and muscle blood flow are, at times, coupled with muscle workload, as in terrestrial exercise. Heart rates, however, are not elevated above resting rates, and the distribution of such increased blood flow is not necessarily selective to exercising muscle. In longer duration and usually deeper dives, heart rates are even slower, but still moderate in range, and the linkage with stroke rate is less evident. In addition, extreme bradycardias (the classic, forced submersion dive reflex) can occur, such as in the long-duration dives of gray seals and during the late descent/early bottom phase of deep dives of California sea lions.

As examined in Weddell seals, gastrointestinal, hepatic, and renal function appear to be maintained during routine, aerobic dives. Maintenance of digestive, hepatic, and renal function during dives is probably more important in "surfacers" such as elephant seals and beaked whales than in "divers" such as sea lions. Hence, the intensity of their cardiovascular responses to diving may vary. However, the exact partitioning of cardiac output into regional blood flow to various organs, although frequently modeled, remains unknown.

5.2 Cardiovascular physiology in seabirds

5.2.1 Forced submersions of birds: bradycardia and peripheral vasoconstriction

Although the heart rate response of ducks to submersion or asphyxia had continued to be investigated by a series of researchers after Bert's original observations in 1870 (Andersen, 1966), it was not until Scholander's monograph in 1940 that a bradycardic response was documented in macaroni (*Eudyptes chrysolophus*) and gentoo (*Pygoscelis papua*) penguins during forced submersions (Scholander, 1940). Although heart rate profiles of complete submersions were not reported, heart rate decreased gradually from pre-submersion rates of 160–200 bpm to rates of 40–60 bpm over the course of 1 min, blood and muscle oxygen decreased to near zero by the end of 5-min submersions, and a wash-out of lactate into the blood occurred after the submersions (Fig. 5.9) (Millard *et al.*, 1973, Scholander, 1940).

However, significant elevations in blood lactate concentration also occurred during the forced submersions of penguins (Fig. 5.9), suggesting that peripheral vasoconstriction in the penguins may not have been as complete as in the seals (Scholander, 1940). In addition, muscle lactate concentrations of penguins did not consistently increase during the submersion. In later studies, femoral artery blood flow, an index of both peripheral blood flow and the degree of vasoconstriction, was reduced to about 25% of the pre-submersion rate in restrained submersions of gentoo penguins (Millard *et al.*, 1973). Complete heart rate profiles during submersions of penguins in this latter study demonstrated considerable variation in heart rate, dependent on the degree of struggling (Millard *et al.*, 1973). It is possible that minor variation in heart rate and vasoconstriction may have also occurred in Scholander's studies and contributed to the elevation in blood lactate during the submersion. Scholander also noted that muscle did not immediately re-oxygenate after the submersion, and that muscle lactate concentration also continued to rise during the early post-submersion period (Fig. 5.9).

During forced submersions of ducks, Scholander again found a moderate decrease in heart rate, near complete depletion of blood oxygen, an elevation in muscle lactate concentration during the submersion, and a wash-out of lactate into the blood after

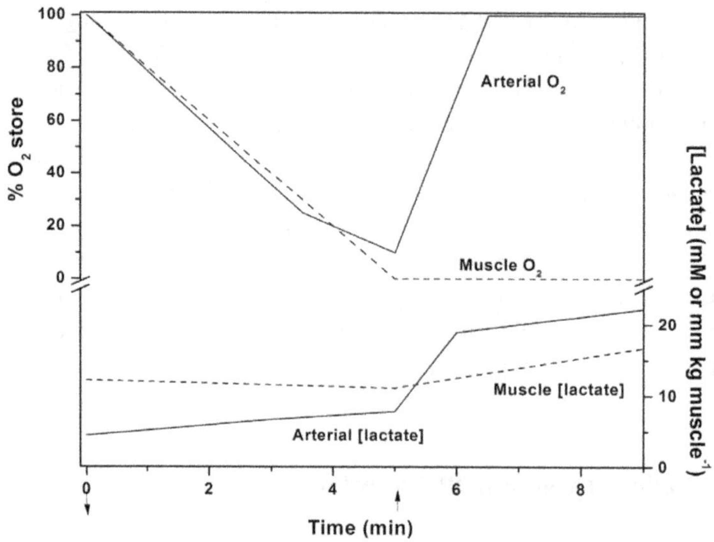

Figure 5.9 Arterial and muscle oxygen contents and lactate concentrations during a forced submersion of a gentoo penguin (*Pygoscelis papua*). The rise in lactate concentration after the submersion provided evidence for peripheral vasoconstriction even during the relatively moderate heart rates of 40–60 bpm reported for the penguin. Notably, muscle oxygen content remained near zero and muscle lactate continued to increase during the early post-submersion period, suggesting that return of muscle blood flow was incomplete or delayed. Arrows indicate start and end of submergence. Adapted from Scholander (1940).

submersion (Scholander, 1940). The decline in heart rate was not as rapid as in seals. Nor were the absolute values of the lowest heart rates as low as in seals (Andersen, 1966). For example, heart rate in "domestic" ducks gradually decreased from a pre-dive rate of 110 bpm to 20 bpm by three minutes of submersion (Andersen, 1959). In mallard or pekin ducks (*Anas platyrhnchos*), forced submersion elicited a gradual decrease from pre-submersion rates of 100 to 150 bpm to a rate of 30–40 bpm over the course of about a minute (Fig. 5.10) (Butler and Jones, 1968, Hudson and Jones, 1986). Such heart rate patterns typified the response of "dabbling ducks" to forced submersion (Furilla and Jones, 1987b). In contrast to dabblers, "diving ducks" such as the pochard, tufted duck, and redhead duck (*Aythya ferina*, *A. fuligula*, and *A. americana*, respectively) had more rapid decreases in heart rate during forced submersions (Fig. 5.10). As an example, heart rate declined from 100 bpm to 40 bpm within two seconds of submersion in the redhead duck (Furilla and Jones, 1986). Similar rapid declines in initial heart rate were originally noted in spontaneous dives of tufted ducks and pochards (Butler and Woakes, 1979).

As in seals, an increase in peripheral vasoconstriction maintained blood pressure despite the decline in heart rate in forcibly submerged ducks (Folkow *et al.*, 1967, Johansen and Aakhus, 1963, Johansen and Krog, 1959). Decreased muscle blood flow due to widespread vasoconstriction was inferred from experiments with use of a hot

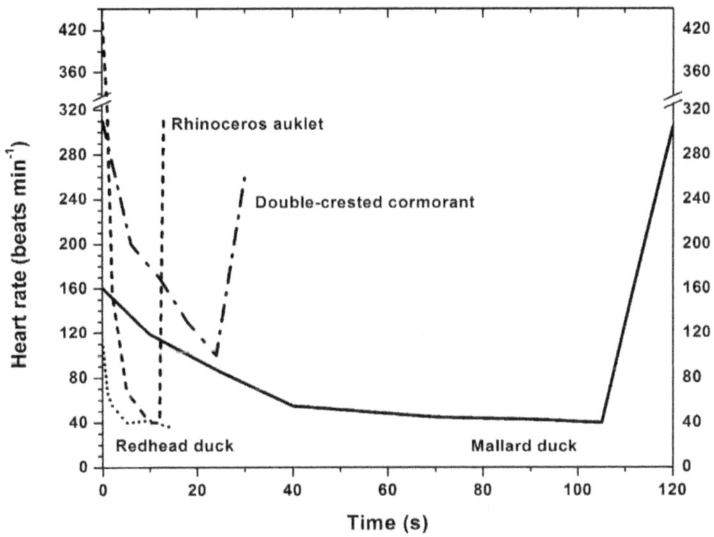

Figure 5.10 Heart rate responses during forced submersions of a mallard duck (*Anas platyrhnchos*), redhead duck (*Aythya americana*), double-crested cormorant (*Phalacrocorax auritus*), and rhinoceros auklet (*Cerorhinca monocerata*). Graphs illustrate generalized pattern of heart rate profiles. Submersions end with the onset of the tachycardias. Heart rates at rest were reported as 90 (restrained), 100 (restrained), 143, and 132–168 bpm in mallards, redhead, cormorants, and auklets, respectively. The gradual decline in heart rate of the mallard, a dabbling duck, is distinct from the more rapid decline in diving ducks and other diving birds. Adapted from Butler and Jones (1968), Enstipp et al. (1999), Furilla and Jones (1986), Stephenson et al. (1992).

wire anemometer as well as with measurements of venous effluent from muscle (Andersen, 1959, Djojosugito *et al.*, 1969).

A bradycardia during forced submersion was also later observed in other diving birds, including cormorants, guillemots, puffins, and rhinoceros auklets (Eliassen, 1960, Stephenson *et al.*, 1992). In these animals, the decline in heart rate after submersion was rapid. In rhinoceros auklets (*Cerorhinca monocerata*), for example (Fig. 5.10), heart rate declined from pre-dive rates of 450 bpm to 80 bpm within 5 sec of submersion (Stephenson *et al.*, 1992). Historical surveys of early cardiovascular research in birds are available in two excellent reviews (Andersen, 1966, Jones and Furilla, 1987).

The pattern of blood flow distribution to organs during the bradycardia and peripheral vasoconstriction of forced submersions of ducks has been examined with both radioisotope tracers and radio-labeled microspheres (Johansen, 1964, Jones and Furilla, 1987, Jones *et al.*, 1979, Stephenson and Jones, 1992a). Similar to findings in seals, these studies confirmed the redistribution of blood flow to essentially the brain, heart, and adrenal glands during forced submersions. Maintenance of blood flow to the adrenal glands was consistent with documented elevations in blood catecholamine levels (Lacombe and Jones, 1990, 1991a, 1991b).

As also already discussed for seals, differences in the sympathetic innervation patterns of major arteries probably contributed to the organ blood flow distribution pattern during forced submersions of birds. For example, sympathetic innervation of the duck's femoral artery, which supplies the propulsive leg muscles, was much more extensive than that in either the turkey or cat (Folkow *et al.*, 1966). This again supported the concept that the distribution of sympathetic fibers to the large, extramuscular arteries of the duck allowed for the maintenance of vasoconstriction even in the presence of muscle hypoxia and accumulation of metabolic vasodilators around intramuscular blood vessels (Folkow *et al.*, 1966).

Interestingly, there are two other exceptions to the widespread vasoconstriction in forcibly submerged ducks. Blood flow to muscles of the head did not decline, presumably because these muscles are supplied by branches of the carotid artery. Blood flow through the carotid artery to the brain was maintained during forced submersion, and, presumably, its arterial branches also did not constrict (Johansen, 1964). The second exception is that of the web of the foot (Djojosugito *et al.*, 1969). Provided the ducks were not alarmed or excited during the forced submersion, blood flow through the A-V anastomoses of the web was maintained. The authors speculated that flow through such A-V shunts allowed blood with high O_2 content to enter the venous pool and return to the heart. However, unlike in seals, increased microsphere deposition in the lungs has not been reported during forced submersions of ducks.

5.2.2 Simulated dives: birds

Simulated dives of birds in a pressure chamber have only been conducted in two studies, one of Adélie and gentoo penguins, and the other of king penguins (*Aptenodytes patagonicus*) (Kooyman *et al.*, 1973c, Ponganis *et al.*, 1999a). In simulated dives of king penguins, heart rate averaged 30 bpm, about 20% of pre- and post-submersion values (Ponganis *et al.*, 1999a). Although not published, similar decreases in heart rate occurred in the other two species.

5.2.3 Surface swimming: birds

Cardiovascular responses during surface swimming in birds have been examined in most detail in tufted ducks (Bevan and Butler, 1992a, Butler *et al.*, 1988, Woakes and Butler, 1983). In tufted ducks, heart rate increased to two-fold above the resting value when the duck was swimming at a metabolic rate almost four times the resting value. Cardiac output and heart rate increased in parallel, and the increase in blood flow was primarily distributed to the working hind limb muscles. Blood flow to the flight muscles, gastrointestinal tract, and kidneys declined while that to the heart increased slightly. Hepatic blood flow was unchanged. At a swimming metabolic rate twice the pre-exercise rate, hind limb muscle blood flow increased three-fold. Such a classic exercise response should be expected in a duck breathing continuously while swimming at the surface.

In graded swimming exercise of emperor penguins to a maximum metabolic rate eight times the value at rest, both submerged and surface heart rates increased with work load (Kooyman and Ponganis, 1994). Surface heart rates were consistently greater than submerged values. Submerged heart rates even at the lowest workloads were above values at rest, and maximum heart rates at the surface were about three times greater than values at rest. Thus, these heart rate responses to exercise were similar to those seen in flume-swimming sea lions. Again, presumably in this situation, even during submergence, there was an increase in blood flow to the primary underwater locomotory muscles. Similar heart rate patterns have also been reported in other surface-swimming penguins. In perhaps the earliest investigation of cardiovascular responses of an unrestrained penguin, average heart rate in a gentoo penguin slowly swimming at the sea surface was 227 bpm, about twice the value while standing upright (Millard *et al.*, 1973). Both carotid and femoral artery blood flow were similarly increased. In gentoo penguins actively swimming in a flume, heart rate was also elevated, near 180 bpm (Bevan *et al.*, 1995b). In Adélie penguins swimming spontaneously in a flume, submerged heart rates were also almost twice the value at rest, and surface heart rates were even higher (Culik, 1992). Therefore, during surface swimming of three species of penguin, an exercise response predominated during the submerged phase. And, of course, submergence durations during surface swimming were quite short.

5.2.4 Free dives: birds

Investigations of cardiovascular function in free-diving birds have primarily focused on heart rate regulation. As will be seen in most avian species, although heart rate during a free dive decreases from the pre-dive level, it usually does not go below the heart rate level at rest. Usually, heart rate only declines to levels observed in birds swimming at the surface or in birds at rest in the water. Technically, therefore, the heart rates of most free-diving birds are not bradycardias, although they are often described as such in the literature (especially when referenced to heart rate at rest in the water or during surface swimming).

5.2.5 Free dives: ducks

One of the earliest papers to describe heart rate patterns in free-diving ducks examined heart rate responses in pochards and tufted ducks during spontaneous and feeding dives of 3–24 sec duration (Butler and Woakes, 1979). Heart rates at rest and while swimming at the surface were near 110 bpm, and 160–220 bpm, respectively. Pre- and post-dive heart rates were very high, ranging between 280 and 473 bpm. During dives, heart rates decreased to about 100 bpm and then progressively increased throughout the dive to final heart rates of 181–249 bpm (Fig. 5.11). Thus, in contrast to seals, heart rates were well above resting rates throughout most of the dive. Similar heart rate responses were also seen in redhead ducks and even in tufted ducks diving as deep as 5.5 m (de Leeuw, 1996, Furilla and Jones, 1986, 1987b). Heart rates below resting levels were only later observed in tufted ducks making extended dives (35-sec mean duration) to obtain food; at 27 seconds into such dives, heart rates were approximately 95 bpm (Stephenson *et al.*, 1986). In even longer dives when surfacing through an exit hole was blocked, heart rate at 27 seconds was 45 bpm, similar to that during forced submersion

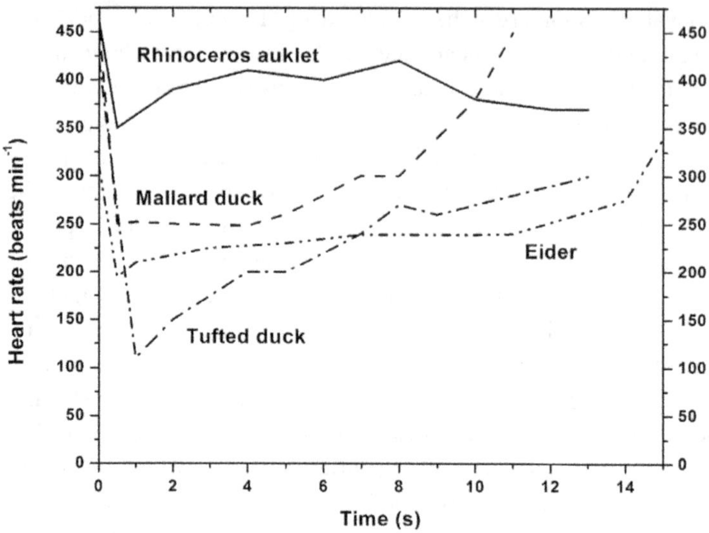

Figure 5.11 Heart rate responses during free dives of a mallard duck (*Anas platyrhnchos*), tufted duck (*Aythya fuligula*), common eider (*Somateria mollissima*), and a wing-propelled, short-duration diver, the rhinoceros auklet (*Cerorhinca monocerata*). Graphs illustrate generalized pattern of heart rate profiles. In each case, dive heart rate is above the reported heart rate at rest (110, 132, and 177 bpm in the tufted duck, eider, and auklet, respectively). Adapted from Butler and Woakes (1979), Furilla and Jones (1987a), Hawkins et al. (2000), Stephenson et al. (1992).

(Stephenson *et al.*, 1986). In the eider, during voluntary dives of up to 26-sec duration, dive heart rate, although declining below pre-dive rates, averaged 239 bpm, well above the rate of 132 bpm while resting in water (Hawkins *et al.*, 2000). Thus, only in rare long dives of ducks were heart rates below resting levels.

In 15-sec dives of tufted ducks, ischiadic artery blood flow to the leg musculature increased five-fold, and constituted 57% of cardiac output by the end of a dive (Bevan and Butler, 1992a). At the same time, brachial artery flow to the inactive wing muscles decreased by about 40%, and carotid artery blood flow to the brain and head/neck muscles more than doubled. Thus, during the high heart rates of free-diving tufted ducks, blood flow to the active leg muscles and to the head/neck increased, while flow to the inactive chest musculature and most other organs of the body was restricted by vasoconstriction. More than 99% of cardiac output during diving was directed toward the leg muscles and the head and neck. In contrast to the situation during forced submersion of ducks, perfusion to leg muscles was increased in free-diving ducks. The cardiovascular response of free-diving ducks does not isolate working muscle from the circulation. It is only during extreme dives that the dive response of forced submersions is invoked in ducks.

5.2.6 Free dives: other flighted birds

In free-diving shags/cormorants, heart rate was high prior to the dive, declined at the start of the dive, usually stabilized at depth, and then did not increase until the start of

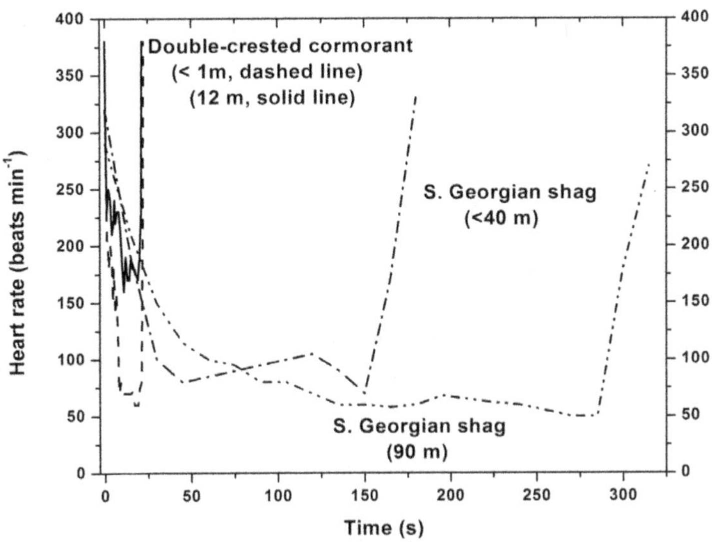

Figure 5.12 Heart rate responses of a double-crested cormorant (*Phalacrocorax auritus*) and South Georgian shags (*Phalacrocorax atriceps georgianus*) during free dives to different depth ranges. Graphs illustrate generalized pattern of heart rate profiles. Heart rate at rest in the double-crested cormorant was 138 bpm, while those of the shallow- and deep-diving shags were 85 and 95 bpm, respectively. Adapted from Bevan et al. (1997, Enstipp et al. (2001).

ascent (Fig. 5.12) (Bevan *et al.*, 1997, Enstipp *et al.*, 1999, 2001, Kanwisher *et al.*, 1981). However, mean dive heart rates were not significantly different from values at rest. The magnitude of the decrease in heart rate and, presumably, the increase in vasoconstriction varied among species. In deep-diving South Georgian shags (*Phalacrocorax atriceps georgianus*), mean minimum heart rates were 65 bpm, about 40 bpm less than the average resting heart rate (Bevan *et al.*, 1997), while in double-crested cormorants (*Phalacrocorax auritus*), heart rate did not decline below 100 bpm in dives up to 12 m in depth (Enstipp *et al.*, 2001). Despite low minimum heart rates, mean dive heart rates were still equivalent to or greater than those at rest, regardless of species.

In free-diving rhinoceros auklets (*Cerorhinca monocerata*), heart rate did not even decrease from pre-dive levels (Stephenson *et al.*, 1992). Thus, during unrestrained dives, the auklet does not exhibit a dive response; rather, it has heart rates in the 350–450 bpm range, approximately twice the 177 bpm rate at rest (Fig. 5.11). However, as demonstrated in "trapped" dives, the auklet can exhibit a dramatic dive response with heart rates as low as 20 bpm.

5.2.7 Free dives: penguins

Heart rates of diving penguins were first reported in tethered 1-min dives of gentoo penguins, in which heart rate declined to about 75 bpm during dives and rose to 180 bpm during surface intervals (Millard *et al.*, 1973). During 30-sec dives of

Humboldt penguins (*Spheniscus humboldti*) (Butler and Woakes, 1984), heart rate declined from elevated pre-dive values, but diving heart rate remained about 120 bpm, which was above or near resting values (Fig. 5.13). It was not until nine years later that diving heart rates of penguins were reported for dives of longer duration. In a 10-min dive of an emperor penguin (Kooyman *et al.*, 1992b), minimum heart rate was 30 bpm, about 50% of the resting value, and the overall dive heart rate was about 15% less than the resting value. This was the first report of an overall dive heart rate in a free-diving bird that was less than the resting value.

In contrast to the bradycardia observed in the emperor penguin, later studies of gentoo, king and macaroni penguins at sea revealed that overall dive heart rates in these species were not below resting values, even for their longest duration dives (Fig. 5.13) (Butler, 2000, Froget *et al.*, 2004, Green *et al.*, 2003). More recently, heart rate profiles in emperor penguins were documented in further detail with use of a digital ECG recorder (Meir *et al.*, 2008). In short dives of less than 6-min duration, diving heart rate, although less than at the surface, was greater than or equivalent to values at rest. However, a "true bradycardia" occurred in longer dives. In an 18-min dive of an emperor penguin, overall heart rate was 23 bpm, minimum heart rate was 3 bpm, and heart rate was 6 bpm for over 5 min (Fig. 5.13).

In summary, for routine dives among avian divers, a true bradycardia (diving heart rate less than resting value) has only been demonstrated in longer dives of emperor penguins. Minimum heart rates of diving South Georgian shags also decreased below rest values, but overall dive heart rate was not significantly different from that at rest. Reviews of avian diving heart rates have primarily focused on implications for peripheral blood flow, especially muscle blood flow, and whether an exercise response or dive response predominates during dives of most birds (Butler, 1991, Butler and Jones, 1997). If similar to the duck, blood flow during these higher heart rates of diving birds would be primarily directed toward the underwater locomotory muscles, with less flow to non-swimming muscles, splanchnic organs, and the kidneys. As suggested by Bevan and co-workers for the South Georgian shag, only when heart rate declines below resting values in the latter segments of dives would blood flow to all tissues, including the primary locomotory muscle, presumably also decrease.

In this model proposed for the South Georgian shag, an exercise response (selective increase in primary swim muscle blood flow) in the initial portion of a long dive is followed by a classic dive response (decreased muscle and organ blood flow) later in the dive. Under such conditions in the early part of the dive, myoglobin desaturation rates would be minimized, while venous O_2 content should decline due to blood O_2 extraction by exercising muscle. Such a cardiovascular response during dives of flighted avian divers is consistent with the large respiratory O_2 store and relatively small muscle O_2 store in these birds (see Chapter 4: respiratory system – about 50% – and muscle – less than 11% of total body O_2 stores). But what of organ and muscle blood flow in birds such as king and emperor penguins, in which respiratory O_2 comprises only one-third of the total body O_2 store and muscle myoglobin concentrations are 4–6 times the concentration in flighted divers?

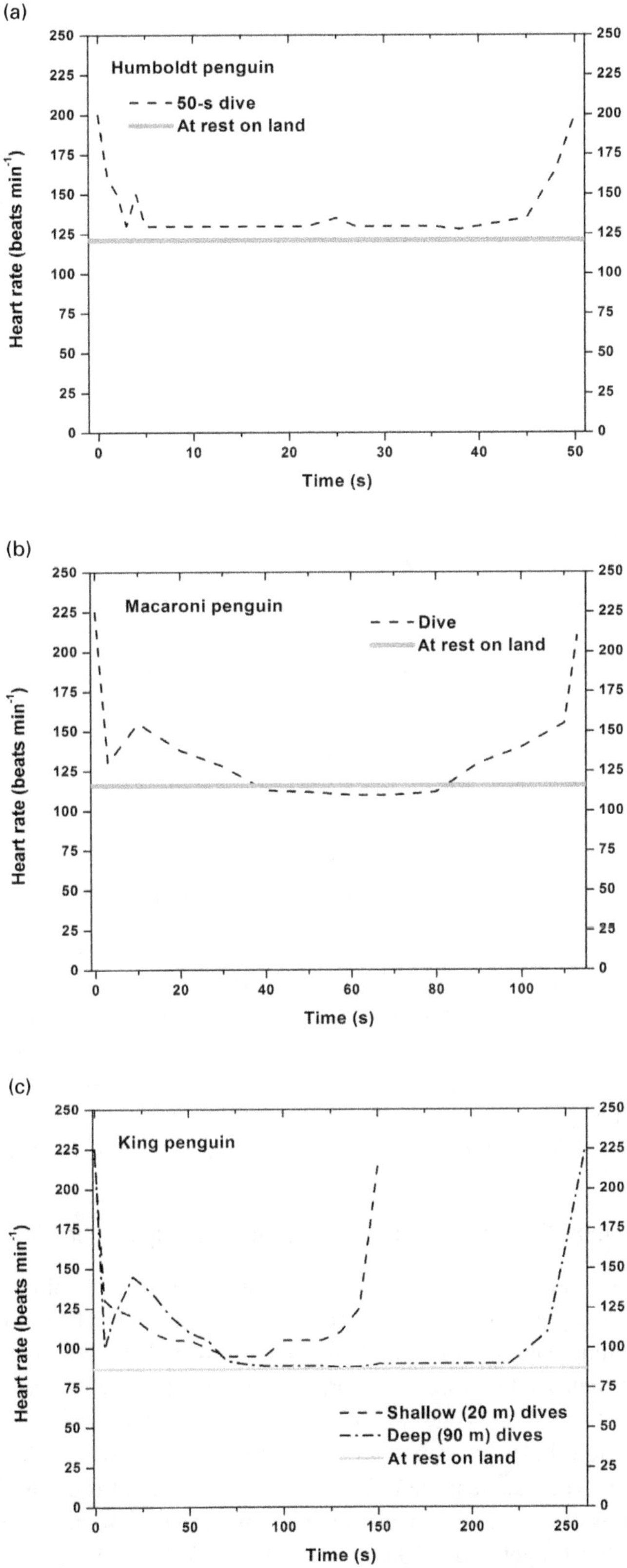

Figure 5.13 Heart rate responses during free dives of four penguin species to different depth ranges. Graphs illustrate generalized pattern of heart rate profiles in the Humboldt penguin

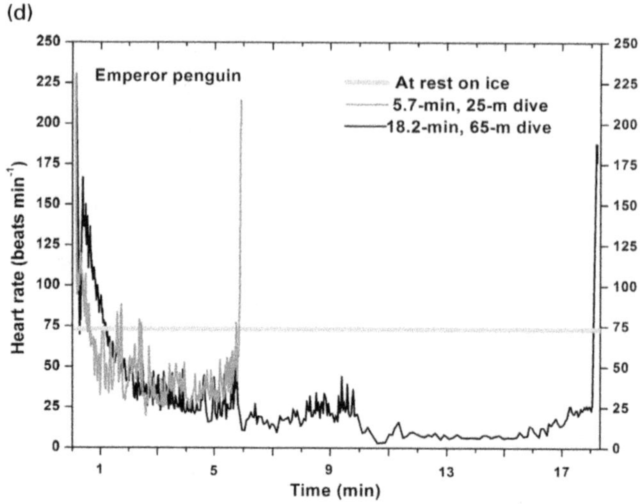

Figure 5.13 (*cont.*) (*Spheniscus humboldti*), macaroni penguin (*Eudyptes chrysolophus*), king penguin (*Aptenodytes patagonicus*), and beat-to-beat profiles in the emperor penguin (*A. forsteri*). Data for the king and macarconi penguins were collected at sea, for the Humboldt penguin in a captive tank, and for the emperor penguin at an experimental dive hole in McMurdo Sound, Antarctica. Adapted from Butler and Woakes (1984), Froget et al. (2004), Green et al. (2003), Meir et al. (2008).

5.2.8 Free dives: recent studies of emperor penguins at an isolated dive hole

Although Millard and co-workers found that femoral artery flow decreased by 75% from surface values in tethered dives of gentoo penguins (Millard *et al.*, 1973), direct measurements of organ and muscle blood flows have not been feasible in other free-diving penguins. Insight into the potential cardiovascular responses of diving penguins can be gained, however, from recent studies of heart rate patterns, blood O_2 profiles, and Mb saturation profiles of free-diving emperor penguins making relatively shallow (<100 m) dives at an isolated dive hole (Meir and Ponganis, 2009, Meir *et al.*, 2008, Williams *et al.*, 2011a).

5.2.9 Heart rate and blood O_2 profiles of emperor penguins: implications for blood flow

Heart rate profiles during dives of emperor penguins reveal that initial heart rates, though far less than pre-dive heart rates, are still relatively high (100–140 bpm) and may not reach resting values until 2–3 min into the dive (see Fig. 5.13). Once at resting levels, heart rate then gradually declines further throughout the remainder of the dive, with ever lower heart rate values in the latter portions of dives as dive duration increases. These low heart rates continue until the ascent, at which time heart rate begins to increase. The maintenance of high initial heart rates also occurs in king and macaroni penguins (Fig. 5.13), and, to some extent, even in South Georgian shags (Bevan *et al.*, 1997, Froget *et al.*, 2004, Green *et al.*, 2003). At least one function of these relatively high initial heart rates is optimization of gas exchange early in the dive.

The significance of this lung-to-blood O_2 transfer was demonstrated by increases in arterial P_{O2} and maintenance of arterial Hb saturations near 100% during much of the dive of the emperor penguin (Meir and Ponganis, 2009). Such gas exchange would also be optimized by increased movement of air through the lung secondary to differential air-sac pressures generated by the high wing stroke rates documented at the start of dives (Boggs *et al.*, 2001, van Dam *et al.*, 2002, Williams *et al.*, 2012).

In addition to O_2 uptake from the lung, another important function of such high heart rates early in the dives of emperor penguins is delivery of O_2 to tissues. If muscle blood flow increases and blood O_2 is extracted by working muscle, one would predict that venous O_2 content would decline or at most remain the same. However, venous P_{O2} and Hb saturation often increased early in the dive; in fact, venous blood could even become arterialized (Meir and Ponganis, 2009, Ponganis *et al.*, 2009). Such high values were not consistent with muscle blood flow and muscle O_2 extraction in these situations. Muscle temperature profiles in diving penguins were also not consistent with muscle blood flow (Ponganis *et al.*, 2003b, Schmidt *et al.*, 2006). In addition, overall dive heart rate and dive stroke frequency did not correlate in emperor penguins, suggesting that heart rate and muscle blood flow were not linked with muscle workload (Meir *et al.*, 2008).

Rather, the observed arterialization of venous blood early in the dive would argue for the occurrence of arterio-venous shunting during these periods. This might occur through A-V anastomoses in the extremities, as observed by the Folkow team in forced submersions studies of ducks, or, as Ponganis and associates suggested, through the wing vasculature (Djojosugito *et al.*, 1969, Meir and Ponganis, 2009, Ponganis *et al.*, 2009). Further support for the maintenance of blood flow through the wing early in the dive was also found in Kooyman's report that a nicked, bleeding wing of an emperor penguin continued to bleed underwater during the initial portion of a dive (Kooyman *et al.*, 1971a). Because of these observations, it has been suggested that one role of the early high heart rates during dives of emperor penguins is to transfer O_2 from the respiratory O_2 store via the arterial system into the venous O_2 store.

However, venous P_{O2} and Hb saturation profiles are highly variable in emperor penguins. The postulated A-V shunting may not always occur. At times, venous P_{O2} and Hb saturation progressively decline, and/or they intermittently increase during the dive (Meir and Ponganis, 2009, Ponganis *et al.*, 2009). Such patterns suggest a very plastic peripheral vascular response with periods of continuous versus intermittent blood flow and/or A-V shunting. The question of muscle blood flow during dives of emperor penguins was further addressed with near-infrared spectroscopy determinations of myoglobin (Mb) saturation in the pectoralis muscle, the primary underwater propulsive muscle of the penguin (Williams *et al.*, 2011a).

5.2.10 Muscle O_2 profiles in emperor penguins: implications for muscle blood flow

Myoglobin saturation profiles of diving emperor penguins revealed two patterns of desaturation. As illustrated in Fig. 5.14, one pattern was a monotonic decline to near 0%

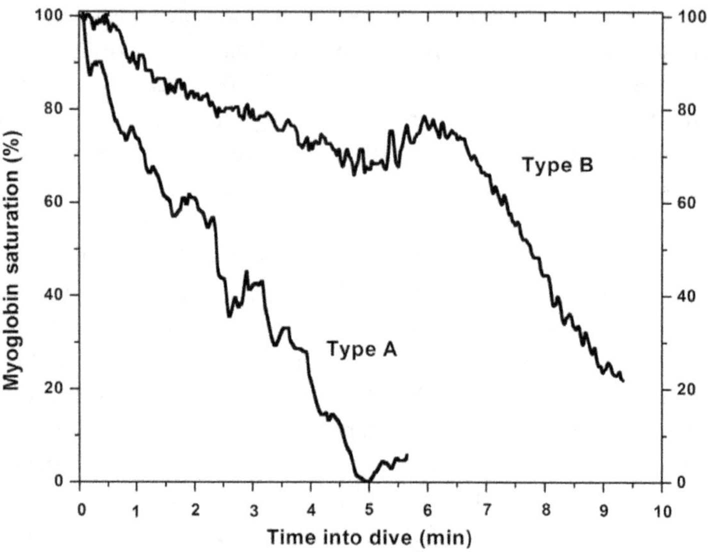

Figure 5.14 Myoglobin desaturation patterns in diving emperor penguins (*Aptenodytes forsteri*). During dives with relatively constant stroke rates, the two distinct patterns are consistent with different muscle blood flow patterns. There is probable complete ischemia in Type A, whereas muscle blood flow is probably intermittent in Type B, where myoglobin saturation is maintained throughout the mid-segment of the dive. Adapted from Williams et al. (2011a).

at about 6-min dive duration, while the other pattern was remarkable for a mid-dive plateau in desaturation followed by resumption of a monotonic decline toward 0% in the final segment of the dive. The first profile, reminiscent of the monotonic declines in Mb saturation in Scholander's forced submersion experiments in seals, was considered consistent with complete muscle ischemia (lack of blood flow) during the dive. In contrast, the plateau in the second pattern was interpreted as evidence for temporary resumption of muscle blood flow with blood-to-muscle O_2 transfer supporting aerobic metabolism in muscle during that period.

The existence of two patterns of Mb desaturation and muscle blood flow responses during dives of emperor penguins was also consistent with the highly variable venous P_{O2} and Hb saturation profiles during dives. Thus, although direct muscle blood flow measurements during dives have not been made, it appears that the peripheral vascular response during dives of emperor penguins can be highly variable both between dives and within a dive – hence, the lack of correlation between heart rate and stroke rate patterns reported in dives of emperor penguins by Meir *et al.* Cardiovascular management during a dive is not a typical exercise response in which heart rate and muscle blood flow are linked with muscle workload. There is certainly little muscle blood flow when heart rate is only 5 bpm in the latter portions of an 18-minute dive. However, when heart rate is high (100–140 bpm) in the early portions of a dive, it appears that emperor penguins have the option to perfuse muscle and supplement muscle metabolism with O_2 or to completely stop muscle blood flow, isolating muscle from the

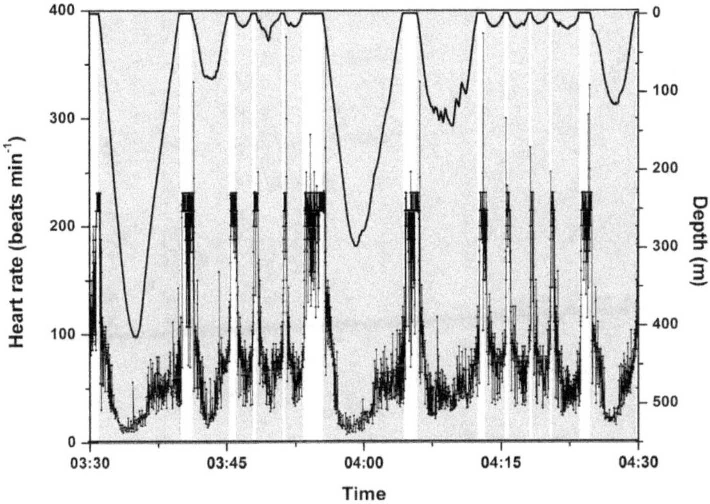

Figure 5.15 Beat-to-beat heart rate profiles of dives of an emperor penguin (*Aptenodytes forsteri*) at sea. Dives ranged from two to nine minutes and were as deep as 415 m. Mean heart rate at rest in this study was 56 bpm. Adapted from data of Wright et al. (2014).

circulation. In the latter situation during complete muscle ischemia, blood flow may be shunted through A-V shunts to increase the magnitude of the blood O_2 store for use later in the dive.

5.2.11 Heart rate profiles of emperor penguins at sea

In order to evaluate heart rate responses during deep dives of emperor penguins, heart rates have been evaluated from the dives of their foraging trips to sea during the chick-rearing period. Heart rate profiles of dives of emperor penguins at sea were characterized by (a) an initial decline from pre-dive rates; (b) a plateau at relatively high rates early in descent; (c) a further decline to lower levels during latter descent and the bottom phase of the dive; and (d) a gradual increase in heart rate during ascent (Wright *et al.*, 2014) (Fig. 5.15). As dives became deeper and longer, it was notable that (a) initial heart rates were higher; (b) the plateau heart rates were higher; and (c) the minimum heart rates were lower. Overall dive heart rate declined as dive duration increased, and was distinctly less than resting heart rate in deeper dives. Heart rates were as low as 10 bpm during the bottom phase of the deepest dives of emperor penguins. Less-than-resting heart rates in dives of emperor penguins at sea again contrasted with heart rates of the longest dives of macaroni and king penguins at sea (which were always near or above resting values). This suggested greater dependence of muscle metabolism on myoglobin-bound O_2 stores in emperor penguins than in other penguin species.

The heart rate profiles of deep dives of emperor penguins were consistent with greater gas exchange at shallow depths early in the dive, and minimization of both pulmonary

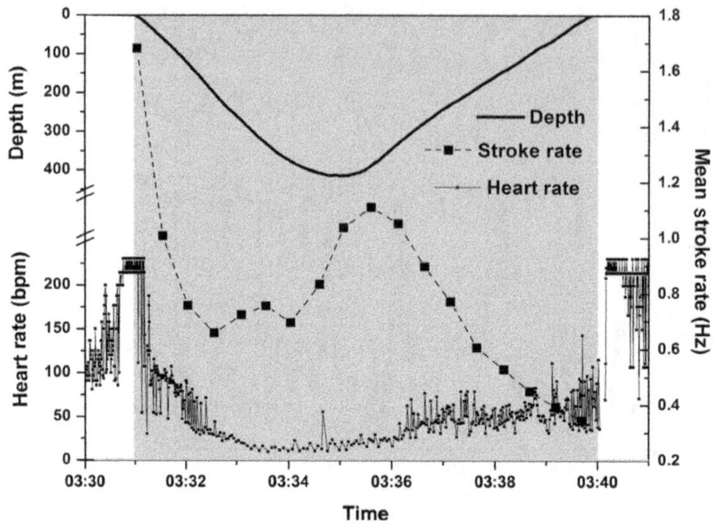

Figure 5.16 Beat-to-beat heart rate profile of a 415-m deep dive of an emperor penguin (*Aptenodytes forsteri*) in comparison to the mean wing stroke rate during dives greater than 400 m. During bottom phase of the dive, the lowest heart rates of the dive do not appear coupled to the typically high stroke rates of such dives. Data are from two different studies. Adapted from Williams et al. (2012), Wright et al. (2014).

gas exchange and peripheral O_2 delivery during the deep phases of the dive. As previously reviewed in this chapter for the similar heart rate profiles of deep-diving sea lions, this heart rate pattern in the emperor penguin should minimize N_2 absorption at depth, and conserve blood O_2 during the deepest portions of the dive. The similarity of heart rate profiles of deep-diving emperor penguins and sea lions suggest that such a heart rate pattern may be characteristic of deep dives of higher vertebrates that dive on inspiration with a relatively large respiratory O_2 store.

Lastly, the heart rate profiles of deep dives of emperor penguins at sea were not consistent with the hypothesis that exercise modifies the dive response and increases heart rate. Although the data have not yet been collected simultaneously, heart rate in the bottom phase of deep dives appears to be lowest when stroke rate is highest (Fig. 5.16) (Williams *et al.*, 2011a, Wright *et al.*, 2014). Further research with simultaneous records of heart rate and stroke rate is required to further resolve the potential mechanisms of heart rate regulation in emperor penguins.

5.2.12 Avian cardiovascular responses during free dives: summary

In summary, cardiovascular responses during free dives of birds are complex. In flighted avian divers with large respiratory O_2 stores but small muscle Mb concentrations, heart rates are usually high during dives, and an exercise response with muscle perfusion is assumed to predominate. Only in more extreme or longer dives does a more classic diving response occur with lower heart rates and presumably more peripheral vasoconstriction.

In contrast, in penguins, birds that have smaller respiratory O_2 stores but much higher Mb concentrations in muscle, heart rates are high initially during dives, but then progressively decline as dive duration increases. Such high initial heart rates early in the dive would support continued gas exchange with the respiratory O_2 reservoir. Peripheral vascular responses in emperor penguins appear to be quite variable early in the dive. Muscle may or may not be perfused during these periods of high cardiac output, and, in addition, A-V shunts may be utilized to enhance venous blood O_2 storage. During deep dives of emperor penguins, extremely low heart rates at the bottom phase of the dive should limit N_2 absorption, conserve blood O_2, and increase dependence of aerobic muscle metabolism on myoglobin-bound O_2 stores.

5.3 Neuroregulation of the dive response in mammals and birds

The neuroregulatory control of the bradycardia and vasoconstriction of the dive response during forced submersion was the subject of extensive investigation from the 1950s through the 1980s. The findings were thoroughly examined in several excellent reviews, to which interested readers are referred (Andersen, 1966, Blix and Folkow, 1983, Butler, 1982, Butler and Jones, 1997, Jones *et al.*, 1988). In both mammals and diving ducks the dive reflex of forced submersion (apnea, bradycardia, and vasoconstriction) could be elicited by facial immersion, wetting of the nostrils and glottis, or stimulation of trigeminal and glossopharyngeal nerves (Drummond and Jones, 1970, Dykes, 1974, Furilla and Jones, 1986).

5.3.1 Neuroanatomical pathways

Such stimulation of the face and nasopharynx is considered to result in inhibition of the respiratory center in the medulla with subsequent increased parasympathetic activity via the vagus nerve to slow the heart, and also with increased sympathetic nerve activity to increase peripheral vascular resistance. Paranasal stimulation results in activation of the anterior ethmoidal nerve (Panneton, 2013). As early as 1965, it was known that stimulation of the periventricular region of seals resulted in a reflex bradycardia (Van Citters et al., 1965). Trigeminal stimulation of medullary centers in the medulla of the muskrat also resulted in bradycardias (Panneton and Yavari, 1995).

In the rat, the anterior ethmoidal nerve has extensive input into the medullary dorsal horn of the spinal trigeminal nucleus (Panneton, 2013). Such linkage of naso-trigeminal stimulation with apnea, bradycardias, and elevated blood pressure in the rat has been shown to be mediated via the neurotransmitter N-methyl-D-aspartate in the pontine neurotaxic center (Dutschmann and Herbert, 1998), and via glutamate, another neurotransmitter, in the spinal trigeminal nucleus interpolaris (McCulloch *et al.*, 1995). Yet another neurotransmitter, serotonin, appears to mediate transmission from trigeminal afferent fibers to cardiac vagal neurons in the nucleus ambiguus (Gorini *et al.*, 2009). Increased sympathetic discharge during the dive reflex is associated with activation of the rostral ventrolateral medulla (Panneton, 2013). It has been proposed that the laboratory rat is an excellent model for further investigation of the central neural regulation of the dive response (McCulloch, 2012, Panneton *et al.*, 2010). A detailed review of these neural pathways in the rat is highly recommended to readers (Panneton, 2013).

5.3.2 Chemoreceptors in dabbling ducks

In dabbling ducks, which have a slow decline in heart rate in contrast to the rapid decrease of diving ducks during forced submersion, the development of the bradycardia was unaffected by nasal or upper airway stimulation, and appeared dependent on a decline in blood oxygen and subsequent carotid body chemoreceptor stimulation (Furilla and Jones, 1986, Jones and Purves, 1970, Jones *et al.*, 1982). The arterial chemoreceptor stimulation was, in turn, associated with increased vagal activity.

5.3.3 Higher cortical input

Despite the demonstrations of the reflex response to forced submersion that occurred even in decerebrate and brain-transected animals (Andersen, 1963, Drummond and Jones, 1970, Gabbott and Jones, 1991), it has always been noted and emphasized that cortical or suprabulbar input can influence the response. The variability in heart rate profiles during trained breath holds and free dives already reviewed in this chapter attest to the ability of these animals to modify their cardiovascular responses during diving. And, as reviewed in Section 5.1.11, seals are capable of altering their cardiovascular responses even during forced or trained submersions (Grinnell *et al.*, 1942, Jobsis *et al.*, 2001).

5.3.4 Reflex pathways in seals and diving ducks

Contributions from other reflex pathways to the dive response have been investigated in both seals and ducks. In seals, Elsner, Angell-James, and de Burgh Daly examined the roles of carotid chemoreceptors, lung inflation, and baroreceptors in a series of studies. Stimulation of chemoreceptors of the carotid bodies by hypoxic blood was required for maintenance of the bradycardia during submersion; isolated perfusion of the carotid bodies with oxygenated blood resulted in a tachycardia which could be reversed by reperfusion with hypoxic blood (de Burgh Daly *et al.*, 1977, Elsner *et al.*, 1977). It was also found that there was an increase in the gain of the baroreceptor reflex toward bradycardias; in other words, for a given change in blood pressure, there was a larger increase in heart beat interval (i.e., a slower heart rate) (Angell-James *et al.*, 1978). The effect of lung inflation on pulmonary stretch receptors and the Herring–Breuer reflex (inflation – tachycardia, deflation – bradycardias) was as expected as in other mammals (Angell-James *et al.*, 1981). Inflation of the lungs interrupted the bradycardia of the experimental submersion. Thus, ambient depth and lung compression have potential input into the intensity of bradycardia just as has been noted for human breath-hold divers (Ferrigno and Lundgren, 2003). For other potential effects of pressure, see Section 2.1.8. In conclusion from all these experiments, reflex input into the dive response is significant from (a) the trigeminal nerve afferents into the respiratory center; (b) carotid body stimulation by hypoxic blood; (c) lung deflation; and (d) adjustment of the gain of the baroreceptor reflex.

In tufted ducks (i.e. a diver, not a dabbling duck), chemoreceptor stimulation did not appear to play a role in the bradycardia of forced submersion (Butler and Woakes, 1982). The role of baroreceptors in regulation of the bradycardia of forced submersion has been debated (Butler and Jones, 1997). Acute baroreceptor denervation did not play a role in the forced submersion response in ducks (Jones *et al.*, 1983). Baroreceptor denervation also had no effect on the heart rate response during free dives of redhead ducks(*Aythya americana*, divers), but it abolished the free dive response in mallards (*Anas platyrhynchos*, dabblers) (Furilla and Jones, 1987a, 1987b).

5.3.5 Autonomic nervous system

In addition to investigations of the role of various reflex pathways in regulation of the dive response, contributions of the sympathetic and parasympathetic nervous system to the dive response have been investigated with the use of various pharmacological blockers (Blix and Folkow, 1983, Butler and Jones, 1997). For readers unfamiliar with the autonomic nervous system and the pharmacology of these drugs, a brief review follows. Effects of the parasympathetic nervous system, mediated by acetylcholine, can be blocked by atropine, which binds to the muscarinic acetylcholine receptor, thereby preventing the physiological response to acetylcholine. Sympathetic responses are mediated by epinephrine and norepinephrine via binding to alpha (α) and beta (β) adrenoreceptors. Alpha receptors are classified into α_1 and α_2 receptors, and it is the α_1 receptors, distributed in the peripheral vasculature, that contribute to vasoconstriction. The primary β receptors involved in cardiovascular regulation are β_1 receptors, the activation of which will increase heart rate and cardiac contractility, and β_2 receptors, which cause smooth muscle relaxation. Alpha receptor blockade can be induced with phentolamine, and selective α_1 blockade with prazocin, while beta blockade can be achieved with use of propranolol or nadolol, and selective β_1 blockade with a drug such as metoprolol.

Atropine has long been known to block the bradycardia of the dive reflex in both ducks and diving mammals (muskrats, seals), thus establishing that it is the parasympathetic nervous system, via its vagal innervation of the heart, that slows the heart rate during diving (Blix and Folkow, 1983, Butler and Jones, 1997). Alpha blockade does not affect the bradycardia of forced submersion in the muskrat (Signore and Jones, 1995), although it does block the reinforcement of the initial bradycardia in forcibly submerged ducks (Blix *et al.*, 1974). Beta blockade with nadolol does not affect the degree of bradycardia in either muskrats or ducks (Furilla and Jones, 1987b, Signore and Jones, 1995). However, consistent with activation of cardiac sympathetic fibers, propranolol does decrease ventricular contractility in the atropinized heart of nutria (*Myocastor coypus*) during forced submersion (Ferrante and Opdyke, 1969). It should also be noted that peripheral vasoconstriction occurs in the absence of bradycardias during forced submersions of atropinized ducks (Blix *et al.*, 1974) and of seals whose hearts were electronically paced at fast rates (Murdaugh *et al.*, 1968). Thus, the development of bradycardia during forced submersion is vagally mediated and not dependent on an increase in peripheral blood pressure. Furthermore, the onset of vasoconstriction is not dependent on the prior development of a bradycardia. In addition, sympathetic nerve fibers to both the heart and the peripheral vasculature are activated during forced submersion. However, the parasympathetic response dominates over activation of the cardiac sympathetic fibers and slows heart rate.

In the 1990s, Elliott *et al.* investigated heart rate responses in free-diving harbor seals which had been selectively blocked either with a muscarinic (acetylcholine) blocker, α_1 blocker, β_1 blocker, or a combination thereof (Elliott *et al.*, 2002). As would be predicted, muscarinic blockade resulted in higher diving heart rates (about twice the control value), but no change in surface heart rates. α_1 blockade resulted in a slight but significant increase in diving heart rate, and no change in surface heart rate. β_1 blockade had no significant effect on diving heart rate, but significantly decreased surface heart rate, supporting a role for sympathetic cardiac activation during the surface interval. During combined muscarinic and β_1 blockade, dive heart rates were lower than in muscarinic blockade alone, suggesting that there was also increased sympathetic cardiac nerve activity during the dive, just as had been documented during forced submersion in nutria. The study also provided evidence to support the role of vagal withdrawal during the pre-surfacing tachycardia of the seals. This increase in heart rate prior to surfacing was unaffected by either alpha or beta blockade, but was decreased by muscarinic blockade. Interestingly, although confirming insights into the autonomic regulation of heart rate in seals, none of the pharmacological blockade regimens had an effect on dive behavior (dive durations) of the freely diving, captive seals.

5.3.6 Neuroregulation of the exercise response

The review of the regulation of the cardiovascular response during diving would not be complete without consideration of neural mechanisms underlying the cardiovascular response to exercise. As already reviewed in this chapter, heart rates are variable during dives and several authors have hypothesized that at least some changes in heart rate during diving are linked to exercise. The coupling of heart rate and stroke rate in a diving mammal or bird could be potentially achieved through the same neural mechanisms responsible for the initiation of locomotion and the cardiorespiratory responses to exercise in non-diving animals. Such processes may primarily occur during the submerged phases of surface swimming or even during short-duration dives. A link between heart rate and activity may especially occur in species that dive on inspiration with a large respiratory O_2 store. These animals include otariids, dolphins, and seabirds, all of which have high heart rates (above resting levels) during sub-surface swimming and short-duration dives. However, during longer dives with heart rates well below resting levels, the neuroregulatory mechanisms that would link heart rate to stroke effort remain unclear.

The exercise response is generally considered secondary to (a) a central command mechanism which leads to the activation of muscle motor units for locomotion and to the activation of cardiorespiratory centers; (b) peripheral feedback of physical activity to higher centers from contracting muscle (exercise pressor reflex); and (c) the arterial baroreceptor reflex (blood pressure regulation: ↑blood pressure → ↓heart rate) (Kaufman and Forster, 1996, Mitchell *et al.*, 1983, Smith *et al.*, 2006, Waldrop and Iwamoto, 2006, Waldrop *et al.*, 1996). Similar to the dive response, cerebral cortical input can influence and contribute to the response. Stimulation of cell bodies in the hypothalamic locomotory region has been found to initiate both locomotory and cardiorespiratory responses; the neurotransmitter, γ-aminobutyric acid (GABA), inhibited this response (Waldrop *et al.*, 1988). In addition, isolated muscle contractions resulted in stimulation of the cells in this hypothalamic region (Waldrop and Stremel, 1989).

Classically, the increase in heart rate with exercise had been considered to be secondary to a decrease in vagal (parasympathetic) nerve activity to the heart (Rowell and O'Leary, 1990). However, vagal nerve activity does not decrease at the start of exercise in experimental models (Kadowaki *et al.*, 2011, Matsukawa, 2012). Rather, muscle stimulation has been shown to increase cardiac sympathetic nerve activity as well as renal sympathetic nerve activity, thus demonstrating a role of the sympathetic nervous system in increasing heart rate and decreasing blood flow to renal and other regional vascular beds during and at the start of exercise (Matsukawa *et al.*, 1990, 1991, 1992, Tsuchimochi *et al.*, 2002, 2009). Therefore, potential neuroregulatory pathways exist whereby stroking activity in a diving mammal or bird could activate sympathetic cardiac accelerator fibers and elicit an increase in heart rate.

5.3.7 Cardiovascular neuroregulation during dives

As reviewed by Butler (1982) in birds, and as recently suggested by Davis and Williams (2012) in mammals, the heart rate response in a diver would then be dependent on whether the dive response or the exercise response predominated. And, of course, both responses are not isolated reflexes, but can be modified by higher cortical input. Certainly, this potential interaction of neuroregulatory pathways may contribute to the variability in heart rate profiles during dives.

However, it must also be remembered that during the classic dive response of a forced submersion, there is both extreme bradycardia (via the vagus nerve – parasympathetic) and extreme vasoconstriction (via sympathetic nerves) at the same time. Pharmacological blockade studies indicate that both the cardiac accelerator fibers and peripheral vascular fibers of the sympathetic nervous system are activated. In terms of heart rate control during forced submersion, the parasympathetic system predominates over activation of the sympathetic nervous system. In breath holds with less intense bradycardias, muscle blood flow persists presumably because vasoconstriction is less due to decreased sympathetic nerve activity (Jobsis *et al.*, 2001, Ponganis *et al.*, 2006b). Because of this decrease in sympathetic activity, the higher heart rate in these situations should be secondary to decreased vagal afferent activity to the heart and not increased cardiac sympathetic activity. Otherwise, the sympathetic vascular response would have to decrease while the activity of sympathetic cardiac accelerator fibers increased.

A key question is whether differential activation of cardiac and vasomotor sympathetic fibers occurs during the dive response. Although differential activation of cardiac and peripheral vascular sympathetic fibers have been demonstrated in human and experimental animal studies (Lovick, 1987, Morrison, 2001), the previously reviewed pharmacological blockade data from seals and nutria indicated that both cardiac and peripheral vascular sympathetic activities were increased during diving (Elliott *et al.*, 2002, Ferrante and Opdyke, 1969). In addition, increased parasympathetic activity and acetylcholine concentrations are known to override both cardiac sympathetic activity and the effects of elevated concentrations of epinephrine and norepinephrine (Levy, 1971, O'Leary, 1993). Lastly, during the bradycardia induced by nasopharyngeal stimulation in the rabbit, simultaneous activation of both cardiac sympathetic activity and vagal nerve activity occurred (Paton *et al.*, 2005, 2006). Therefore, variation in heart rate during dives is probably most dependent on parasympathetic control.

If increases in heart rate during diving were secondary to a further increase in sympathetic nerve activity, a simultaneous withdrawal of vagal activity would presumably still be required. Furthermore, if increased sympathetic nerve activity beyond that already induced by the dive response were to induce elevations in heart rate during dives, increased activation of the densely distributed sympathetic fibers in the extramuscular, proximal branches of the aorta would constrict those vessels, preclude an increase in blood flow to muscle, and result in an increase in blood pressure. Consequently, if the sympathetic nervous system links muscle workload to heart rate in order to increase muscle blood flow in diving marine mammals, it should primarily involve sole activation of sympathetic cardiac accelerator fibers. Thus, the exact mechanisms by which an exercise response might modify the dive response remain unclear.

Based on the extensive work by Butler and co-workers on birds (see Sections 5.2.5–5.2.6), and the recent observations and suggestions of Hindle *et al.* (2010) and Davis and Willliams (2012) for diving mammals, perhaps a simpler conceptual model for neuroregulation of heart rate during dives of most marine mammals is that an exercise response always occurs during active dives but that it can be reduced and even overwhelmed by activation of the dive response. Partial activation of the dive response would (a) limit the increase in heart rate; (b) only partially constrict the proximal arterial tree (remember the dense proximal sympathetic innervation of those vessels); (c) allow for a limited exercise response with small elevations in heart rate and increased blood flow to active muscle via release of local vasodilators; (d) result in continued or decreased flow to splanchnic organs dependent on the intensity of exercise; and (e) allow for heat dissipation/conservation via peripheral vascular thermoregulatory responses. Complete activation of the dive response would result in the dive reflex of forced submersion, with (a) intense vagal stimulation overwhelming the cardiac sympathetic responses associated with exercise as well as the dive reflex; and (b) intense peripheral sympathetic activation and vasoconstriction in the proximal aortic tree, thus inducing widespread ischemia to most major organs and muscle, even despite the release of local vasodilators (except for possible A-V shunts and thermoregulation). Such a model is consistent with findings during forced submersions, sleep apnea, short-duration dives, and long-duration dives. And, of course, many factors, including nasopharyngeal reflexes, venous return, atrial stretch receptors, depth, lung inflation/deflation, blood gases and pH, exercise intensity, fright, and suprabulbar control could affect the response, dependent on the species and the nature of a given dive.

More research is clearly needed to further define the mechanisms by which cardiovascular responses are regulated during diving. Although laboratory models can further define anatomical and molecular pathways, well-designed investigations of cardiovascular responses during free diving are also needed because of the many factors that can influence the nature of the response.

6 Adaptations in cardiovascular anatomy and hemodynamics

This chapter will review the cardiovascular anatomy and hemodynamics of marine mammals and seabirds. Anatomical features will be considered first as many of these adaptations underlie hemodynamic stability during periods of decreased heart rate and peripheral vasoconstriction. And, as will be seen, the functions of some unique vascular structures are still incompletely understood. The latter section of the chapter will address hemodynamics, with particular attention given to the balance of myocardial oxygen supply and demand in diving mammals.

6.1 Marine mammals: the heart

The four-chambered hearts of marine mammals are typical of mammals. The heart is 0.5–1.3% of body mass in most pinnipeds and small cetaceans, but slightly less, about 0.3–0.5% of body mass, in the great whales (Drabek, 1975, Drabek and Burns, 2002, King, 1983, Ridgway and Johnston, 1966, Ridgway and Kohin, 1995, Slijper, 1962). Slight differences in heart to body mass ratios may be secondary to variations in physiological demands between similarly sized species (Drabek and Burns, 2002, Ridgway and Kohin, 1995). Chamber size, stroke volume, and resting cardiac output and heart rate (where measured) are also in the general mammalian range, and in agreement with mammalian allometric equations (Drabek, 1975, Ponganis and Kooyman, 1999, Ponganis *et al.*, 1990, 1991, 2006b, Smodlaka *et al.*, 2008, Sommer *et al.*, 1968).

It is notable that the hearts of diving mammals are dorso-ventrally flattened, the right ventricular chambers are enlarged, and, dependent on the species, the relative thickness of the right ventricular wall may or may not be increased (Bisaillon *et al.*, 1987, Drabek, 1975, Drabek and Burns, 2002, Ochrymowych and Lambertsen, 1984, Rowlatt, 1981, Rowlatt and Gaskin, 1975, Smodlaka *et al.*, 2008, Truex *et al.*, 1961). The dorso-ventral flattening results in re-positioning of the right ventricle so that its filling is not compromised by chest compression during dives. The changes in chamber size and wall thickness of the right ventricle are considered to improve its function during chest compression, lung collapse, and probable increased pulmonary vascular resistance during dives.

Although delayed closure has been reported in some young seals, both the foramen ovale and ductus arteriosus are closed in adult seals and cetaceans as in other mammals (Dennison *et al.*, 2011, Lydersen *et al.*, 2002, Slijper, 1961, 1962, van Nie and van der Kamp, 1988). Therefore, utilization of an intermittent fetal circulatory pathway does not appear to be a mechanism to bypass a potential increase in pulmonary vascular resistance during diving. Furthermore, a patent ductus arteriosus would interfere with the windkessel functions of the ascending aorta and aorbic bulbs in various species (see descriptions below).

6.2 Marine mammals: general vascular features and the spleen

General aspects of the arterial and venous systems in marine mammals are remarkable for several features (Butler and Jones, 1997, Elsner and Gooden, 1983). First, dense sympathetic nerve innervation of proximal as well as distal arteries in seals may represent a mechanism by which the intense sympathetic vasoconstriction of the dive response can be maintained independent of local tissue metabolite induced vasodilation in the periphery (White *et al.*, 1973). Second, venous capacitance is highly developed, especially in phocid seals and whales (Galantsev, 1991, Harrison and Tomlinson, 1956, Slijper, 1962). This includes a large hepatic sinus and posterior vena cava, the latter of which in seals has been estimated to be capable of storing one-fifth of the seal's blood volume (Elsner *et al.*, 1964b). Presumably, this large venous capacitance is related to the large blood volume of seals, already discussed in Chapter 4. In addition, contraction of the large spleen in seals (see next paragraph) also injects blood into the hepatic sinus (Thornton *et al.*, 2001).

A third anatomical feature of the circulatory system in some marine mammals is a large spleen that appears to be a significant storage organ for red blood cells. Increased splenic volumes in several pinniped species, and extensive sympathetic nerve innervation and smooth muscle development in the splenic capsule, are consistent with this storage role (Bryden and Lim, 1969, Cabanac *et al.*, 1997, 1998, 1999, Cabanac, 2002, Castellini and Castellini, 1993, Hurford *et al.*, 1996, Ponganis *et al.*, 1992b, Qvist *et al.*, 1986, Schumacher and Welsch, 1987). As reviewed in these papers, estimated masses of fully dilated spleens in pinnipeds (1–7% of body mass in most animals; up to 14% in some juveniles and end-of-fast adult females) are greater than the 0.2–0.4% values in humans, and are beyond or in the upper range of relative splenic masses found in terrestrial mammals with large splenic storage capacities (0.6–1.9% in horses, goats, sheep, and dogs). Fluctuations in hematocrit between resting and diving states, or anesthetized and stressed states also support such a role for the spleen in seals (Castellini *et al.*, 1986, Hurford *et al.*, 1996, Ponganis *et al.*, 1992b, Qvist *et al.*, 1986). It has been estimated that 30% of the blood volume or about 50% of red cell mass can be stored in the spleen in Weddell seals (Hurford *et al.*, 1996). Indeed, agonal contractions of the spleen can result in extremely high hematocrits (90%) in the hepatic sinus of Weddell seals (Kooyman, 1968). Depending on rheological properties of a given species' blood (Chapter 4), storage of red cells in a large spleen during periods of inactivity and low oxygen demand could lower hematocrit, decrease blood viscosity, and ultimately decrease the work of the heart (Elsner and Meiselman, 1995, Meiselman *et al.*, 1992, Wickham *et al.*, 1989, 1990a).

As in the regulation of muscle blood flow during dives, modulation of sympathetic tone is probably key to the regulation of the frequency and magnitude of splenic contraction and relaxation during dive/surface activity. It was originally proposed that the spleen dilated completely with each surface interval and then contracted fully during each dive, essentially acting as a "splenic scuba tank" to inject oxygenated red blood cells into the circulation (Hochachka, 1986a, Qvist *et al.*, 1986, Zapol, 1987). Subsequent investigations have revealed that although the spleen could contract rapidly with sympathetic stimulation, splenic relaxation times were too slow to allow for complete dilation of the spleen during short surface intervals (Hurford *et al.*, 1996, Thornton *et al.*, 2001). For example, in the elephant seal, return of splenic volume to its pre-submersion value required 18–22 minutes. Incomplete dilation of the spleen during surface intervals or eupneic intervals on land was also consistent with maintenance of high or only slightly reduced hematocrits during those periods (Castellini *et al.*, 1986, 1988). Thus, during serial dives with short surface intervals, it appears that there would only be minor oscillations in splenic volumes and hematocrits. Large changes in splenic volume and hematocrit would probably only occur during and after prolonged surface intervals. As already pointed out, a primary effect of a lowered hematocrit during these prolonged rest intervals would be a decrease in blood viscosity, which could contribute to a lower myocardial workload (Elsner and Meiselman, 1995). It is also notable that the magnitude of the calculated blood oxygen store is not significantly different whether the blood volume is assumed to be distributed between the arterial and venous system vs. the spleen and arterial/venous systems (Hurford *et al.*, 1996, Ponganis *et al.*, 1992b, 1993a,).

6.3 Marine mammals: the extradural venous system of phocid seals

A fourth aspect of the circulatory system, again well developed in both seals and whales, is the extradural venous system (Harrison and Tomlinson, 1956). This vein is located within the vertebral canal and above the spinal cord; it receives blood flow from the brain, back, and pelvic regions (Fig. 6.1). It is linked with both the posterior and anterior vena cava via paravertebral communicating veins. In seals and cetaceans, the extradural vein is the primary venous drainage of the brain; the internal jugular vein is poorly developed or absent. The function of such a prominent vertebral venous system in these animals is unclear. It has been noted in humans that extradural vein flow may contribute to brain temperature regulation, and that, in the upright posture, the vertebral veins, kept open by attachment to the bony walls of the vertebral canal, are the primary cerebral venous drainage because blood flow decreases in the jugular veins due to venous collapse in the upright posture (Gauer and Thron, 1965, Zenker and Kubik, 1996).

In seals, the direction and magnitude of flow within the extradural vein vary with the respiratory cycle (Nordgarden *et al.*, 2000, Ronald *et al.*, 1977). During a breath hold (apnea), extradural vein flow is low and has been reported to vary in direction (i.e., rostrally or caudally). However, during breathing (eupnea), extradural vein flow is increased and rostral in direction. It has been estimated that the extradural vein may contribute as much as 20% of eupneic venous return to the heart via its intrathoracic connections to intercostal veins, the enlarged right azygous vein, and the anterior vena cava (Ponganis *et al.*, 2006a). These intrathoracic connections allow transmission of

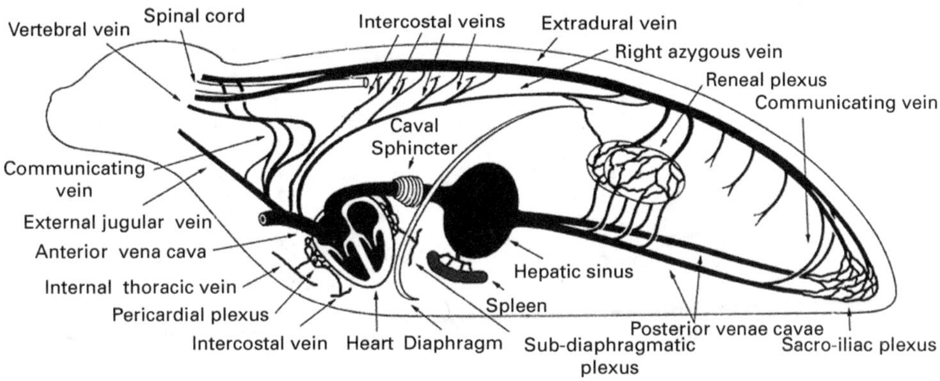

Figure 6.1 The extradural vein and its connections to other venous vessels in the phocid seal. Not all vascular plexuses are depicted, and structures are not drawn to anatomical scale. Adapted from King (1983).

negative inspiratory pressures within the chest to the extradural vein. This enhances rostral blood flow during the inspiration since the ligamentous attachments of this vein to the vertebral canal prevent collapse of the vein. Therefore, at least one important function of the prominent extradural vein in seals may be its contribution to eupneic venous return to the heart, especially during inspiration. It has also been postulated that, in phocid seals, the extradural vein provides a route of venous return to the heart when the posterior vena cava and hepatic sinus are compressed when a seal squeezes in and out of an ice hole (Blix, 2011).

6.4 Marine mammals: thermoregulatory structures

Vascular structures involved in thermoregulation represent the fifth remarkable feature of the circulation in marine mammals. The parallel pattern of counter-flowing arteries and veins, characteristic of countercurrent exchange units, is present in the dorsal fins (Fig. 6.2), flukes, and flippers of cetaceans (Scholander and Schevill, 1955). These arrangements, characteristic of blood vessel groupings in the limbs of many animals, are considered to conserve body heat by transferring heat from warm, out-going arterial blood to cool venous blood returning from the limb. A superficial venous system, which does not return in conjunction with out-going arteries, also occurs in the skin. These veins, which have well-developed muscular walls, represent a route by which heat can be dissipated to the environment during periods of thermal stress.

An additional structural adaptation observed in pinnipeds is the presence of numerous arterio-venous (A-V) anastomoses in the skin (Bryden and Molyneux, 1978). These structures represent a mechanism by which blood bypasses tissue capillary networks, and instead shunts directly from the arterial to venous system. The A-V anastomoses are distributed uniformly over the body surface of phocid seals, but in otariids they are

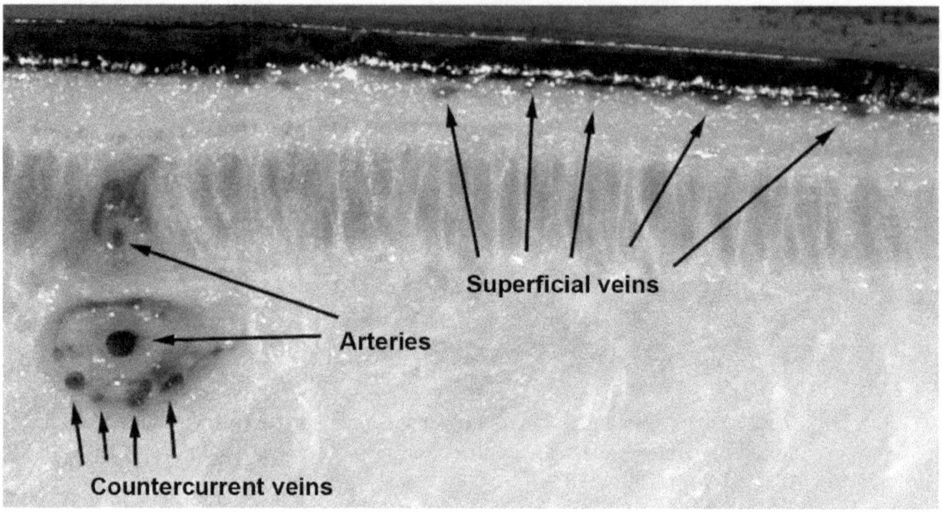

Figure 6.2 Countercurrent arrangement of central artery and surrounding veins in cross-section of the tail fluke of a bottlenose dolphin, *Tursiops truncatus*. Return of blood through the central veins conserves heat, while return through the superficial veins near the skin provides a route for heat loss.

found in greater densities in the flippers. It is presumed that flow through these vessels allows heat exchange at the skin surface. In addition, they represent a blood flow route through which venous blood can become arterialized (as reviewed in Chapter 5).

Countercurrent anatomy has also been observed in the reproductive organs of dolphins and pinnipeds (Rommell *et al.*, 1995). It has been proposed that return of blood from the skin via vascular anastomoses allows relatively cool venous blood to prevent overheating of these organs. Temperature patterns along the length of the colon in the dolphin have been consistent with this hypothesis.

Countercurrent vascular anatomy in the tongues of mysticete whales also apparently contributes to heat conservation during feeding (Heyning and Mead, 1997, Werth, 2007). Another vascular structure that may contribute to thermoregulation in baleen whales has been named the corpus cavernosum maxillaris (Ford and Kraus, 1992, Ford *et al.*, 2013, Heyning *et al.*, 1993). This structure has been described as extending along the midline of the entire length of the palate in both right and bowhead whales (*Eubalaena glacialis*, *Balaena mysticetus*). It has been proposed that engorgement of the structure allows excess heat to be lost into the water engulfed into the mouth during feeding.

6.5 Marine mammals: the aorta

A sixth important characteristic of the circulatory system in marine mammals involves the ascending aorta. In pinnipeds, again particularly in phocid seals, the aortic root (ascending aorta) is dilated, forming the so-called aortic bulb (Drabek, 1975, Rhode *et al.*, 1986). The bulb can accommodate the stroke volume ejected by the heart, and it is

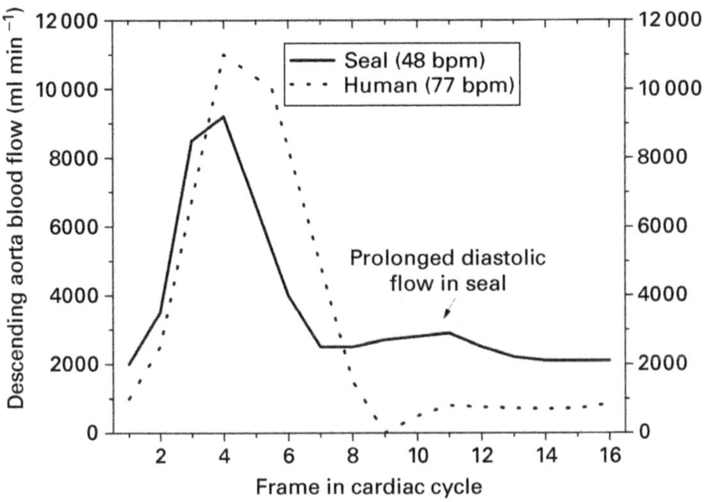

Figure 6.3 Blood flow in the descending aorta during a single cardiac cycle in humans and seals, at heart rates of 78 and 48 bpm. Maintenance of higher blood flow during the diastolic (relaxation) phase of the cardiac cycle in the seal was considered secondary to slow contraction of the distended aortic bulb during diastole in the seal. Adapted from Hochachka (2000), based upon Thornton et al. (1997).

more distensible than the distal aorta. It has been proposed that the aortic bulb acts as a windkessel: gradual contraction of the bulb due to elastic fibers within its wall contributes to maintenance of blood flow, especially to the brain and heart during diastole (relaxation phase of the cardiac cycle) (see Fig. 6.3).

The ascending aorta, aortic arch, and proximal carotid arteries of whales are also very compliant, and have also been hypothesized to act as a windkessel and preserve blood flow during diastole (Shadwick and Gosline, 1994). This is especially important in whales because long diastoles accompany slow heart rates. Low heart rates can occur in whales due to both their large body masses and the cardiovascular responses which occur during diving.

Maintenance of blood flow and pressure during a long diastole is, of course, critical to the brain, but also to the heart. This is because coronary perfusion occurs during diastole when the heart is relaxed. Myocardial flow is dependent on the diastolic blood pressure as the driving pressure. Thus, species which are either large or have more profound diving responses are likely to have some form of an aortic windkessel. Such maintenance of aortic blood flow during the slow heart rates and long diastoles of trained submersions was confirmed in Thornton's MRI studies of elephant seals (Hochachka, 2000, Thornton *et al.*, 1997).

A compliant ascending aorta may also contribute to a reduction in the impedance that the left ventricle must pump against during the peripheral vasoconstriction of the diving response. This reduction in afterload will decrease the work and oxygen consumption of the heart, which is of course beneficial to a diver with a limited oxygen supply.

A non-compliant descending aorta, an essential component of the windkessel, has also been demonstrated in fin whales (Gosline and Shadwick, 1996, Lillie *et al.*, 2013). The stiffness of the aorta has been attributed to a high collagen content. A non-compliant descending thoracic aorta has also been postulated to prevent aortic collapse in the event of depth-induced changes in transmural pressure of the thoracic aorta (Lillie *et al.*, 2013).

6.6 Marine mammals: retia mirabilia

The retia mirabilia (wonderful nets) of cetaceans are another feature of vascular anatomy that have long been noted by anatomists (McFarland *et al.*, 1979, Melnikov, 1997, Slijper, 1962). These plexuses of anastomosing arteries and veins occur along the vertebrae and base of skull, and are especially prominent in the thorax (Fig. 6.4). The vascular retia are well developed in dolphins; they are also found in sirenians. Excellent descriptions of the retia of the narwhal and beluga whale emphasize that the thoracic and cervical retia originate from the carotid and intercostal arteries to supply the spinal cord and brain, and that these retia are (a) primarily composed of anastomotic arterial vessels; (b) embedded in fatty connective tissue (white fat); and (c) poorly innervated (Vogl and Fisher, 1982, Vogl *et al.*, 1981). Notably, the thoracic retia are larger in the better diver (narwhal) of these two closely related cetaceans. The thoracic retia are also well developed in deep-diving kogiid whales, comprising almost 10% of thoracic volume in comparison to 5% in the bottlenose dolphin (Piscitelli *et al.*, 2010). These retia also extend ventrolaterally, and are reported to envelop the apex of the lung in kogiid whales. Retia mirabilia are also extensively developed in the sperm whale (*Physeter macrocephalus*) (Melnikov, 1997). The vertebral rete of the sperm whale extends to the posterior lumbar vertebrae, whereas the rete in dolphins often ends anterior to the lumbar spine.

The thoracic rete in the dolphin is supplied by vessels from the aorta, which anastomose to form a complex, spongiform structure beneath the dorsal thoracic wall. This vascular tissue extends around the vertebrae into the vertebral canal, and forms the primary arterial blood supply to the brain in dolphins (Galliano *et al.*, 1966, McFarland *et al.*, 1979, Viamonte *et al.*, 1968). The carotid arteries are vestigial or absent. The spinal meningeal artery in dolphins extends from the rete to the brain. Although mean blood pressure in the spinal meningeal artery of dolphins is equal to the aortic pressure, it is notable that the pressure is non-pulsatile; there is no systolic peak or diastolic trough. Thus, the cetacean brain appears to receive non-pulsatile blood flow. The significance of such a flow pattern as well as the function of the retia are unknown.

Slijper also reported the presence of large venous retia in the abdomens of whales. Hypotheses about the role of the retia have included windkessel functions, intrathoracic vascular engorgement to prevent "lung squeeze" during diving, thermoregulation, and modification of composition of the blood, including the trapping of air bubbles (Hui, 1975, Vogl and Fisher, 1982). Ophthalmic retia also occur in marine mammals (Ninomiya *et al.*, 2014). These structures are considered to warm the eye.

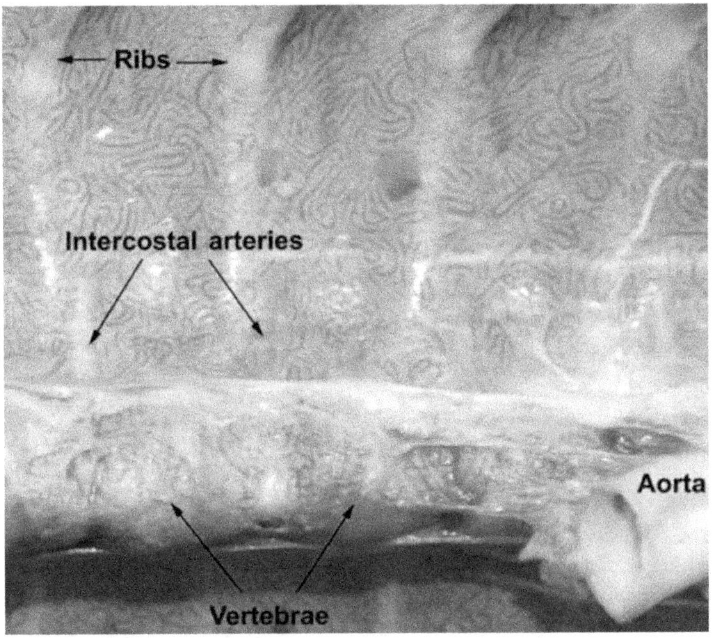

Figure 6.4 Ventral view of the dorsal thoracic rete of the common dolphin, *Delphinus delphis*. Intercostal arteries extend from the aorta, over the vertebrae, and into the rete vessels.

6.7 Marine mammals: vena caval and portal vein sphincters and the pericardial plexus

This review of vascular anatomy of marine mammals will conclude with three more remarkable modifications to the vascular system. In most pinnipeds, the posterior vena cava is associated with a striated muscle sphincter at the level of the diaphragm (Figs 6.1, 6.5, 6.6). Again, this is most well developed in phocid seals (Harrison and Tomlinson, 1956, King, 1983). The sphincter is innervated by the right phrenic nerve, and is located cranial to the large hepatic sinus and inferior vena cava. Relaxation/contraction of the sphincter has been observed angiographically during forced submersions, and it is assumed that this is a mechanism to regulate venous return to the heart during diving bradycardias (Elsner and Gooden, 1983, Harrison and Tomlinson, 1956, Ronald *et al.*, 1977). Vena caval sphincters are also described in whales (Slijper, 1962); they presumably regulate blood return from the large venous capacitance vessels in the abdomen (posterior vena cava and venous rete).

Smooth muscle sphincters have also been observed surrounding intrahepatic segments of the portal vein in some, but not all, cetaceans (Hilton and Gaskin, 1978, Simpson and Gardner, 1972). The primary function of such sphincters is unknown, but their contraction presumably regulates blood flow from the gastrointestinal tract into the liver.

Another venous structure, especially developed in phocid seals, is the pericardial venous plexus (Harrison and Tomlinson, 1956). This extensive venous network (Fig. 6.6) is connected to intercostal veins, internal thoracic veins, and a sub-diaphragmatic venous

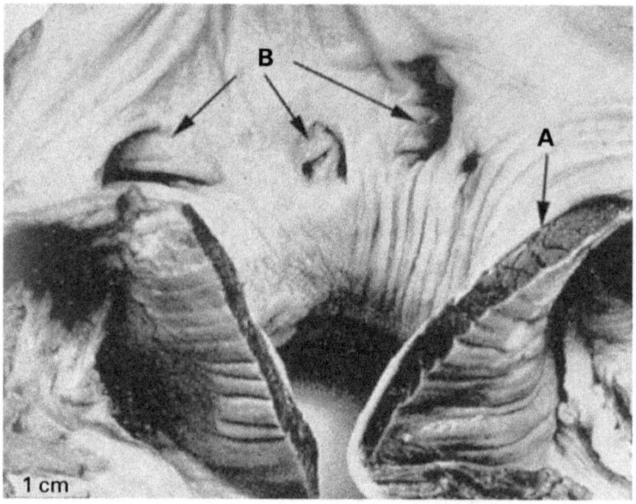

Figure 6.5 The posterior vena caval sphincter of an elephant seal. **A:** transected muscle of the caval sphincter. **B:** Opening of the pericardial venous plexus into the thoracic segment of the posterior vena cava. Adapted from Harrison and Tomlinson (1956).

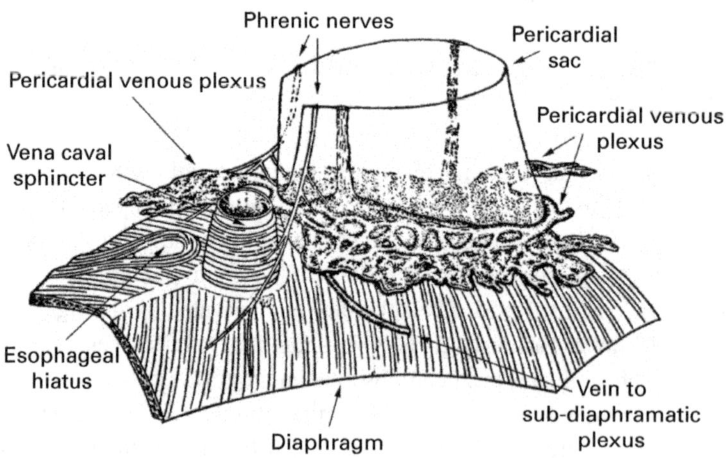

Figure 6.6 Pericardial venous plexus in the phocid seal. Adapted from Harrison and Tomlinson (1956).

plexus; it empties into the posterior vena cava just cranial to the vena caval sphincter (Fig. 6.5). This venous plexus is especially developed in the better-diving seals such as elephant seals. Its significance is unknown, although it has been reported to be associated with brown fat and has been hypothesized to function in thermoregulation (Blix *et al.*, 1975). Blood return to the heart via venous plexuses has also been observed angiographically during forced submersions (Hol *et al.*, 1975). In addition, the pericardial venous plexus may be part of the intrathoracic pathway for eupneic venous return to the heart. Engorgement of this plexus may also contribute to the prevention of barotrauma during deep dives.

6.8 Diving birds: the heart

Details of the anatomy of the circulation are far less documented in diving birds than in marine mammals. Heart size scales allometrically with body mass in birds (Schmidt-Nielsen, 1984), and, on a mass-specific basis, birds, in general, have larger hearts than mammals (Hartman, 1955). In a comparison of geese, dabbling ducks, and diving ducks, relative heart mass was greatest in the diving ducks (Bethke and Thomas, 1988). Among the smaller penguin species, cardiac mass comprised 0.7–1.1% of body mass, and both chinstrap and rockhopper penguins (*Pygoscelis antarctica*, *Eudyptes chrysocome moseleyi*) had larger relative heart masses than predicted by allometric equations (Drabek, 1989, 1997, Drabek and Tremblay, 2000). The right ventricle comprised 21% of heart mass in emperor penguins (*Aptenodytes forsteri*), but only 7% in the little penguin (*Eudyptula minor*). In general, the relative size of the right ventricle in penguins increased with dive capacity.

6.9 Diving birds: vascular anatomy

The vascular anatomy in diving birds is notable for several features, but, in general, is similar to that of non-diving birds. As in many other birds, these notable characteristics include the ophthalmic rete mirabile, and countercurrent heat exchange arrangements of the vessels in the wings and legs. The ophthalmic rete mirabile in penguins is considered thermoregulatory in nature, decreasing heat loss from the eye and conserving cranial temperature during diving (Frost *et al.*, 1975, Midtgård, 1983). The ophthalmic retia of flighted birds also contribute to maintenance of a constant brain temperature, but via a different mechanism. In flighted birds, brain temperature does not increase despite elevations in arterial temperature, because relatively cool venous blood from the evaporative surfaces of the cornea and upper airway lowers the temperature of arterial blood in the retia before the blood reaches the brain (Bernstein *et al.*, 1979, Midtgård, 1983). In addition, these vascular structures form a non-pulmonary gas exchanger, enhancing brain oxygenation due to venous gas exchange via the vascularized surface of the upper airway during the hyperventilation of flight (Bernstein *et al.*, 1984). However, during the breath hold of a diving penguin, neither evaporative heat loss nor gas exchange over the mucosa of the upper airway will occur.

The other notable vascular features of penguins, again also found in other birds, are countercurrent arrangements of arteries and veins in the wings and legs (Arad *et al.*, 1989, Frost *et al.*, 1975, Midtgård, 1980b, 1981). In penguins, the brachial artery forms multiple branches in the axilla, forming a humeral plexus in which each arterial vessel is associated in parallel with small veins (Frost *et al.*, 1975, Thomas and Fordyce, 2007,

2012, Trawa, 1970). This classic countercurrent arrangement allows for the transfer of heat from warm, out-going arterial blood to cooler, returning venous blood, and, thus, acts to conserve body heat (Scholander, 1955). The marginal vein, a large vein in the proximal wing but not in contact with the humeral plexus, appears to be a route by which the axillary heat exchanger can be bypassed (Frost *et al.*, 1975, Thomas and Fordyce, 2007). Countercurrent arrangements of blood vessels also occur in the legs of penguins and other birds, again allowing for heat conservation in cold surroundings (Frost *et al.*, 1975, Midtgård, 1980b, 1981).

Arterio-venous anastomoses (AVAs) occur in the skin of birds, especially in the eyelids and webs of the feet (Midtgård, 1980a, 1984). These structures are considered to play a role in thermoregulation as heat loss through the feet is considerable during flight (Baudinette *et al.*, 1976). Increased blood flow through AVAs in the brood patch also contributed to maintenance of egg temperature during incubation (Midtgård, 1984, 1985). The role of AVAs in thermoregulation during diving is undetermined, although, as will be reviewed in Chapter 8, subcutaneous brood patch temperature did occasionally increase in diving king penguins (Schmidt *et al.*, 2006). In addition, as already discussed, blood flow through the AVAs in the webs of the feet of ducks can be maintained during forced submersions, and may contribute to elevated venous oxygen contents (Djojosugito *et al.*, 1969).

6.10 Hemodynamics

Hemodynamic function during the "dive response," and, in particular, regulation of myocardial oxygen supply and demand, will be examined in this section. The function and associated metabolic costs of the heart as a pump are primarily dependent on four parameters. These factors are heart rate, contractility, preload (the venous filling of the heart), and afterload (the impedance or resistance the heart must pump against) (Berne and Levy, 1998, Braunwald, 1971, Colin *et al.*, 2003, Warltier *et al.*, 2000). Each of these factors will be reviewed in this section in relation to the dive response and to the cardiovascular anatomical adaptations in diving mammals and birds. Particular attention will be paid to phocid seals, in which both the dive response and cardiovascular anatomy have been extensively investigated.

6.10.1 Bradycardia and myocardial oxygen supply/demand

Provided venous return to the heart is adequate and stroke volume (the volume of blood ejected per heart beat) is constant, higher heart rates will proportionately increase the heart's output of blood to the periphery (cardiac output = heart rate and stroke volume). An increased heart rate will also obviously increase myocardial oxygen demand because of the increased rate of cardiac contractions. At the same time, increased heart rates potentially worsen myocardial oxygen supply because of shortened diastolic intervals (the time between heart beats when the heart is relaxed and, importantly, when blood oxygen delivery to the heart occurs). In fact, heart rate is probably the most critical determinant of myocardial oxygen supply and demand. Hence, the treatment of patients with coronary artery disease with beta sympathomimetic blockers to slow heart rate and prevent episodes of tachycardia and chest pain (angina). Thus, the slow heart rate of the seal during a dive not only decreases peripheral blood flow and oxygen delivery, but also optimizes myocardial oxygen supply and demand by increasing diastolic perfusion time for the heart while at the same time decreasing myocardial workload.

6.10.2 Contractility and myocardial oxygen supply/demand

Enhanced myocardial contractility can also increase cardiac output and myocardial oxygen demand. Myocardial contractility, the force or wall tension with which the heart contracts (the rate of myocardial fiber shortening) is increased by sympathetic stimulation and is often measured as the maximum rate of pressure change (dp/dt_{max}) of the left ventricle. Left ventricular dp/dt_{max} as well as left ventricular wall segment shortening are decreased during the bradycardia of forced submersion in seals (Elsner *et al.*, 1985, Kjekshus *et al.*, 1982). Thus, myocardial contractility and its associated oxygen demand are decreased during the dive response, even despite the well-documented increase in sympathetic tone. In this regard, it is notable that vagal (parasympathetic) stimulation, which slows heart rate during the dive response, is known to decrease the effect of sympathetic stimulation on myocardial contractility (Casadei, 2001, Ferrante and Opdyke, 1969).

6.10.3 The caval sphincter, cardiac preload, and myocardial oxygen supply/demand

Cardiac preload also affects the performance and oxygen consumption of the heart. Ventricular filling pressure, central venous pressure (for the right heart), and left atrial pressure (for the left heart) have often been used as indices of preload. Preload is more accurately defined as ventricular end diastolic volume, and, perhaps ultimately, is best considered as ventricular end diastolic wall stress, a function of ventricular volume, filling pressure, and wall thickness (Norton, 2001). The variables are all related by the Law of LaPlace, where the wall tension of a sphere = (chamber pressure × chamber radius) / 2, or wall stress = (chamber pressure × chamber radius) / (2 × wall thickness). Increases in wall tension are associated with increased myocardial oxygen consumption, and, if the heart becomes over-distended, this can result in decreased myocardial performance. In this light, the regulation of preload during the dive response is critical to both the performance of the heart and to its oxygen consumption.

In phocid seals, the vena caval sphincter, which has been observed to open and contract with each cardiac cycle during forced submersions, represents a unique mechanism to regulate venous return and optimize preload of the heart (Elsner *et al.*, 1971, Ronald *et al.*, 1977). During the severe bradycardia of forced submersion, decreased left ventricular wall segment dimensions, decreased left ventricular dp/dt_{max}, unchanged left ventricular diastolic pressures, unchanged central venous pressures, and decreased stroke volumes were all consistent with decreased ventricular filling, decreased wall tension, and decreased contractility (Blix *et al.*, 1983, Elsner *et al.*, 1985, Hol *et al.*,

1975, Kjekshus *et al.*, 1982). All these factors would minimize myocardial oxygen demand. In addition to forced submersions, stroke volumes in harbor seals are also decreased during rest apneas, underwater flume swimming, and simulated dives in pressure chambers (Ponganis *et al.*, 1990, Sinnett *et al.*, 1978). There is only one report of ventricular dilation, in the right ventricle, during the forced submersion of a seal (Blix and Hol, 1973). Presumably, such dilation is minimized through action of the caval sphincter (Blix, 2011). Lastly, in terms of optimizing myocardial perfusion, the lack of any increase in left ventricular diastolic pressure will aid in maintenance of myocardial perfusion because the driving pressure for coronary perfusion during diastole is the difference between the arterial pressure and the intraventricular pressure.

6.10.4 The aortic bulb, afterload, and myocardial oxygen supply/demand

The fourth major determinant of myocardial oxygen consumption is afterload. Afterload has often been equated with the resistance or impedance that the heart must pump against. Blood pressure and systemic vascular resistance have both been used as indices of afterload. In addition, blood viscosity can contribute to impedance. Ultimately, however, afterload is perhaps best defined by the stress or tension in the wall of the ventricle (Norton, 2001); in terms of the Law of LaPlace, afterload = (systolic ventricular pressure × systolic ventricular radius) / (2 × wall thickness). There is usually no gradient between systolic ventricular pressure and aortic pressure, hence the frequent use of blood pressure or vascular resistance as an index of afterload. The greater the wall stress of the ventricle, the greater the oxygen consumption of the heart.

In seals, the effects of widespread vasoconstriction and increased systemic vascular resistance on the afterload and oxygen consumption of the left ventricle are minimized by the aortic bulb (Drabek, 1975, Rhode *et al.*, 1986). The distensible walls of the aortic bulb decrease the afterload the ventricle would otherwise face. Thus, the workload of the heart during the dive response is decreased by yet another mechanism. In addition, as previously described in this chapter, the elastic wall of the expanded bulb gradually contracts during diastole, maintaining both blood flow and diastolic blood pressure which is necessary for coronary perfusion. It is remarkable that the physiological effects of the aortic bulb in seals are exactly analogous to the therapeutic effects of the intra-aortic balloon pump, a medical device used to treat severe heart failure in human patients (Papaioannou and Stefanadis, 2005). Inflation of the pump's balloon within the aorta during diastole maintains blood pressure and coronary perfusion as the heart relaxes. Deflation of the balloon at the start of systole decreases afterload and improves the output of the failing heart. The aortic bulb in seals accomplishes the same optimization of myocardial oxygen supply/demand.

6.10.5 Pulmonary vascular resistance and the right ventricle

As will be discussed in Chapter 13, the right ventricle also faces a potential increase in afterload during dives due to hypoxia, hypercarbia, and alveolar collapse, all factors that increase pulmonary vascular resistance in most mammals (Barer and Shaw, 1971, Hyman and Kadowitz, 1975, Sommer *et al.*, 2008, Viles and Shepherd, 1968). It is notable, however, that pulmonary artery pressures do not increase during sleep apnea in seals (Ponganis *et al.*, 2006a). Evidence for mechanisms that potentially decrease pulmonary vascular resistance and the workload of the right ventricle include (a) increased production of nitric oxide, a vasodilator, due to elevated concentrations of substrate precursors in the blood (Soegaard *et al.*, 2012), and (b) hypoxia-induced pulmonary vasodilation (Olson *et al.*, 2010). In addition, the thickened right ventricular walls of some marine mammals should allow for development of higher ventricular pressures to accommodate any increases in pulmonary vascular resistance. However, a hypertrophied ventricle increases myocardial oxygen consumption, a potential limitation during times of limited oxygen supply.

6.10.6 Summary: hemodynamics and myocardial oxygen supply/demand

In summary, the performance of the heart during the dive response of marine mammals is not just remarkable for the degree of bradycardia. The regulation of preload, afterload, and contractility in the optimization of myocardial oxygen supply/demand are just as significant. Indeed, the dive response of the seal provides the clinician with a textbook example of therapeutic principles for the treatment of the ischemic, failing human heart. In addition, from the perspective of an experienced cardiac anesthesiologist, perhaps the most notable performance of the seal's heart is during the "ascent tachycardia" and the surface transition to breathing, especially after a long dive. Here, the heart increases its workload under all the worst conditions for myocardial oxygen supply/demand – hypoxia, hypercarbia, respiratory acidosis, elevated hematocrit (and potentially increased blood viscosity), lung collapse with possibly increased pulmonary vascular resistance, and widespread peripheral vasoconstriction.

6.11 Avian hemodynamics

Hemodynamic investigations of diving birds have primarily been conducted on ducks during forced submersion. In penguins, measurements of cardiac output and stroke volume have only been determined in emperor penguins at rest (Kooyman *et al.*, 1992b).

During the bradycardia of forced submersion in ducks, cardiac output was reduced, and stroke volume was either reduced or unchanged (Folkow *et al.*, 1967, Jones and Holeton, 1972). The right atrium and ventricle were found to be distended, and central venous pressures were elevated (Aakhus and Johansen, 1964, Folkow *et al.*, 1967, Johansen and Aakhus, 1963). Angiographic evidence was consistent with constriction of distal pulmonary arterial vessels and an increase in pulmonary vascular resistance (Aakhus and Johansen, 1964). The left ventricle was also slightly dilated, even at the end of systole, and left ventricular end diastolic pressure was elevated; such findings were all consistent with a reduced stroke volume despite the increase in left ventricular end diastolic volume (Aakhus and Johansen, 1964, Johansen and Aakhus, 1963). The reduction in stroke volume was attributed to decreased left ventricular contractility secondary to parasympathetic stimulation. This was demonstrated by a decrease in left ventricular dp/dt during forced submersion of the duck (Folkow and Yonce, 1967).

These hemodynamic findings in forcibly submerged ducks indicate that decreased heart rate and contractility are the primary mechanisms which serve to optimize myocardial oxygen supply and demand. In contrast to the seal, preload does not appear to be regulated during submersion of the duck; both filling pressures and ventricular volumes are increased. These factors potentially lead to increased wall tension and decreased gradients for coronary perfusion in the duck. There also do not appear to be any biomechanical adaptations, such as the aortic bulb in the seal, that would reduce afterload during the widespread vasoconstriction and elevated systemic vascular resistance during forced submersion of the duck. Hemodynamics during the bradycardias of emperor penguins may be different, but have not been investigated. Notably, preliminary studies indicate that the pulmonary artery of emperor penguins demonstrates hypoxic pulmonary vasodilation similar to that of the sea lion (K. Olson, personal communication).

7 Muscle and locomotory work

The biochemistry and energy metabolism of muscle is an essential component of the physiological basis of an animal's dive capacity. Despite the development of significant blood hypoxemia and muscle ischemia during a dive, muscle must continue to function to provide propulsion. And the work that muscle must perform during a dive is, in turn, dependent on hydrodynamics, drag, buoyancy, swim speed, and the pattern/efficiency of locomotion. Ultimately, hydrodynamics, swim patterns, and cost of transport are important because they increase the efficiency of diving and extend the duration of aerobic metabolism during a dive by decreasing the rate at which body oxygen stores are consumed. Therefore, this chapter will first review muscle fiber types and biochemistry in diving mammals and birds, and then outline the various factors that contribute to the locomotory workload of muscle during a dive.

In evaluating potential muscle adaptations and their significance to diving performance, it is most relevant to examine the muscles that are primarily responsible for underwater movement (see Table 7.1). An excellent review and reference source for the anatomy underlying the various locomotory styles of marine mammals is chapter 8 in Berta *et al.*'s 2006 textbook, as well as in Howell's classic text on the anatomy of marine mammals (Howell, 1930).

7.1 Muscle fiber types

The contractile properties and metabolic characteristics of a muscle are determined by its fiber type composition. Fibers are typically classified on the basis of myosin heavy chain isoforms and oxidative/glycolytic capacities into slow twitch oxidative (SO, Type I), fast twitch oxidative, glycolytic (FOG, Type IIa), and fast twitch glycolytic (FG, Type IIb) fibers (Zierath and Hawley, 2004). However, there are multiple heavy chain isoforms that can be expressed at different levels and even in different combinations in various muscles and species (Flück and Hoppeler, 2003, Rivero *et al.*, 1999). Hence, there were at least as many as five different Type II fiber subtypes in one study (Rivero *et al.*, 1999). It should also be remembered that myosin heavy chain isoforms are reflected in the contractile properties of muscle. Hence, myosin heavy chain immunohistochemistry classifies contractile fiber types, but not necessarily the metabolic capacities of the fibers. The expression of genes for enzymes in the oxidative and glycolytic pathways are under the influence of multiple signaling pathways (Flück and Hoppeler, 2003).

Table 7.1 Underwater locomotory strokes and primary propulsive muscles of various divers.

Diver	Locomotory stroke	Primary muscles	References
Cetaceans	Dorso-ventral oscillation of tail fluke	Epaxial/hypaxial muscles of the spine	A
Manatees	Dorso-ventral oscillation of trunk and tail fluke	Epaxial/hypaxial muscles of the spine	B
Otariids	Stroke of fore-limb flippers	Pectoralis, latissimus dorsi, shoulder muscles	C
Phocids	Lateral stroke of hind flippers	Longissimus dorsi-iliocostalis muscle complex	D
Walruses	Lateral stroke of hind limb flippers (primary)	Longissimus dorsi-iliocostalis muscle complex	E
Sea otters	Dorso-ventral undulation of hind trunk	Para-spinal muscles	F
Foot-propelled avian divers	Stroke of hind limbs	Leg muscles	G
Wing-propelled avian divers	Stroke of wings	Pectoralis and supracoracoideus muscles	G

A: Fish and Hui, 1991, Fish *et al.*, 1988, Pabst, 1990, Strickler, 1980; B: Berta *et al.*, 2006, Kojeszewski and Fish, 2007; C: English, 1976, 1977, Feldkamp, 1987a; D: Bryden and Felts, 1974, Fish *et al.*, 1988, Pierard, 1971; E: Berta *et al.*, 2006, Gordon, 1981; F: Berta *et al.*, 2006, Williams, 1989, Tarasoff, 1972; G: Baudinette and Gill, 1985; Butler, 1991.

Classically, slow twitch fibers have been considered to underlie slow, prolonged activities such as marathon running, whereas fast twitch fibers contribute to rapid and/or powerful motions such as those in flight, sprinting, or weight lifting. Consistent with this concept, a larger percentage of slow twitch fibers predominates in the locomotory muscle of a large cetacean, such as the narwhal, in comparison to smaller dolphin species; based on size alone, the larger animal will probably have a lower stroke rate than its smaller counterparts (Sato *et al.*, 2007, Williams *et al.*, 2011b). However, this increase in slow twitch fiber composition is not the case in all large cetaceans; beaked whales, and fin whales have 50–75% fast twitch fibers (see Table 7.2).

With the exception of pilot whales and beaked whales (Velten *et al.*, 2013), fiber types in the primary locomotory muscles of diving mammals and birds are not unusual in comparison to non-diving species (Table 7.2). In slow-swimming beaked whales, the epaxial muscles had high myoglobin content, but had only 20% slow twitch oxidative fibers of small fiber diameter that were uniformly distributed among the more numerous large-diameter, fast twitch fibers with high glycolytic capacity (Velten *et al.*, 2013). The authors postulated that the slow twitch fibers provided routine stroke power and that these fibers utilized their own myoglobin-bound oxygen as well as that from the uniformly distributed surrounding fast twitch fibers. They also hypothesized that the 80% fast twitch fibers were utilized (a) during faster stroking rates and (b) after the myoglobin-bound oxygen depot was exhausted. The authors postulated that larger-diameter fibers allowed for relatively less metabolic demand due to a lower surface-to-volume relationship, fewer membrane ion pumps, and an overall slower muscle metabolic rate.

Table 7.2 Muscle fiber type distribution in muscles of marine mammals and diving birds.

Species	Muscle	%SO	%FT	%FOG	%FG	Reference
Dolphins (3 species)	Longissimus dorsi	~50	~50			A
Narwhal *Monodon monoceros*	Longissimus dorsi	79	21			B
Pilot whale *Globicephala macrorhynchus*	Epaxial muscle	62	38			C
Beaked whales (*Mesoplodon*, 4 species)	Epaxial muscle	17–23	77–83			C
Pygmy sperm whale *Kogia breviceps*	Longissimus dorsi	53	47			D
Fin whale *Balaenoptera physalus*	Longissimus dorsi	46	54			E
Antarctic fur seal *Arctocephalus gazella*	Pectoralis	10	90	28	61	F
California sea lion *Zalophus californianus*	Neck/shoulder muscle	44	56	18	38	G
Grey seal *Halichoerus grypus*	Longissimus dorsi	55	42	6	36	F
Harbor seal *Phoca vitulina*	Longissimus dorsi	47	53	53		F, H
Weddell seal *Leptonychotes weddellii*	Longissimus dorsi	67	33	33		I
Northern elephant seal *Mirounga angustirostris*	Longissimus dorsi	100	0			J
Sea otter *Enhydra lutris*	Gastrocnemius	56	44	2	42	G
Tufted duck *Aythya fuligula*	Gastrocnemius – red	15	85	85	0	K
	Gastrocnemius – white	0	100	46	54	K
Atlantic puffin *Fratercula arctica*	Pectoralis	0	100	100		L
Cassin's auklet *Ptycoramphus aaleuticus*	Pectoralis	0	100	100		L
Emperor penguin *Aptenodytes forsteri*	Pectoralis	0	100	100		M

References: A: Bello *et al.*, 1985, Ponganis and Pierce, 1978, Suzuki *et al.*, 1983; B: Williams *et al.*, 2011b; C: Velten *et al.*, 2013; D: Kielhorn *et al.*, 2013; E: Hochachka and Foreman, 1993; F: Reed *et al.*, 1994a; G: Ponganis and Pierce, 1978; H: Watson *et al.*, 2003; I: Kanatous *et al.*, 2002; J: Moore *et al.*, 2014; K: Turner and Butler, 1988; L: Kovacs and Meyers, 2000; M: Ponganis *et al.*, 1997a.

The percentage fast twitch fibers are further subdivided into component fiber types where available. Abbreviations: SO: slow twitch oxidative fibers, FT: fast twitch fibers, FOG: fast twitch oxidative, glycolytic fibers, FG: fast twitch glycolytic fibers.

In contrast to the beaked whales, in pilot whales, which exhibit "underwater sprints" with high stroke rates, one-third of fibers were fast twitch, and two-thirds slow twitch, of which half were slow twitch oxidative fibers and half were a unique fiber type – slow twitch oxidative and glycolytic (Velten *et al.*, 2013). Fiber diameters were not enlarged as in the beaked whales. The authors postulated that slow twitch fibers were utilized for routine swimming, and that the fast twitch and more glycolytic fibers were for underwater sprints. And, of course, glycolysis was also possible in both fiber types when myoglobin-oxygen stores were depleted.

Among birds, the 100% FOG fiber composition of pectoral muscle in wing-propelled divers is characteristic of birds with flapping flight (Butler, 1991). Similarly, the fiber type distribution of the central, white portion of the gastrocnemius muscle and the lateral, red portion of that muscle in the legs of tufted ducks are also characteristic of other birds (Butler, 1991, Turner and Butler, 1988). However, differences in fiber type distribution can be associated with different locomotory patterns. For example, in contrast to a hummingbird or finch, which have 100% FOG fibers in their flight muscles, the pelican and albatross also have SO fibers in addition to FOG fibers (Meyers and Stakebake, 2005, Rosser *et al.*, 1994, Welch Jr. and Altshuler, 2009). The slow twitch fibers are considered to help maintain wing position during soaring of the pelican and albatross.

It is also worth noting that different distributions of fast twitch fibers into FOG fibers and FG fibers have been reported in harbor seal muscle (47% FG vs no FG fibers) (Reed *et al.*, 1994a, Watson *et al.*, 2003). Both studies, however, found the same proportion of total FT fibers (53%). This difference may, in part, be due to differences in histochemical staining techniques, and the probable range of oxidative capacities within FOG fibers. The earlier work utilized a myofibrillar ATPase stain dependent on acid/alkaline inhibition of ATPase isoforms in combination with serial oxidative enzyme stains, which allowed assessment of oxidative capacity based on subjective interpretation of the relative stain intensity. The latter study relied on immunohistochemical stains for the different myosin heavy chains, and therefore was specific for the contractile properties of the fibers (i.e., there were Type IIa, but not Type IIb fibers in the harbor seal). The fiber-specific oxidative and glycolytic capacities of those Type IIa fibers were not independently measured, but were assumed to be high (i.e., a FOG fiber) on the basis of enzyme activities in whole-muscle homogenates and on fiber type-specific findings in other species. These findings raise the issue of whether the terms, I, IIa, and IIb, should be used interchangeably with SO, FOG, and FG, especially in species in which characterization of individual fiber contractile properties and oxidative/glycolytic capacities have not been conducted. This issue was again exemplified in Weddell seal muscle (Kanatous *et al.*, 2002). By immunohistochemistry, 67% of fibers were Type I, and 33% Type IIa; there were no Type IIb fibers. By classic fiber type terminology, the Type I fibers would correspond to SO fibers, and Type IIa fibers to FOG fibers. One would predict high oxidative capacities in the muscle. Yet, muscle mitochondrial density and citrate synthase activities, both indices of oxidative capacity, were low in Weddell seal muscle, presumably as an adaptation for low aerobic metabolic rates during hypoxic conditions (Kanatous *et al.*, 2002). Therefore, readers must remember the complexities and limitations of fiber-typing techniques in interpretation of fiber type composition, especially in relatively exotic, non-standard species such as marine mammals and seabirds.

7.2 Muscle enzyme activities, energy substrates, and mitochondria

The oxidative and glycolytic enzyme activities measured in the muscles of these diving animals are consistent with the large percentages of oxidative fibers (SO and FOG) in their muscles, and, in general, are characteristic of non-divers (Baldwin, 1988, Baldwin *et al.*, 1984, Blix and From, 1971, Blix *et al.*, 1970, Butler, 1991, Castellini and Somero, 1981, Castellini *et al.*, 1981, Davis and Guderley, 1987, 1990, Hochachka and Foreman, 1993, Kanatous *et al.*, 1999, 2002, 2008, Mill and Baldwin, 1983, Polasek *et al.*, 2006, Ponganis and Pierce, 1978, Ponganis *et al.*, 1997a, Reed *et al.*, 1994a, Turner and Butler, 1988).

The findings of normal glycolytic enzyme activities in these animals contrasts with earlier views that an enhanced glycolytic capacity underlied the diving abilities of marine mammals (George and Ronald, 1973, Simon *et al.*, 1974, Storey and Hochachka, 1974). It was even postulated that anaerobic pathways generating succinate and alanine contributed to diving ability (Hochachka *et al.*, 1975). Subsequent studies with documentation of non-elevated glycolytic enzyme activities in multiple tissues as well as demonstration of a lack of elevation of alanine in blood of diving seals (Castellini *et al.*, 1981, Guppy *et al.*, 1986) did not support these views.

This is not to say that glycolysis does not play a role in diving, but muscle of non-diving animals is already well-adapted for anaerobic energy production via glycolysis (Zierath and Hawley, 2004); hence, the lack of further elevation of glycolytic enzyme activities. Skeletal muscle glycogen concentrations are also not elevated (Groscolas, 1990, Kerem *et al.*, 1973, Williams *et al.*, 2012). However, there is enhanced buffering capacity in muscle of most marine mammals, up to 60% greater than terrestrial controls (Castellini and Somero, 1981, Lestyk *et al.*, 2009, Noren, 2004). Buffering capacity is also elevated in most penguin species, but in puffins and murres it lies between values for pigeons and pheasants (Castellini and Somero, 1981, Davis and Guderley, 1990, Mill and Baldwin, 1983). This buffering is considered secondary to histidine-containing dipeptides in the sarcoplasm (Castellini and Somero, 1981, Crush, 1970, Harris *et al.*, 1990), and it undoubtedly contributes to a more stable intracellular pH, especially during long dives.

Lastly, in regard to anaerobic energy sources, phosphocreatine concentrations are also not elevated in diving mammals and birds (Blix, 1971, Stephenson and Jones, 1992b, Stephenson *et al.*, 1997, Williams *et al.*, 2012). It is notable that despite the high myoglobin concentration in emperor penguin muscle (see Chapter 4), the "normal" glycogen and phosphocreatine stores of the emperor penguin can provide about six times more ATP than the myoglobin-bound O_2 (Williams *et al.*, 2012). However, due to the low metabolic rate of muscle during diving of emperor penguins, significant glycolysis and post-dive wash-out of lactate do not occur until after dives of about 6-min duration (Ponganis *et al.*, 1997b). In addition, after an 11-min dive with no muscle blood flow, it has been estimated that phosphocreatine would be only 56% depleted, and muscle glycogen less than 10% depleted (Williams *et al.*, 2012). Thus, there appears to be a large anaerobic energy reserve in emperor penguin muscle even without elevations in glycolytic enzyme activities, glycogen content, or phosphocreatine concentrations. The same argument has also been applied to Weddell seal muscle (Butler and Jones, 1997).

Muscle oxidative capacity has been assessed with examination of both enzyme activities and mitochondrial volume densities. It is notable that citrate synthase activity and mitochondrial volume densities were elevated in the upper mammalian range in two smaller pinnipeds (harbor seal, northern fur seal) and in the intermediate-sized Steller sea lion (Kanatous *et al.*, 1999, Watson *et al.*, 2007). However, in the Weddell seal, neither of these parameters were increased (Kanatous *et al.*, 2002, 2008). Differences may be secondary to a lower rate of energy demand in the Weddell seal, a larger, slower-stroking, longer-duration diver. Relatively low citrate synthase activities were also reported in two other large phocids, crabeater seals and leopard seals (Hochachka and Foreman, 1993). This latter reference also emphasized that higher pyruvate kinase/lactate dehydrogenase ratios favored pyruvate oxidation in large phocids, whereas lower ratios in dolphins and fin whales favored glycolysis (Hochachka and Foreman, 1993).

Remarkable among all the pinnipeds examined, however, is the relatively high activity of hydroxylacyl-CoA dehydrogenase, a key enzyme in fatty acid oxidation (Kanatous *et al.*, 1999, 2002, Polasek *et al.*, 2006, Reed *et al.*, 1994a). This enzyme pattern is consistent with the high free fatty acid turnover rate in flume-swimming seals (Davis *et al.*, 1991). In addition, muscle mitochondria from adult elephant seals demonstrated enhanced fatty acid utilization over pyruvate and increased phosphorylation control (Chicco *et al.*, 2014). Thus, these animals appear geared toward fat metabolism. Increased numbers of intramuscular lipid droplets have also been reported in several species of marine mammals (Kanatous *et al.*, 1999, 2002, Tulsi, 1975, Watson *et al.*, 2007), and in slow twitch versus fast twitch fibers of beaked whales (Velten *et al.*, 2013). Intramuscular triglycerides may serve as an important free fatty acid source because both the ischemia of diving (Chapter 5) and the relatively low capillary density in marine mammal muscle would decrease the potential for delivery of blood-borne fatty acids to muscle during dives (Kanatous *et al.*, 2001, 2002, Reed *et al.*, 1994a).

Much less is known about the potential for fatty acid metabolism in diving birds. In the tufted duck, capillarity of the leg muscle fibers is 70–80% that of the pectoralis (Turner and Butler, 1988), but, as reviewed in Chapter 5, most of the cardiac output during diving is delivered to the leg musculature. The ratio of hydroxyacyl-CoA dehydrogenase to citrate synthase in the leg is similar to that in pectoral muscle, indicating similar propensity for fatty acid oxidation in the two muscles (Turner and Butler, 1988). In emperor penguins, the density of lipid droplets in the pectoral muscle was about three times that in the leg muscles (Ponganis *et al.*, 1997a).

Another ultrastructural observation on muscle mitochondria in marine mammals is that most of the mitochondria (greater than 80%) are interfibrillar (Kanatous *et al.*, 1999, 2002, Velten *et al.*, 2013, Watson *et al.*, 2007). In contrast, 20% or more of mitochondria are subsarcolemmal in terrestrial mammals. The authors have proposed that the increased mitochondrial volume and interfibrillar distribution of mitochondria increase mitochondrial surface area and decrease distances for diffusion of O_2 from myoglobin, thus maintaining aerobic metabolism under hypoxic conditions. High mitochondrial volumes and homogeneous mitochondrial distributions have also been found in the liver, kidneys, and stomach of seals, which the authors again propose supports aerobic metabolism under hypoxic conditions (Fuson *et al.*, 2003). In diving birds, much less is known about mitochondrial distribution, although 75–90% of mitochondria were interfibrillar in both the chest and leg muscles of emperor penguins (Ponganis *et al.*, 1997a).

The potential role of hypometabolism in prolongation of muscle function during periods of hypoxia and ischemia (low or no blood flow) has been considered since the forced submersion experiments of Scholander (Scholander, 1940, Scholander *et al.*, 1942a). Oxygen consumption of muscle does decrease during muscle ischemia in many mammals (Duran and Renkin, 1974, Gutierrez *et al.*, 1988, Mizuno *et al.*, 2003). Such a decrease in muscle oxygen consumption may account for the maintenance of myoglobin oxygen saturation during sleep apnea of seals despite a calculated decrease in muscle O_2 extraction (Ponganis *et al.*, 2008).

Hypothermia and various hypoxia-linked mechanisms (ion channel arrest, a reversed Pasteur effect (lack of increased glucose consumption and lactate production during hypoxia), and decreased protein synthesis; see Chapter 8) have been proposed to decrease metabolic rate in muscle of diving mammals and birds (Butler, 2004, Hochachka, 1988, Hochachka and Guppy, 1987, Scholander *et al.*, 1942b). These hypoxia-linked mechanisms of metabolic suppression have been documented extensively in turtle tissues and, to some extent, in seal hepatocytes (Hochachka and Lutz, 2001, Hochachka *et al.*, 1988). However, as regards hypothermia, muscle temperature remained near 37 °C or even increased during dives in free-diving Weddell seals, emperor penguins, and king penguins (Ponganis *et al.*, 1993b, 2003b, Schmidt *et al.*, 2006). In regard to metabolic suppression, the muscle ATP turnover rate did not decline during forced submersion of ducks, but was maintained through depletion of myoglobin-bound oxygen, and breakdown of phosphocreatine and glycolysis (Stephenson and Jones, 1992b). In addition, the contribution of phosphocreatine breakdown could account for the energetic shortfall calculated in Scholander's forced submersion experiments (Kooyman and Ponganis, 1998). Thus, there is still no evidence of a unique mechanism of metabolic suppression in muscle of marine mammals and diving birds.

After reviewing all these investigations into muscle fiber type distribution, enzymes activities, energy substrate concentrations, buffering capacities, and mitochondrial distributions, the most consistent, outstanding feature of muscle in divers is the elevation of myoglobin concentration (Chapter 4). By far, the increase in myoglobin content is the predominant adaptation in the muscle of diving mammals and birds. The duration of aerobic metabolism in muscle is extended not only by that increase in myoglobin concentration, but also by a low muscle metabolic rate, which is due to the low energy cost of underwater locomotion in these animals.

7.3 Locomotory work

In contrast to a terrestrial mammal, which works primarily against gravity during locomotion, a marine mammal or deep-diving seabird primarily works against drag (resistance in water to movement) (Williams, 1999). In this respect, the cost of underwater locomotion is potentially increased because the density and viscosity of water are 800 and 60 times greater than those values in air, respectively (Dejours, 1987). Thus, efficient locomotion is essential to the success of diving mammals and birds. Minimization of the cost of underwater locomotion is achieved through several mechanisms. Perhaps most significant is hydrodynamic design and drag reduction. Other cost-saving measures include stroke and glide locomotory patterns, and utilization of buoyancy changes to decrease stroke effort.

7.3.1 Hydrodynamics and drag

The hydrodynamics of diving mammals and birds are based on their streamlined body shapes and the consequent reduction in drag. Excellent articles and reviews on hydrodynamics are available for cetaceans (Fish, 1994, Fish and Hui, 1991, Fish *et al.*, 2008), pinnipeds (Feldkamp, 1987b, Stelle *et al.*, 2000, Williams and Kooyman, 1985), sea otters (Williams, 1989), and diving birds (Lovvorn and Liggins, 2002, Lovvorn *et al.*, 2001). Readers with particular interest in hydrodynamics are referred to these citations.

Key components of body design relative to hydrodynamics include a fusiform shape, the location of maximum width relative to body length, and the design of limb appendages. These factors contribute to drag reduction through maintenance of a boundary layer (a layer of water extending from the body surface to a point at which it is moving at 99% of full speed; see Fish and Hui, 1991 for full details). Flow in the boundary layer may either be laminar (thereby reducing the frictional component of drag), turbulent (which is more stable, causing boundary layer separation to occur further down the length of the body and decreasing the size of the pressure wake, thus decreasing the pressure component of drag), or transitional (Fish, 1994).

Flow properties within the boundary layer vary according to the non-dimensional Reynolds number (R_e). $R_e = vl\rho/\mu$, where v is swim speed, l is body length, ρ is water density, and μ is water viscosity (Webb, 1988). Boundary flow is laminar, turbulent or transitional at $R_e < 5 \times 10^5$, $> 5 \times 10^6$, or between those two values, respectively (Fish, 1994). As of 2008, there was still no conclusive evidence that laminar flow occurs in the boundary layer of dolphins (Fish *et al.*, 2008). Therefore, as emphasized by those authors, it is the maintenance of a boundary layer over as much of the body length as possible, and not necessarily laminar flow, which contributes most to drag reduction.

The fineness ratio (ratio of body length to maximum width) serves as an index of the fusiform shape; a value of 4.5 has been considered optimal in that it represents maximal volume, minimum surface area, and, thus, the lowest drag for a given volume. However, optimal fineness ratios and drag reduction may vary with swim speed, swim style and body size; most values of fineness ratios in marine mammals have ranged between 3 and 8 (Fish *et al.*, 2008). It has been proposed that a fineness ratio of 6.3 contributes to the ability of spinner dolphins (*Stenella longirostris*) to perform their aerial spins (Fish *et al.*, 2006).

The location of maximum width along body length (usually the shoulder) is also important because it is at this point that laminar flow is most likely to become turbulent, or where boundary layer separation will occur. The shoulder occurs at 35–40%, 40%, and 50–60% of body length from the tip of the rostrum or nose in cetaceans, otariids, and phocids, respectively (Fish *et al.*, 2008). The appendages, the movement of which has been estimated to contribute to 28% of drag in a dolphin (Lang and Pryor, 1966), are also streamlined, promoting maintenance of a boundary layer over their surface. One of the most remarkable appendage modifications are the foreflippers of the humpback whale. The tubercles along the scalloped leading edge of the flipper contribute to increased lift and prevention or delay of stall during high-speed underwater turns (Fish and Lauder, 2006, Fish *et al.*, 2008).

Lastly, the bare skin of cetaceans and the hair properties (flat fur pelage) of other marine mammals also contribute to streamlining (Fish *et al.*, 2008). However, drag reduction through turbulence dampening by dolphin skin properties has never been demonstrated (Fish and Hui, 1991). Presumably the flat feather layer of penguins also contributes to streamlining. Notably, it has been recently proposed that release of air microbubbles from the feather layer of penguins contributes to a reduction in drag during high-speed exits onto sea ice (Davenport *et al.*, 2011).

Drag is proportional to body size and swim speed, and its relationship to these and other parameters is determined by the drag equation: $D = ½(\rho v^2 A)C_D$, where D is drag, ρ is density of the fluid, v is swim speed, A is the reference area, and C_D is the drag coefficient, a dimensionless constant related to the body geometry and flow around it. Because of this relationship, drag is proportional to body size and it increases curvilinearly with swim speed. Drag force can be measured by towing frozen carcasses, models, or even trained, live animals, or it can be calculated from deceleration rates of gliding animals (Feldkamp, 1987b, Lovvorn and Liggins, 2002, Lovvorn *et al.*, 2001, Williams and Kooyman, 1985).

Two additional features of drag are especially relevant to diving. First, the drag of an object at the surface is greater than that at depth due to the creation of wave drag at the surface (Fish, 1994, Williams and Kooyman, 1985). Drag of a towed harbor seal was 2.5 times greater at the surface than at depth (Williams and Kooyman, 1985). At depth, the sea otter can swim at higher speeds and with up to 41% less oxygen consumption than at the surface (Williams, 1989). And, in penguins, oxygen consumption during submerged swimming is less than that during surface swimming (Hui, 1988b). Thus, the metabolic demand and physiological responses of an animal swimming at the surface in a flume may be quite different from those of an animal during a dive. Second, drag measurements are usually determined when an animal is gliding. Drag during the stroking phase of locomotion is increased due to resistance against the propulsive limbs and early disruption of the boundary layer (Fish, 1994, Williams and Kooyman, 1985). In the swimming harbor seal, it was estimated that drag increased nearly two-fold during the stroke phase versus the glide phase of swimming (Williams and Kooyman, 1985). Drag also increased three-fold during the stroke phase of the cormorant (Ribak *et al.*, 2005a). Hence, the advantage of a stroke–glide pattern of swimming and the importance of the design and thrust development of locomotory appendages.

The design and function of locomotory appendages in the production of thrust in diving mammals and birds have been examined in multiple species and described in several reviews (Feldkamp, 1987a, 1987b, Fish, 1994, 1998, Fish and Lauder, 2006, Fish *et al.*, 1988, 2006, 2008). A recent novel analysis and thorough review of thrust production in the dolphin is highly recommended (Fish *et al.*, 2014).

The cetacean tail fluke acts as a flexible hydrofoil and has propulsive efficiencies (0.75 to 0.90 in different species), which are greater than those (<0.70) of propellers of commercial ships (Fish *et al.*, 2008). Further increasing the efficiency of locomotion in the dolphin are (a) the anatomical levering by long spinous and transverse processes and chevron bones on their vertebrae, and (b) the attachment of epaxial muscles to a subdermal connective tissue sheath, allowing transmission of force to caudal vertebrae (Fish and Hui, 1991, Pabst, 1990, 1993, Pabst *et al.*, 1999).

The hind flippers of seals also act as hydrofoils (Fish *et al.*, 1988). In the sea lion, due to the design of the foreflipper and anatomy of the shoulder, the foreflipper acts as both a hydrofoil and a paddle, generating thrust in all phases of the stroke cycle (English, 1976, 1977, Feldkamp, 1987a). Penguins, guillemots, and puffins fly through water with their wings acting as hydrofoils with thrust production in both phases of the stroke (Clark and Bemis, 1979, Hui, 1988a, Johansson and Aldrin, 2002, Lovvorn *et al.*, 2004). In foot-propelled avian divers such as grebes and cormorants, the motion of the webbed feet is considered to transit from drag-based to lift-based propulsion throughout the stroke cycle (Johansson and Lindhe Norberg, 2001, Johansson and Norberg, 2003, Ribak *et al.*, 2004). The power output of a stroke in all these animals is also potentially affected by the amplitude and velocity of the stroke as well as the angle of attack of the appendage.

Lastly, the behavior of an animal during a dive may also affect drag. An excellent example is the lunge feeding of rorqual whales. Although these animals have a streamlined body shape, feeding involves opening the mouth to about 80°, engulfment of a large mass of water, distention of the buccal cavity, and creation of an "engulfment" drag as water is pushed forward in the mouth (Goldbogen *et al.*, 2011, Potvin *et al.*, 2009). This engulfment drag forms about 90% of the increase in drag during the mouth-opening phase. The increased drag and cost of lunge feeding is thought to limit the dive durations of these large whales (Croll *et al.*, 2001, Goldbogen *et al.*, 2007, 2008, 2011, 2012, Simon *et al.*, 2012).

7.3.2 Locomotory patterns and costs

The basic locomotory pattern of all diving mammals and seabirds is a stroke and glide pattern (Clark and Bemis, 1979, Johansson and Aldrin, 2002, Johansson and Lindhe Norberg, 2001, Johansson and Norberg, 2003, Lovvorn, 2001, Lovvorn *et al.*, 1999, 2001, 2004, Miller *et al.*, 2004b , Ribak *et al.*, 2004, 2005a, Sato *et al.*, 2003, van Dam *et al.*, 2002, Watanuki *et al.*, 2003, 2005, 2006, Williams, 1989, 2001, Williams *et al.*, 2000, Williams *et al.*, 2012, Wilson and Liebsch, 2003). As reviewed earlier in this chapter, this pattern of locomotion is optimal in that there is decreased drag during the glide phase. Although the thrust and amplitude of a stroke have not been examined as closely as stroke rate in these divers, it appears that change in stroke rate primarily contributes to changes in propulsive effort and swim speed.

Stroke rates vary considerably during diving and among divers and are probably governed by (a) minimization of drag (body size, drag coefficient, swim velocity), (b) swim speed requirements for prey pursuit, and (c) buoyancy (see references in prior paragraph). During diving, changes in buoyancy secondary to air volume compression play a particularly significant role in determination of the rate and pattern of stroking (Enstipp *et al.*, 2006, Lovvorn *et al.*, 1991, Lovvorn *et al.*, 1999, 2004, Skrovan *et al.*, 1999, Watanuki and Sato, 2008, Watanuki *et al.*, 2003, 2005, 2006, Williams, 2001, Williams *et al.*, 2000, Wilson *et al.*, 2010).

Stroke rates are usually highest at the start of a dive when the animal is most shallow and air spaces are least compressed. As the animal descends, air volume in the respiratory system and the air layer in the outer pelage (if present, as in birds, fur seals, sea otters) are compressed and buoyancy decreases. The decrease in buoyancy allows many animals to either decrease stroke rate or begin prolonged gliding during descent. For example, in deep dives of elephant seals, Weddell seals, and dolphins, glides began at a mean depth of 86 m and comprised 80% of descent time (Williams *et al.*, 2000). In blue whales, prolonged glides began at 18-m depth (Williams *et al.*, 2000). Prolonged gliding during descent has not been observed in diving birds, although most species decrease their stroke rate below 20-m depth (Watanuki *et al.*, 2006, Wilson *et al.*, 2010). Another species which does not glide during descent is the right whale (*Eubalaena glacialis*), which is especially buoyant due to its large percentage fat composition (Nowacek *et al.*, 2001).

During ascent, however, most species must increase stroke rate and even stroke amplitude to overcome negative buoyancy at depth (Williams, 2001). As air within the body re-expands later during ascent, many animals can again take advantage of increased buoyancy to decrease stroke rate and eventually glide to the surface during the final phases of the dive (Nowacek *et al.*, 2001, Sato *et al.*, 2002, 2011, Watanuki and Sato, 2008, Watanuki *et al.*, 2006, Williams, 2001, Williams *et al.*, 2012, Wilson *et al.*, 2010). In shallow-diving cormorants, the entire ascent is a passive glide (Kato *et al.*, 2006, Watanuki *et al.*, 2005).

Prolonged gliding and lower stroke rates secondary to buoyancy changes in diving animals translate into lower metabolic costs and potentially into lower rates of oxygen store depletion. For equidistant dives of Weddell seals, deep dives with prolonged gliding had 35% less recovery oxygen consumption in comparison to shallow dives with continuous stroking (Williams, 2001, Williams *et al.*, 2000). Thus, decreased buoyancy secondary to air compression provides an advantage to deeper divers. During the long foraging trips of pregnant elephant seals, increased body weight and fat deposition throughout the trip resulted in the achievement of neutral buoyancy, with decreased stroke rates, more ascent glide time, and increased bottom time during dives (Adachi *et al.*, 2014). In comparison to negative buoyancy, neutral buoyancy may result in decreased locomotory effort and metabolic rate, thus contributing to the long-duration dives of these seals.

The increased buoyancy of a flying bird versus that of a neutrally buoyant penguin may also account for lower metabolic costs of penguins even during shallow dives (Butler and Woakes, 1984, Culik *et al.*, 1991, 1994, Schmid *et al.*, 1995). This is especially evident in shallow-diving ducks, in which oxygen costs are 21% greater than that of ducks surface swimming at maximum speed; most of that work is to overcome buoyancy (Butler, 2000, Lovvorn and Jones, 1991, Lovvorn *et al.*, 1991, Stephenson, 1993, 1994, 1995, Stephenson *et al.*, 1989a, Woakes and Butler, 1983). Decreased buoyancy may be achieved by reduced plumage air in deeper diving, flighted birds (Wilson *et al.*, 1992b), as well as by differences in the pneumatization of bones in penguins and various diving birds (Fajardo *et al.*, 2007, O'Connor, 2004, 2009). However, loss of plumage air due to wetting of feathers in cormorants leads to increased thermoregulatory and transport costs (Schmid *et al.*, 1995).

Overall, hydrodynamic design, propulsive efficiency, and stroke/glide patterns have made marine mammals efficient swimmers. The cost of transport (mass-specific metabolic rate/speed) for swimming marine mammals scales negatively with body mass and is indistinguishable from values for running terrestrial mammals, and flying bats (Williams, 1999). In other words, on a mass-specific basis, it costs the same for a marine mammal to swim a fixed distance as it does a terrestrial mammal to run or a bat to fly the same distance. Although the total costs of transport for swimming marine mammals and terrestrial runners are indistinguishable, the relative contributions of locomotion and body maintenance functions differ between the two groups (Williams, 1999, 2001). In marine mammals, 22–77% of the cost of transport is attributed to maintenance processes, whereas that value is only 12% in terrestrial mammals. This large maintenance component of the cost of transport in marine mammals is attributable to the cost of endothermy in the aquatic environment (see Chapter 8). Thus, the negative relationship of cost of transport in marine mammals to body mass should be primarily dependent on differences in maintenance costs and resting metabolic rates (Noren *et al.*, 2012b, Noren and Williams, 2000, Williams, 1999, 2001). In addition, a component of this negative relationship is probably also attributable to increased storage of elastic energy in tendons of larger mammals, and the lower rates of acto-myosin crossbridge cycling associated with the lower stroke rates of larger animals (Heglund *et al.*, 1982a, 1982b).

For semi-aquatic mammals such as muskrats, mink, and humans, the cost of transport is three to four times higher than that in marine mammals and running/flying mammals (Williams, 1999, 2001). This difference has been attributed to the increased drag of near-surface swimming and less aquatic specialization (increased hydrodynamic drag and decreased propulsive efficiency). Minimal cost of transport data for many penguin species are in the same range as that for swimming marine mammals (Baudinette and Gill, 1985, Culik *et al.*, 1994, 1996a, Hui, 1988b). Minimum costs of transport for foot-propelled avian divers such as cormorants are generally greater than those for penguins (Ancel *et al.*, 2000, Enstipp *et al.*, 2005, Schmid *et al.*, 1995). This increase in cost for cormorants in comparison to penguins has been attributed to differences in buoyancy, stroke style, and heat loss.

8 Thermoregulation

Thermoregulation during diving is an important physiological process in endotherms such as marine mammals and diving birds because the heat conductance and heat capacity of water are 24 and 3400 times greater, respectively, than those of air (Dejours, 1989). Thus, both convection and conduction are poised to remove heat from the body of a mammal or bird moving through water. Heat loss is decreased by (a) insulation in the form of subcutaneous fat (i.e., blubber) or pelage air (feathers in birds, fur in fur seals and sea otters) (Irving and Hart, 1957, Irving *et al.*, 1962, Kooyman *et al.*, 1976a, Liwanag *et al.*, 2012, Scheffer, 1964); and (b) decreased heat flow to the periphery via vasoconstriction in the skin and appendages (Chapter 5).

However, thermoregulation during dives is more than just insulation; it also involves rates of heat production and heat loss (Irving and Hart, 1957, Scholander *et al.*, 1950a, 1950b). Heat is produced within the body from metabolic processes, especially in active locomotory muscle. Only 20% of the energy released from adenosine triphosphate during muscle contraction is converted into muscular work; 80% is lost as heat (Smith *et al.*, 2005). In addition, specific dynamic action (SDA, the post-feeding increase in resting metabolism; also known as HIF, the heat increment of feeding) may contribute to heat production and maintenance of body temperature. In contrast, a decrease in metabolic rate of tissues secondary to decreased organ perfusion during dives would result in less heat production. Other possible mechanisms of heat production include the metabolism of brown fat and shivering. It has been proposed that the brown fat associated with the pericardial venous plexus of seals has a role in thermoregulation during diving activity in both seals and muskrats (Blix *et al.*, 1975, MacArthur, 1986). Shivering represents another route by which muscle produces heat. Seals do shiver as body temperature decreases (Kvadsheim *et al.*, 2005). However, the shivering response is inhibited or at least greatly reduced during experimental forced submersions (Kvadsheim *et al.*, 2005). Therefore, at least in seals, shivering during a dive is unlikely to occur regardless of how low core temperature is reduced.

Mechanisms of increased heat loss during dives include cold prey ingestion (Wilson and Culik, 1991), potential arterio-venous shunting in the skin and limbs (Chapter 5, Willis *et al.*, 2005), compression of the pelage air layer due to depth (Kooyman *et al.*, 1976a), water penetration of the feather–air layer in some birds (Elowson, 1984, Ribak *et al.*, 2005b, Rijke, 1968), and potential loss of heat through the poorly insulated brood patch of penguins (Handrich *et al.*, 1997, Schmidt *et al.*, 2006).

This chapter focuses primarily on thermoregulation during diving. The many thermoregulatory challenges that are faced on shore by pinnipeds or penguins will not be addressed. The role of metabolic processes in rates of energy/heat production, and early observations of temperature responses during forced submersions will be reviewed first. Then, basic anatomical and physiological thermoregulatory adaptations will be considered. These reviews will conclude with observed temperature profiles during diving in several species and assessment of various hypotheses of thermoregulatory mechanisms during diving.

8.1 Metabolism and heat production

Since Scholander's earliest forced submersion experiments with seals, it has been known that temperature decreased variably in tissues, and as much as 2.5 °C in the brain and blubber (Scholander, 1940, Scholander *et al.*, 1942b). These temperature declines were considered secondary to a decrease in metabolic rate during forced submersions. In forcibly submerged ducks, although deep body temperatures increased during submersion, temperatures decreased during the recovery period (Scholander, 1940, Scholander *et al.*, 1942b). Scholander interpreted these changes as evidence for both decreased metabolism during the submersion as well as for conservation of core temperature and cooling of the periphery (i.e., regional heterothermy) during the submersion, with subsequent rewarming of the periphery during the recovery.

Hypothermia, as demonstrated in Scholander's studies, is theoretically beneficial to a diver in two ways. First, the rate of oxygen store depletion could be depressed by a temperature-induced decline in tissue metabolic rate via the Q_{10} effect (Butler, 2004, Geiser, 2004). Second, even mild hypothermia of a few degrees can be beneficial during ischemia; neurological outcomes after cardiac arrest are improved by such therapy (Bernard *et al.*, 2002, Milde, 1992, Sessler, 2009). However, maintenance of body temperature in the normothermic range for an awake mammal or bird is important for optimal tissue function such as neural processing or muscle contraction (Bennett, 1984, Fischbeck and Simon, 1981, Mallet, 2002). Certainly, humans with mild hypothermia (32–35 °C) present with confusion, amnesia, dysarthria, and ataxia (Kempainen and Brunette, 2004, Mallet, 2002). On the other hand, spatial learning is conserved in rats cooled to as low as 32 °C (Moser and Andersen, 1994). Perhaps the temperature tolerance of the central nervous system of diving mammals and birds is more similar to that of the rat than the human.

Given the challenges of thermoregulation in water, Scholander's documented temperature changes during forced submersions, and the potential benefits/risks of hypothermia, maintenance of proper body temperature is an important determinant of an animal's ability to dive. This optimization of body temperature is achieved through regulation of metabolic processes and regulation of heat conservation/heat loss mechanisms.

As already mentioned, heat production is associated with ATP breakdown during muscle contraction. Swimming activity and, presumably, the secondary heat production in muscle due to swimming, have been associated with maintenance of core temperatures in sea otters (Costa and Kooyman, 1984). Similarly, secondary heat production in exercising muscle may well account for the lack of an increase in metabolic rates in sea lions swimming in water temperatures below their thermoneutral zone (Liwanag *et al.*, 2009). It is also notable that muscle mitochondria from elephant seals, despite increased phosphorylation control, have a 50% greater respiratory mitochondrial leak, which may be related to thermogenesis in the diving seal (Chicco *et al.*, 2014). Exercise has also been found to decrease thermoregulatory costs in diving ducks (Kaseloo and Lovvorn, 2006), and wing flapping has been suggested to aid in warming of ingested prey in cormorants (Grémillet, 1995). In addition, the heat increment of feeding (HIF) may also contribute to heat production and maintenance of body temperature. HIF has been demonstrated in sea otters, seals, cormorants, ducks, penguins, and other birds (Bech and Praesteng, 2004, Costa and Kooyman, 1984, Enstipp *et al.*, 2008, Hawkins *et al.*, 1997, Janes and Chappell, 1995, Kaseloo and Lovvorn, 2006, Markussen *et al.*, 1994, Rosen and Trites, 1997).

However, it has also been questioned in some species and circumstances (i.e., low-protein diet) whether HIF can effectively substitute for thermogenesis and decrease the cost of thermoregulation during foraging/diving activity (Kaseloo and Lovvorn, 2003, Rosen and Trites, 1997, Rosen *et al.*, 2007). Indeed, Wilson and Culik have emphasized that the thermoregulatory costs to rewarm cold prey may be much greater in magnitude than HIF (Wilson and Culik, 1991). In addition, the digestive process may even be delayed until periods of relative inactivity so that the oxygen consumption associated with digestion does not compete for oxygen stores during foraging activity (Crocker *et al.*, 1997, Mitani *et al.*, 2010, Sparling *et al.*, 2007). In such cases, any contribution of HIF to thermoregulation might occur well after diving activity has stopped.

In terms of metabolic rate and heat production, mass-specific metabolic rates and thermoregulatory costs are higher in smaller-sized animals due to the well-known increased ratio of body surface area to volume in smaller animals (Kleiber, 1975). This relationship is also true in marine mammals, but in general the metabolic rates of marine mammals resting in water are typically elevated above that predicted by the Kleiber equation for mammals at rest (Lavigne *et al.*, 1986, Williams *et al.*, 2001b). This elevation has been attributed to both thermoregulatory costs and the carnivory of many marine mammals. Hence, the very high metabolic rates and food requirements of the smallest marine mammal, the sea otter (*Enhydra lutris*) (Kenyon, 1969, 1975, Morrison *et al.*, 1974, Yeates *et al.*, 2007).

Although the regulation of metabolic rate during free dives is debated and metabolic rates can be reduced during trained breath holds versus surface rest (Hurley and Costa, 2001), larger animals do appear to have lower rates of blood oxygen depletion, and, hence, lower metabolic rates during breath holds (Hudson and Jones, 1986, Noren *et al.*, 2012b). These lower metabolic rates are consistent with lower thermoregulatory costs in the larger divers, and confer an advantage in terms of the rate of oxygen store depletion. In addition, as discussed in Chapter 7, larger animals have lower locomotory stroke rates, and, thus, lower rates of heat production in muscle. Thus, the potential rate of mass-specific heat production by metabolic processes will be relatively lower in larger animals versus smaller animals. Interestingly, the potential effects of size and body design of minke whales (*Balaenoptera acutorostrata*) swimming in cold waters result in maintenance of an effective blubber insulation, a decrease in deep body temperature (35.1 °C), and a low estimated metabolic rate (Blix and Folkow, 1995, Folkow and Blix, 1992, Kvadsheim *et al.*, 1996).

8.2 Marine mammals: thermoregulatory anatomy and physiology

In marine mammals, blubber and fur serve as insulation during diving. Pinnipeds and sea otters both have fur, but fur is an effective insulator underwater only in the fur seals and sea otter (Kvadsheim and Aarseth, 2002, Liwanag *et al.*, 2012, Scholander *et al.*, 1950c). In adult harp seals (*Phoca groenlandica*) and hooded seals (*Cystophora cristata*), thermal resistance of the fur decreased in water by 92% from that in air and was only 3% of total insulation, while the blubber contributed 85% to total insulation in water (Kvadsheim and Aarseth, 2002). Among phocid and otariid pinnipeds, only the fur seals retain an air layer upon immersion in water; blubber plays the primary role in the other species, even the sea lion, another otariid (Liwanag *et al.*, 2012).

In the Antarctic fur seal (*Arctocephalus gazella*), the difference between skin temperature below the fur and adjacent sea water can be as high as 20 °C at the surface, but that difference ranges between 4 and 0 °C during dives, decreasing with depth and increasing with ascent (Boyd, 2000). In the fur seal and sea otter, high hair density, elongation/flattening of the guard hairs, and interlocking underhairs facilitate air trapping (Kuhn et al., 2010, Liwanag *et al.*, 2012, Williams *et al.*, 1992a). Absent arrector pili muscles in pinnipeds also contribute to flattening of the fur layer in these animals (Kvadsheim and Aarseth, 2002, Liwanag *et al.*, 2012, Scholander *et al.*, 1950c).

The effectiveness of blubber as an insulator is considered dependent on its heat conductivity and thickness. It is important to note that heat conductivity correlates negatively with blubber lipid content, and that these parameters, as well as blubber thickness, can vary throughout regions of the body and with age (Dunkin *et al.*, 2005, Kvadsheim *et al.*, 1996, Ryg *et al.*, 1993, Worthy and Edwards, 1990). In pinniped and cetacean species with over a 4000-fold range in body mass, average blubber thickness ranged from 2 to 9 cm (Ryg *et al.*, 1993). Blubber thickness in different body regions ranged from 7 to 40 cm in one fin whale specimen (Lockyer and Waters, 1986).

Overall, heat loss to the environment through blubber is a function of surface area, blubber heat conductivity and thickness, and the temperature difference between the skin surface and the muscle–blubber interface (see above references). In such calculations, the temperature at the muscle–blubber interface is preferred over core temperature as the temperature gradient between the body core and environment is not necessarily confined to the blubber layer. As emphasized by Kvadsheim *et al.* (1996), use of an incorrect temperature may have caused errors in earlier estimations of metabolic requirements of whales based on calculations of heat loss to the environment. In addition, heat loss through the blubber layer is due to both conduction (the conductivity of the blubber) and also the magnitude of blood flow through the blubber (convective heat loss). These differences were well illustrated in young harp seals, in which blubber heat flux in water at 0 °C was equivalent to the calculated conductive heat loss through blubber, and to that of the blubber of a seal carcass, in which there was obviously no blood flow. However, at higher water temperatures, the calculated conductive heat loss was only 4–43% of the total blubber heat flux, which supports the concept that heat transfer via blood perfusion into the blubber accounted for the remainder (Kvadsheim and Folkow, 1997).

Heat conductivity in blubber tissues with similar total lipid content may also vary with lipid composition (Bagge *et al.*, 2012). The blubber of pygmy sperm whales (*Kogia breviceps*), with a similar total lipid content but higher wax ester content than that of short-finned pilot whales (*Globicephala macrorhynchus*), had 30% greater thermal conductivity.

Sirenians, as represented by manatees, have dense skin that is very low in lipid content; they do not have a true blubber layer (Gallivan *et al.*, 1983, Kipps *et al.*, 2002). Manatees do, however, have subcutaneous fat which presumably aids in insulation. In addition, there is a countercurrent heat exchanger in the tail vasculature (Rommel and Caplan, 2003). Minimum thermal conductance of the manatee has been reported to be similar to that of the seal, but that value is dependent on nutritional status and the quantity of subcutaneous fat (Gallivan *et al.*, 1983). In addition, the low metabolic rate of these animals limits their ability to maintain body temperature in colder water (Gallivan *et al.*, 1983). It has been argued that these factors limit the lower end of their thermoneutral zone and their geographical distribution to a water temperature of 20 °C (Irvine, 1983). Although the dense skin is disadvantageous in terms of heat retention, the high density of the skin as well as the pachyostotic bone of the manatee are considered to provide negative buoyancy in an animal with increased positive buoyancy due to its lung capacity (Kipps *et al.*, 2002).

Heat loss through the less-insulated appendages of diving animals has been estimated to represent 8–28% of total body heat loss in a modeling study of a large series of seals and whales (Ryg *et al.*, 1993). In the killer whale (*Orcinus orca*), 40% of total body heat loss may occur through the appendages, in part due to the larger percentage of body surface area represented by those structures (Kasting *et al.*, 1989). In young harp seals in 0 °C water, heat loss from the flippers was 2–6% of total body heat loss (Kvadsheim and Folkow, 1997). In contrast, in 24 °C water, flipper heat loss accounted for 19–48% of total heat loss in these animals, illustrating the importance of peripheral blood flow regulation under such conditions. A prior estimate of 84% of total body heat loss through the flippers of the free-diving harp seal (Gallivan and Ronald, 1979) was not consistent with these findings, but both studies emphasize the important role of the appendages in increasing heat loss during thermal stress.

In contrast to the role of insulation in conservation of body heat and temperature in an endotherm, blood flow regulation to the skin and appendages is considered the primary route of dissipation of excess heat generated through muscular exercise. It is here where the superficial arterio-venous anastomoses and blood vessels in the skin and appendages (see Chapter 5) potentially play an important role (Bryden and Molyneux, 1978, Molyneux and Bryden, 1975, 1978, Rommel and Caplan, 2003, Scholander and Schevill, 1955). Indeed, infrared imaging has revealed hot spots (skin temperatures above sea temperature) on the trunks of surfacing humpback whales (*Megaptera novaeangliae*), and the pectoral fins and dorsal fins of swimming minke whales and fin whales (*Balaenoptera physalus*) (Cuyler *et al.*, 1992). Localized heat flux from the dorsal fin is highest above superficial veins in the dolphin (Meagher *et al.*, 2002). Heat loss from the dorsal fin of the dolphin is highly variable. Heat flux decreased during diving, increased during ventilatory periods, and also increased as ambient water temperature declined (Meagher *et al.*, 2002, Noren *et al.*, 1999, Westgate *et al.*, 2007, Williams et al., 1999).

Although temperature regulation has been demonstrated to override the dive response in experimental studies (Hammel *et al.*, 1977), post-exercise heat flux at the surface was reduced at depth in diving dolphins (Noren *et al.*, 1999). Heat flux increased only slightly during ascent, but on return to the surface there was an immediate 80–100% increase in heat flux in the dorsal fin and tail fluke, presumably secondary to the surface tachycardia and peripheral vasodilation (Noren *et al.*, 1999). The occurrence of heat loss versus heat conservation responses at depth is most probably a function of the particular circumstances of a dive. It is notable that one exceptionally high measurement of heat flux occurred in the dorsal fin of a diving dolphin during submersion after it had performed an unusually strenuous exercise routine prior to the dive (Noren *et al.*, 1999). This increase in dorsal fin blood flow and heat loss would be consistent with Hammel *et al.*'s forced submersion observations that elevated body temperature may override the bradycardia and vasoconstriction of the dive response.

Besides an overriding of the dive response, there is another mechanism that could allow for external heat loss through the appendages. In two forced submersion studies of seals, Blix, Odden, and colleagues confirmed Scholander's findings of a 2.5–3.3 °C decrease in brain temperature during the submersion period (Blix *et al.*, 2010, Odden *et al.*, 1999). In the 2010 study, Blix also proposed a vascular anatomical basis to selectively cool the brain by examination of the brachial arterial and venous vasculature in the foreflipper. It was hypothesized that brachial artery blood flow through arterio-venous anastomoses in the foreflipper and then through large superficial veins in the flipper bypassed the heat-exchanging rete in the central flipper and allowed for the return of cooled blood to the heart, which would result in the lowering of aortic blood temperature and brain temperature. Such shunting through the hind flippers and pelvic plexus was also consistent with an observed decrease in rectal temperature.

Temperature regulation of organs within the body, especially the reproductive organs, can also be critical. There is both physiological and anatomical evidence for the role of countercurrent vascular heat exchangers in maintaining the temperature of reproductive organs in both cetaceans and pinnipeds (Pabst *et al.*, 1995, Rommel *et al.*, 1992, 1994, 1995).

Other routes of external heat loss in divers include respiration and feeding. Respiratory heat loss in both dolphins and seals is minimized by apneic respiratory patterns and reduction of the temperature and water saturation of exhaled air (Coulombe *et al.*, 1965, Folkow and Blix, 1987, 1989, Kasting *et al.*, 1989). In the seals, the reductions in temperature and water vapor saturation are achieved through a nasal countercurrent heat exchanger. In dolphins, cooler tissue lining the cranial airway sinuses is considered to absorb heat from air prior to exhalation. It has also been proposed that compression of air below the blowhole prior to exhalation leads to water vapor condensation in the airway, thus decreasing the water content of exhaled air of the dolphin (Coulombe *et al.*, 1965).

Heat conservation during prey capture and feeding has not been investigated, although heat loss in the mouth of the mysticete whale is presumably minimized by a countercurrent, potential heat exchanging vasculature in the tongue (Heyning and Mead, 1997, Werth, 2007). Effects of cold prey ingestion on stomach temperatures are exemplified in stomach temperature pill studies of elephant seals and sea lions. After two prey ingestion events (unknown prey mass) in an elephant seal diving at sea, stomach temperature decreased from 36 to 33 °C, and then recovered to baseline within 40 minutes after the last feeding event (Kuhn *et al.*, 2009). In captive sea lions and elephant seals, ingestion of 1 kg of herring at an average temperature of 17.5 °C resulted in a decline of 4 °C in stomach temperature, which recovered to baseline within about 70 minutes in both species (Kuhn and Costa, 2006).

8.3 Marine mammals: body temperatures during dives

Before reviewing body temperatures in diving marine mammals, it is first important to consider the distribution of heat and temperature within the body. Temperature data from harvested whales documented temperature gradients with depth into the body as well as temperature variation within different regions of the body (Folkow and Blix, 1992, Morrison, 1962). The highest temperatures were near the heart and liver. Such regional heterothermy is not unusual, and this variation in temperature raises the question of the definition of core temperature, i.e., where should it be measured?

As an example of regional heterothermy in a living animal, temperature was 35 °C in superficial muscle during sleep apnea of elephant seals, while pulmonary artery temperatures during sleep apnea in the same animals were near 36.5 °C (Ponganis *et al.*, 2006b, 2008,). Even in humans, muscle temperatures at rest averaged 35 °C, with a range of 33.0 to 36.5 °C, and these muscle values were always less than simultaneous rectal temperatures (Saltin *et al.*, 1968). Indeed, it is the loss of vascular regulation during anesthesia and the redistribution of heat to cooler regions of the body that accounts for the initial decreases in core temperature during anesthesia (Sessler, 2000). And, in the other direction, it is probably the warming of the cooler blubber layer and other tissues of the dolphin that allows the animal to transiently move from cool off-shore water to near-shore water with surface water temperatures at or greater than body temperature (Heath and Ridgway, 1999). Because of the regional variation in body temperatures, core temperature is probably best represented by temperatures near the heart and liver, and in the blood of the central aorta, pulmonary artery, and/or vena cava.

Core body temperatures have not been measured during diving in cetaceans. Data from the whaling era revealed a mean post-mortem deep-body temperature of 35.8 °C, although pre-mortem deep muscle temperatures were above 38 °C in sperm (*Physeter macrocephalus*), humpback, and fin whales (Morrison, 1962). Post-mortem core temperature of harvested minke whales was 35.1 °C (Folkow and Blix, 1992). The closest available data from live cetaceans are stomach temperature and abdominal temperature profiles of captive spinner (*Stenella longirostris*) and Pacific bottlenose (*Tursiops gilli*) dolphins collected over the course of the day in the 1970s (Hampton and Whittow, 1976, McGinnis *et al.*, 1972). Abdominal and stomach temperatures in various animals were reported to range from 36 to 38 °C. Similarly, stomach temperatures of a captive manatee remained stable near 35.6 °C during daily activity (Gallivan *et al.*, 1983). Mean rectal temperature of captive sea otters during metabolic studies was 38.1 °C (Morrison *et al.*, 1974). In muskrats (*Ondatra zibethicus*), abdominal cavity temperatures decreased from a baseline of 38 °C to 34 °C after 15 min of diving activity in 10 °C water (Hindle *et al.*, 2006). Notably, in muskrats pre-chilled by 2 °C, diving behavior and dive oxygen consumption were unchanged.

It is in the pinnipeds that temperature profiles have been most successfully recorded during diving. Such profiles are important in that they are the end result of all the cardiovascular, metabolic, and thermoregulatory responses during the dive, and, therefore, provide insight into the nature of these responses during the actual dive. In addition, they address the question of the optimal temperature during diving, and whether lower body temperatures are critical to dive performance. The first evidence of hypothermia during free-diving were immediate post-dive arterial temperatures of 34.9 to 36.3 °C after five dives greater than 25-min duration in Weddell seals (Kooyman *et al.*, 1980). These post-dive temperatures were always lower than the final pre-dive temperature. It was also noted that shivering did not occur even with a temperature of 34.9 °C during the surface recovery period.

The first continuous arterial temperature profiles were reported in a single Weddell seal, and these data revealed mean dive temperatures between 35 and 37 °C for dives of 15–35-min duration and temperatures between 35 and 38.5 °C for shorter-duration dives (Hill *et al.*, 1987). These authors emphasized that temperatures did not change so much during the dive as prior to the dive. Temperatures were always less than 37 °C before dives longer than 15 min, and were less than 36.3 °C before dives longer than 30 min. Thus, Hill *et al.* concluded that the lower temperature prior to and during diving was not the result of a lower metabolic rate during the dive, but rather that the lower temperature was potentially the cause of a lowered metabolic rate. They also postulated that the pre-dive decrease in temperature was secondary to increased blood flow to the skin and flippers.

Continuous muscle temperature profiles in a free-diving Weddell seal revealed that muscle temperatures remained between 36 and 37 °C, with no obvious consistent trends up or down during diving (Ponganis *et al.*, 1993b). As in Hill *et al.*'s arterial temperature profiles, higher temperatures occurred during prolonged surface rest periods. Although hypothermia was not evident, it was still remarkable that these animals could perform muscular exercise during a dive without a significant increase in muscle temperature. Such a temperature profile in the primary locomotory muscle could be consistent with maintenance of some muscle blood flow during the dive. In addition, muscle temperature did decline a few tenths of a degree during the final segment of a 45-min dive, again suggesting that there could be some flow of cooler blood to muscle during the ascent tachycardia.

The range of internal body temperatures in seals at sea was also demonstrated in foraging studies utilizing stomach temperature pills. Although continuous temperature

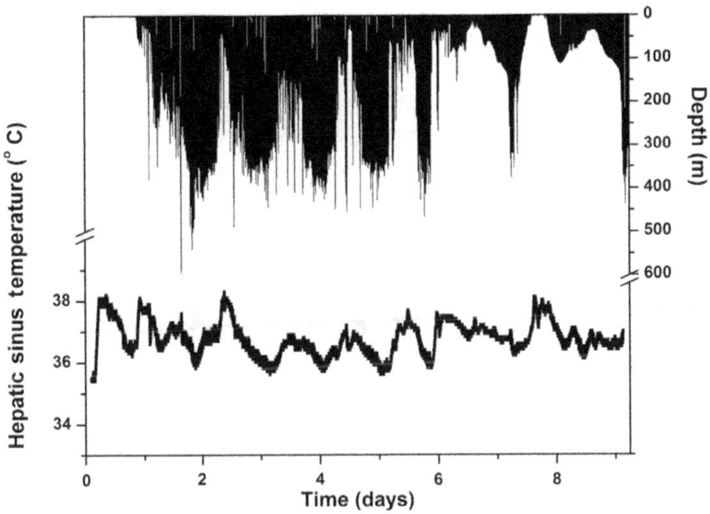

Figure 8.1 Hepatic sinus temperature of a young, translocated elephant seal (*Mirounga angustirostris*) varied between 36 and 38 °C during eight days of diving activity. There was no evidence of significant core hypothermia. Adapted from Meir and Ponganis (2010).

and depth profiles were not analyzed or extensively presented, it is notable that stomach temperatures prior to prey ingestion were near 36 °C in northern elephant (*Mirounga angustirostris*) seals and above 38 °C in gray (*Halichoerus grypus*) seals (Austin *et al.*, 2006, Kuhn *et al.*, 2009).

Almost 15–20 years after the last of the Weddell seal temperature studies, intravascular temperature profiles were obtained in juvenile, translocated northern elephant seals and in adult female California sea lions (*Zalophus californianus*) during maternal foraging trips to sea. In the elephant seal study, temperatures were recorded in the extradural vein, hepatic sinus, and aorta (Meir and Ponganis, 2010). There was a significant, albeit weak, negative correlation of mean dive temperature with dive duration in all but one of 13 seals. However, most temperatures were between 36 and 38 °C (Fig. 8.1). In only one seal were arterial temperatures consistently lower (between 34 and 36 °C). The lack of consistent, significant hypothermia (below 35 °C) in the hepatic sinus and aorta during dives as long as 25 min was taken as evidence that core hypothermia and a subsequent cold-induced decrease in metabolic rate were not required in such dives.

It was also notable that arterial temperature of elephant seals did not increase during what appeared to be routine, repeated foraging dives to 500–600-m depth (Fig. 8.2). However, during bouts of short-duration, shallow dives, temperatures could be more variable and higher (Fig. 8.2). Lastly, longer dives were occasionally associated with abrupt, transient decreases in either aortic and extradural vein temperature to as low as 30 °C (Figs 8.3, 8.4). The rapidity of the onset and recovery of these temperature declines argues for confinement of the temperature change to the blood, and for cooling through superficial arterio-venous shunts as described by Blix *et al.* (2010) in the flippers of the seal as the mechanism responsible for such a change.

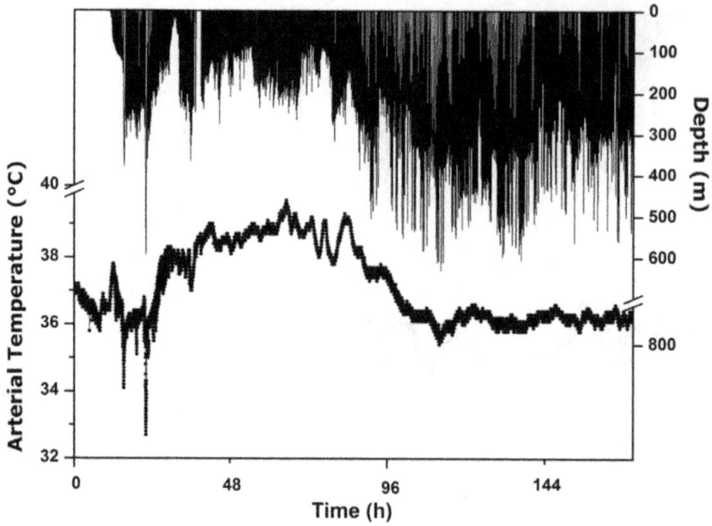

Figure 8.2 Aortic temperature of a young translocated elephant seal (*Mirounga angustirostris*) varied greatly during six days of diving activity. Higher temperatures to near 39 °C predominated during shallow diving activity while temperatures closer to 36 °C occurred during bouts of deeper dives.Adapted from Meir and Ponganis (2010).

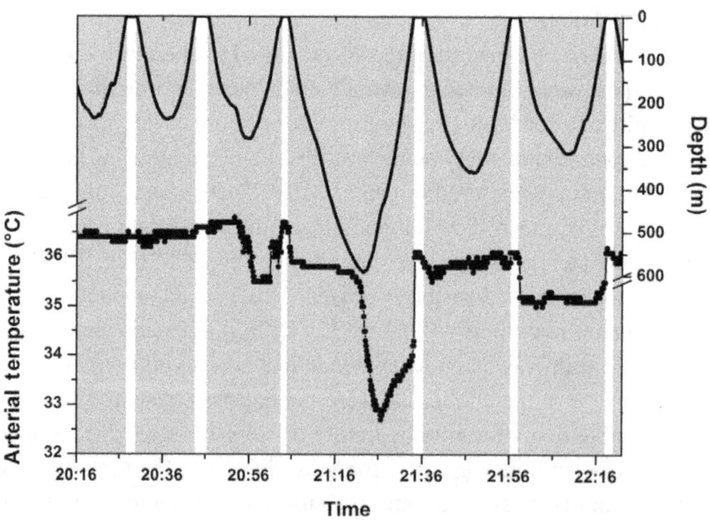

Figure 8.3 Transient decreases in arterial temperature can occur in longer dives of elephant seals (*Mirounga angustirostris*). The rapid decline to 32 °C and quick recovery to 36 °C suggest that such changes are probably confined to the blood and secondary to blood flow patterns. Shaded areas indicate dive periods in this and following figures. Adapted from (Meir and Ponganis, 2010).

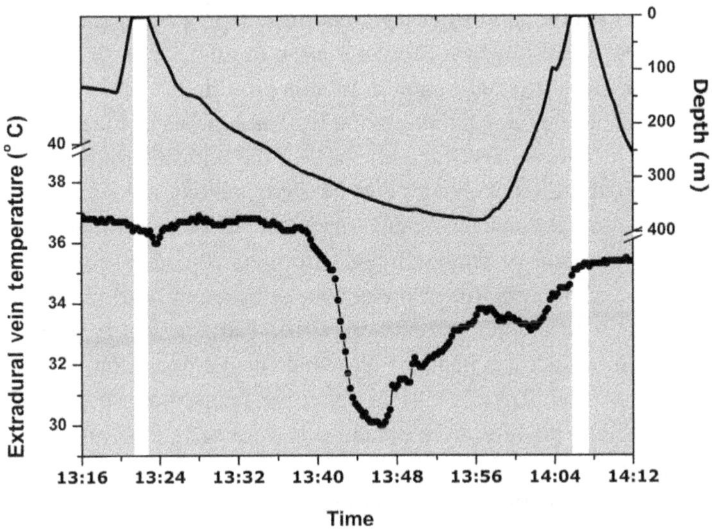

Figure 8.4 Transient decreases can also occur in extradural vein temperature during longer dives of elephant seals (*Mirounga angustirostris*). In this case, temperature declined almost 7 °C over four minutes, and then returned to near 35 °C by the end of the dive over the course of 22 minutes. Adapted from Meir and Ponganis (2010).

In Antarctic fur seals, intraabdominal temperatures have been reported to decline as much as 20 °C during a dive (Woakes *et al.*, 1992). In adult female California sea lions during maternal foraging trips to sea, vena caval temperatures remained between 36 and 38 °C, even during dives as deep as 400 m and long as 8 min (McDonald and Ponganis, 2013). Although temperature was again not elevated during diving activity, there was no evidence that significant core hypothermia was required in the sea lions to perform long dives.

In summary, temperature profiles during dives of Weddell seals, elephant seals, and sea lions reveal that core temperatures in the aortic and central veins primarily remain in the 36 to 38 °C range and that core hypothermia is not essential to perform routine and even long dives. Muscle temperature profiles in the diving Weddell seal and the dive behavior/oxygen consumption of pre-chilled muskrats support the same argument. On occasion, however, marked temperature changes can occur in blood, most probably related to selective shunting of blood to the periphery. Regional heterothermy, which has been demonstrated in marine mammals and terrestrial mammals at rest, most probably also occurs during diving, but regional temperatures have not been extensively examined in free-diving marine mammals. Certainly, however, data from harvested whales and forced submersion experiments support the concept of regional heterothermy.

8.4 Marine birds: thermoregulatory anatomy and physiology

Feathers and the air layer beneath them provide the major component of insulation in birds. Even in the emperor penguin (*Aptenodytes forsteri*) standing on ice, 86% of insulation is provided by the feather layer (Jarman, 1973). Maintenance of the air layer and water repellency are important for insulation in all birds, especially aquatic birds. Feather structure and uropygial gland secretions provide the basis of water repellency (Stettenheim, 2000). The basic structure of a feather consists of the main shaft with side branches (barbs) and secondary side branches off the barbs (barbules). Downy barbs at the base of the shaft are more slender and flexible, creating a thick, fluffy, air-trapping structure. In addition, there can be a small afterfeather associated with the shaft at its base.

A uniform distribution of feathers about the body, broad, flat shafts, and the size, spacing, density, and hooking (interconnection) of barbs and barbules are considered to contribute to water repellency (Chandler, 1916, Dawson *et al.*, 1999, Stettenheim, 2000), although the exact mechanisms allowing for the exclusion of water have been debated (Elowson, 1984, Rijke, 1968). In penguins, feathers are short, stiff, and capable of being "locked" into position by subcutaneous muscles; a long afterfeather and down provide insulation beneath (Dawson *et al.*, 1999). Mathematical modeling suggests that the small radius of the barbule and the barbule's geometric structure contribute most to the insulative properties of the feather layer (Du *et al.*, 2007).

In the cormorant, which is considered to have a partially wettable feather, the middle section of the feather is densely packed while the outer margins are more spaced (Grémillet *et al.*, 2005). In the center, hooklets on adjacent barbules interconnect with one another as in other feathers. This non-homogeneous arrangement in the cormorant feather allows water penetration on the periphery and air-trapping in the center. Water penetration of the feathers can account for observed weight gains in cormorants after diving, and lower feather air volumes than in other aquatic species (Grémillet *et al.*, 1998, 2005, Ribak *et al.*, 2005b, Wilson *et al.*, 1992b). Although water penetration of the feather layer decreases the cormorant's insulation, it also decreases buoyancy, a potential advantage for a diver. It is notable that in cormorant species adapted to colder regions, the feather air layer is larger, and in the deep-diving, sub-Antarctic blue-eyed shag (*Phalacrocorax atriceps*), the feather structure is reported to be homogeneous, like that of other aquatic birds, thus potentially making the feathers in this species water-repellent and preserving more air in the feather layer for insulation (Grémillet *et al.*, 1998). In general, mass-specific feather air volumes of carcasses were lowest in pursuit divers, intermediate in plunge divers, and highest in surface-feeding and flight-feeding birds, with the lowest values in cormorants, penguins, and divers (loons) (Wilson *et al.*, 1992b).

Water penetration of the feather layer has been assessed in ducks and cormorants, with assessment of an experimental index – the critical water penetration pressure (Grémillet *et al.*, 2005, Stephenson and Andrews, 1997). When they first developed this technique, Stephenson and Andrews found that the critical penetration pressure in duck and goose feathers was three times the pressure on feathers of a bird floating at the surface, and two times that of a partially submerged bird, thus providing experimental evidence of feather water repellency and protection of the feather air layer. Notably, critical penetration pressure of the dense, center portion of the cormorant feather was three times greater than that of the duck, again re-enforcing Grémilet *et al.*'s conclusion that the central section of the feather was resistant to water penetration, allowing for trapping of air beneath it.

The thermal conductance of birds in water has been evaluated with both carcasses and live animals. Cooling rates of carcasses of 14 species of aquatic birds revealed that thermal conductance increased two-fold with water contact and 4.8-fold with submergence (de Vries and van Eerden, 1995). Thermal conductances across gentoo (*Pygoscelis papua*) and Adélie (*P. adeliae*) penguin pelts increased 1.1–1.9 times on immersion into water, and 5.2-fold on compression to 10-m depth (Kooyman *et al.*, 1976a). However, during compression, unlike in live penguins, water penetrated the unwashed pelts. In a washed pelt free of oil, conductance on immersion alone increased 3.4-fold, supporting the role of feather oil in water repellency. In live little penguins (*Eudyptula minor*), and king penguin (*Aptenodytes patagonicus*) chicks, thermal conductance in water increased two- to four-fold, all consistent with compression of the air–feather layer (Barre and Roussel, 1986, Stahel and Nicol, 1982). Although total conductance in a live Adélie penguin increased three-fold on immersion in water, it did not increase further with compression to 10-m depth (Kooyman *et al.*, 1976a). In live thick-billed (*Uria lomvia*) and common (*U. aalge*) murres, calculated thermal conductances were low and did not decrease with decreasing water temperature (Croll and McLaren, 1993). The low conductance values of the murre were attributed to (a) buoyancy and minimal contact (including the wings) with water, (b) withdrawal of feet into the feather layer, and (c) maintenance of the feather air layer.

Besides compression of the feather layer, other potential routes of heat loss during diving are the appendages and the brood patch. For example, during flight of the herring gull (*Larus argentatus*), 80% of heat production is lost through the feet (Baudinette *et al.*, 1976). However, in the feet and wings of diving birds, the extensive countercurrent associations of arteries and veins discussed in Chapter 5 (Thomas and Fordyce, 2007, 2012) should promote heat conservation in the body core and cooling of the periphery even in the presence of some peripheral blood flow during diving. In addition, as already pointed out in this chapter, feet can be retracted into the feathers in birds such as murres. Superficial veins that bypass the countercurrent system in the wings of both diving and non-diving birds have been considered by most authors to provide a route for heat dissipation for periods of heat stress (in flight or on land) or during surface intervals. Hence, the flushed pink feet and inner wing surfaces of Adélie penguins exiting from the water.

Increased blood flow and heat loss through the brood patch during dives of king penguins has also been proposed as a mechanism of heat dissipation and thermoregulation (Schmidt *et al.*, 2006). As discussed in Chapter 5, blood flow through the arteriovenous anastomoses in the brood patch provides a mechanism of heat transfer to warm an incubated egg. In diving king penguins, although brood patch subcutaneous temperature always had a net decrease over the course of a dive, transient warming episodes of about 0.3 °C were recorded in 21% of dives. On this basis, Schmidt *et al.* hypothesized that fine peripheral blood flow adjustments could contribute to thermoregulation and observed temperature profiles during dives.

Cold prey ingestion represents another process that can contribute to lower body temperatures. Stomach temperature responses are highly variable and are dependent on the mass and temperature of ingested prey. Ingestion of a single prey item by a bank cormorant resulted in a 2 °C decrease in stomach temperature, with recovery to the baseline temperature of 40.5 °C in ten minutes (Wilson et al., 1995). In contrast, over a 50-minute feeding dive bout by the same species, stomach temperature could decrease as much as 5 °C (Wilson and Gremillet, 1996). In king penguins at sea, stomach temperature declined by as much as 20 °C over a ten-hour period of diving; the return to baseline temperatures of 36 °C took as long as another ten hours (Wilson *et al.*, 1995).

8.5 Marine birds: body temperatures during dives

Body temperature profiles during diving have been examined to both determine the effectiveness of the feather air layer underwater as well as to investigate the potential role of temperature in regulation of tissue oxygen demand during diving. In tufted ducks (*Aythya fuligula*), abdominal temperature increased during dive bouts in summer but decreased to about 40.5 °C during diving in colder winter waters, resulting in about a 1 °C difference in end-of-bout abdominal temperature between seasons (with a 15.5 °C lower ambient water temperature in winter) (Bevan and Butler, 1992b).

Further development of temperature loggers allowed application to diving birds in the wild, and documented 4–8 °C decreases in abdominal temperatures during dive bouts of gentoo penguins and South Georgian shags (Bevan *et al.*, 1997, Woakes *et al.*, 1992, 1995). In addition, variable but smaller temperature decreases were also reported in a king penguin, with thermistors located below the lower sternum (near the liver and stomach), in the mid-sternal region near the heart, and in the upper abdomen (Culik *et al.*, 1996b). On average, in both gentoo and macaroni (*Eudyptes chrysolophus*) penguins, abdominal temperature declined about 2.4 °C during dive bouts (Bevan *et al.*, 2002, Green *et al.*, 2003).

The most remarkable decreases in abdominal temperature were reported in king penguins, in which lower intraabdominal temperature (between the brood patch wall and stomach) decreased as much as 20 °C during dive bouts (Handrich *et al.*, 1997). These reports all raised issues of regional heterothermy, effects of cold prey ingestion, effectiveness of insulation, and the potential role of diving cardiovascular responses in the distribution and loss of heat, i.e., decreased heat production due to perfusion-related decreases in organ oxygen consumption, decreased muscle blood flow and subsequent transfer of muscle heat to the rest of the body, and possible heat loss through the well-vascularized brood patch and through the wings. In particular, the potential savings in oxygen costs due to hypothermia were highlighted. As an example, it was calculated that, if the temperature of the entire body of a gentoo penguin decreased the average 2.4 °C found in the abdomen, then, based on the Q_{10} effect, the duration of aerobic metabolism would be increased by 29%, and all recorded dives of the gentoo penguin would be within its calculated aerobic limit (Butler, 2004).

The question remained, however, whether temperature did indeed decline in the whole body or, for that matter, at least in the highest O_2-consuming organs (brain, liver, kidneys, muscle). This question was particularly relevant because many of the abdominal temperature sensors in these studies were implanted near the brood patch

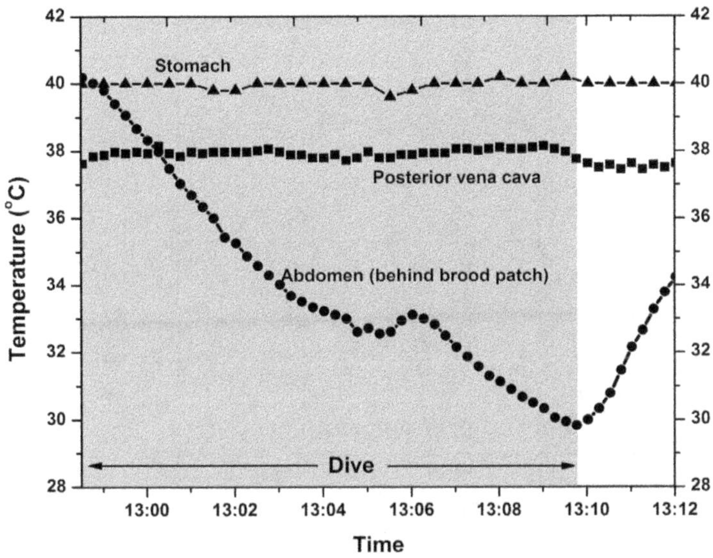

Figure 8.5 Regional heterothermy and preservation of core temperature occurred during this 11.3-min dive of an emperor penguin (*Aptenodytes forsteri*) at an isolated dive hole. Both stomach and posterior vena caval temperature are preserved while the intraabdominal temperature behind the brood patch declines almost 11 °C during this dive. Decreased peripheral perfusion and heat loss through the less-insulated brood patch probably contributed to the decline in abdominal temperature. Adapted from Ponganis et al. (2003b).

wall and/or near the stomach, the former poorly insulated, the latter a reservoir for ingested cold prey items, and neither an organ of high oxygen consumption. This question was addressed in a series of investigations of emperor penguins voluntarily diving at an experimental dive hole (Ponganis *et al.*, 2001, 2003b, 2004).

The dives of emperor penguins at the experimental dive hole were interspersed with surface periods out of the water on the sea ice and were usually only to about 50-m depth due to prey distribution (Ponganis *et al.*, 2000). Nonetheless, despite this behavioral difference with other penguin species that stay in the water and dive continuously at sea, the temperature inside the abdomen behind the brood patch of the emperor penguin declined to as low as 19 °C, confirming the previous findings in king penguins. However, these declines did not correlate with dive duration, and, in addition, simultaneously recorded vena caval temperatures showed only minor fluctuations (Fig. 8.5), and usually increased during the dive period (Ponganis *et al.*, 2001). In addition, during deep dives to 225-m depth, aortic temperatures actually increased about one 1 °C (Fig. 8.6). Lastly, during the second longest dive currently reported for an emperor penguin, vena caval temperature remained near 36.7 to 37.0 °C (Fig. 8.7). Because vena caval and aortic temperatures were considered more representative of both core temperature and the temperature of central organs, it was concluded that central hypothermia and an associated decrease in organ metabolic rate did not occur despite the large changes in abdominal temperatures behind the brood patch. Rather, the temperature

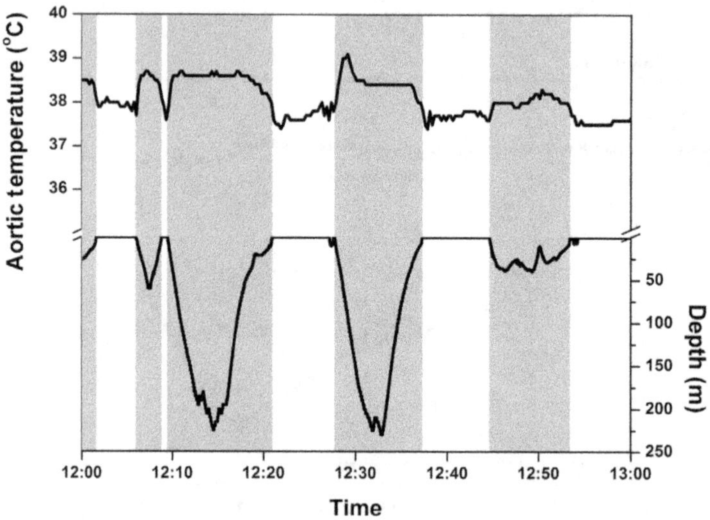

Figure 8.6 Maintenance and even elevation of arterial temperatures occurred during >10-min, 225-m deep dives of an emperor penguin (*Aptenodytes forsteri*) at an isolated dive hole. Thus, core hypothermia is not essential to make deep, long-duration dives. Adapted from Ponganis et al. (2004).

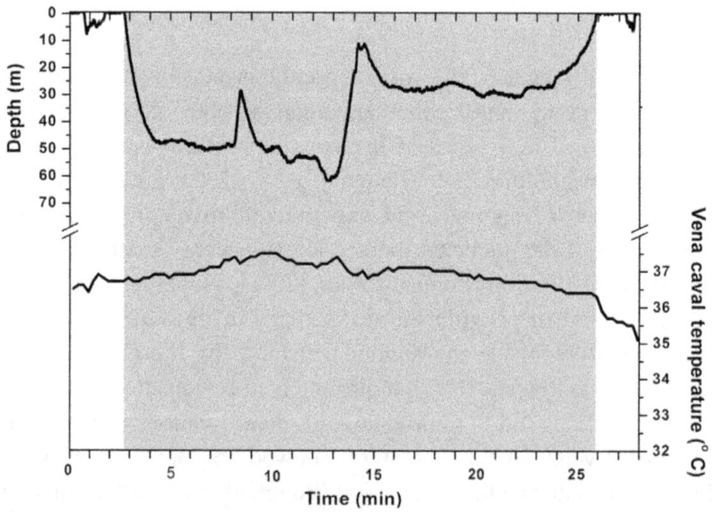

Figure 8.7 Posterior vena caval temperature was preserved at 36.5–37.0 °C during this 23.1-min dive of an emperor penguin (*Aptenodytes forsteri*). Vena caval hypothermia did not occur in one of the longest dives reported for an emperor penguin. Although regional heterothermy may occur, whole-body and core hypothermia were not essential to such dive performance. Adapted from Ponganis et al. (2007, 2010a).

profiles were considered consistent with a model of regional heterothermy based on conservation of core temperature, peripheral vasoconstriction, and cooling of an outer body shell. This model was also consistent with elevations in aortic temperatures and pectoral muscle temperatures during dives as well as with (a) conservation of intradive temperatures in the large central limb veins, and even in the stomach (despite documented cold prey ingestion); and (b) large decreases in intradive temperature in the peripheral wing and foot veins, subcutaneous tissues, and sub-feather space (Ponganis *et al.*, 2003b, 2004).

Although hypothermia in peripheral regions during a dive can decrease the metabolic cost of thermoregulation, in that heat need not be generated to maintain those peripheral temperatures during a dive, it does not appear, at least in emperor penguins at an isolated dive hole, that core hypothermia leads to a decrease in the metabolic rate of central organs. This conclusion was also supported by the lack of correlation of dive temperature and dive duration in these animals, and the maintenance of vena caval temperature even during a 23.1-min dive (Fig. 8.7), one of the longest recorded dives of any emperor penguin (Ponganis *et al.*, 2010a). Although deep body temperatures of other penguin species have not been measured during diving, it is notable that, although pectoral muscle temperature of king penguins at sea also increased during dives, muscle temperature did decrease gradually throughout an entire dive bout due to declines during surface intervals (Schmidt *et al.*, 2006). During dive bouts with mean durations of 4.8–12.4 hours, mean pectoral muscle temperatures declined 0.5–4.0 °C. However, it is unknown whether longer dives were associated with lower muscle temperatures in these king penguins as correlations between dive duration and intradive pectoral muscle temperature were not reported. As previously reviewed in seals, it is also remarkable that penguins were able to perform continuous diving for up to 12 hours without significant elevations in locomotory muscle temperature.

Central temperature profiles have also been recorded in an alcid diver, the guillemot; core hypothermia did not occur during diving (Niizuma *et al.*, 2007). Core temperature (beneath the right lobe of the liver, adjacent to the heart and lungs) was conserved and even increased as much as 1 °C during dives, while peripheral temperature (subcutaneous abdominal wall) decreased as much as 2.5 °C during a dive and 8 °C during a dive bout. Thus, in the guillemot, a bird weighing less than 1 kg, there does not appear to be a hypothermia-associated reduction of metabolic rate in central organs during dives.

9 Diving metabolism

The regulation of diving metabolism has been the subject of much investigation, discussion, and debate. Since the experiments of Irving and Scholander (Irving, 1939, Irving *et al.*, 1941b, Scholander, 1940, Scholander *et al.*, 1942a, 1942b), the significance of a low rate of oxygen consumption and slow depletion of oxygen stores has long been emphasized. It is the rate of metabolism during the breath-hold period that is the critical determinant of breath-hold capacity, the duration of aerobic metabolism, and any need for anaerobic metabolism. Despite many investigations and experimental approaches, however, the metabolic rate during a dive has remained an elusive parameter to measure or even estimate.

There do not appear to be any unique molecular mechanisms of hypometabolism in marine mammals and seabirds (see review below). Low metabolic rate during a dive is most likely secondary to (a) regulation of heart rate and tissue perfusion; (b) regional hypothermia, and (c) a low cost of locomotion during the dive. Actual measurement of that metabolic rate during the dive has been difficult. This chapter will focus on the measurement and interpretation of whole-body metabolic rates in diving animals. Scholander's findings during forced submersions will be summarized first, followed by a review of potential mechanisms of metabolic suppression. Next, the terms *basal metabolic rate* and *resting metabolic rate* will be defined as these are essential control reference values from which to evaluate the level of hypometabolism. Fourth, metabolic rate determination techniques and the difficulty of assigning a value for metabolic rate during a breath hold in intermittently breathing animals will be reviewed. Lastly, estimates of diving metabolic rate from the literature will be summarized.

9.1 Forced submersions and metabolic rate

Diving metabolism is the end result of the many physiological processes and biomechanical adaptations reviewed in previous chapters. Heart rate, organ perfusion, body temperature, muscle workload, buoyancy, hydrodynamics, and the costs of hunting and prey capture/digestion all potentially contribute to the rate of metabolism and oxygen store depletion during a dive. In his 1940 monograph, Scholander reported that the pre-submersion oxygen consumption of the seals in his study was elevated and that the seals incurred an oxygen debt during submersion. That oxygen deficit was not fully accounted for by oxygen store depletion, glycolysis, or excess post-submersion oxygen consumption. He concluded that the submerged seal reduced its metabolism relative to the elevated metabolic rate of the pre-submersion period, i.e., it was hypometabolic. As reviewed in Chapters 5, 7, and 8, that decreased metabolic rate was most probably due to the observed decline in heart rate/organ perfusion and possibly also to decreased body temperatures with hypothermia-related declines in tissue metabolic rate.

In relation to muscle tissue metabolic rate during forced submersion, it should also be noted that rate of decline of muscle O_2 content during the forced submersion of seals was elevated, at five times the typical resting rate of oxygen consumption in mammalian muscle (Blei *et al.*, 1993, Kooyman and Ponganis, 1998, Scholander, 1940). During the anaerobic phase of the forced submersion, the measured muscle lactate accumulation and presumed creatine phosphate depletion could account for maintenance of that elevated muscle metabolic rate (Kooyman and Ponganis, 1998). Consequently, there was no evidence of hypometabolism at the muscle-tissue level during the forced submersion of seals.

Scholander also found that ducks did not repay their oxygen debts, but again, changes in heart rate and possibly body temperature probably contributed to the lower metabolic rate. Notably, in a study 40 years later, there was no evidence for a decrease in ATP turnover rates at the tissue level in muscle (Stephenson and Jones, 1992b). Interestingly, in contrast to seals and ducks, high ventilatory rates in Scholander's forcibly submerged penguins more than accounted for the oxygen shortfall during their submersions. Thus, there was no indirect evidence of a decline in metabolism in the forcibly submerged penguin.

9.2 Biochemical mechanisms of metabolic suppression

At this point, it is worth reviewing the potential role of metabolic suppression in limiting the rate of metabolism during diving. As mentioned briefly in Chapter 7, there is no evidence that the decrease in metabolic rate during the forced submersions of seals is secondary to any unique underlying mechanism at the tissue level. Hypoxia-linked metabolic suppression was proposed to account for low diving metabolic rates and less-than-expected increases in post-dive blood lactate concentrations of Weddell seals (*Leptonychotes weddellii*) in the mid-1980s (Guppy *et al.*, 1986, Hochachka, 1988). This hypothesis of downregulation of metabolism in diving seals was based on the lack of a compensatory increase in lactate concentration (a reverse Pasteur effect) in anoxia-tolerant lower vertebrates (Hochachka, 1986a, 1986b, 1986c, Hochachka and Guppy, 1987, Hochachka and Somero, 1984). In fish, amphibians, and reptiles that tolerate prolonged dormant periods of hypoxia/anoxia, the decrease in metabolic rate is primarily based upon ion channel arrest (decreased Na-K ATPase activity, and density) and arrest of protein synthesis at the level of translation (Bickler and Buck, 2007, Hochachka and Lutz, 2001, Hochachka *et al.*, 1996). Metabolic rate is also decreased in hibernating mammals (Malan, 2014, Quinones *et al.*, 2014), but again, these animals are not travelling and foraging during hibernation.

In mammalian tissues, hypoxic metabolic suppression has been demonstrated, usually in cell cultures, but only after exposures to extremely low concentration of oxygen (i.e., P_{O2}: 10 mm Hg or 1.33 kPa) for prolonged periods of time (30 min to hours) (Wheaton and Chandel, 2011). Cellular mechanisms of such metabolic suppression in mammalian cells include: (a) decreased Na-K-ATPase activity (probably through endocytosis of the enzyme); (b) decreased sensitivity of contractile myofilaments in "hibernating" heart muscle; (c) stabilization of hypoxia-inducible factor 1 (HIF 1), which (1) limits electron transport chain activity through several mechanisms including alteration of the expression of cytochrome c oxidase subunits, and (2) upregulates expression of glycolytic enzymes and also expression of pyruvate dehydrogenase kinase which decreases the activity of pyruvate dehydrogenase, decreasing conversion of pyruvate to acetyl-coA; (d) inhibition of protein synthesis at the level of mRNA translation via activation of the mTOR (mammalian target of rapamycin) pathway; and (e) activation of adenosine monophosphate protein kinase (AMPK) (promotes endocytosis of Na-K-ATPase and activates mTOR pathway) (Arai *et al.*, 1991, 1995, Arsham *et al.*, 2003, Comellas *et al.*, 2006, Emerling *et al.*, 2009, Hudson *et al.*, 2002, Iyer *et al.*, 1998, Liu *et al.*, 2006, Semenza, 2004, 2007).

Such metabolic suppression may also occur in the cells of seals on exposure of cell cultures to prolonged, severe hypoxia, but whether this explains the low metabolic rate, hypoxic tolerance, and diving ability of seals is another question for several reasons. First, the tissues of most mammals are not tolerant of extreme hypoxia, and, therefore, despite the observation of these mechanisms of hypoxic metabolic suppression in mammalian cell cultures, they do not afford additional protection to severe ischemia and hypoxemia in the typical mammal, and, therefore, do not yet account for the hypoxic tolerance of seals. Second, most of these responses require hours of exposure to extreme hypoxemia (P_{O2} of 10 mm Hg or 1.33 kPa), far longer and lower than the hypoxemia in dives of most seals, even elephant seals. Third, metabolic suppression, especially as demonstrated in lower vertebrates, occurs during periods of dormancy or hibernation, not during diving and hunting. Although various physiological indices of renal, hepatic, and neural function are better preserved in seals than terrestrial controls during hypoxic/anoxic exposure (Folkow *et al.*, 2008, Halasz *et al.*, 1974, Hochachka *et al.*, 1988, Hong *et al.*, 1982, Koschier *et al.*, 1978), the underlying mechanisms of such protection and their contribution to metabolic rate reduction are still unknown.

Two additional possible mechanisms of metabolic suppression under hypoxic conditions involve the gasotransmitters hydrogen sulfide (H_2S) and nitric oxide (NO). H_2S is known to bind to complex IV (cytochrome c) of the mitochondrial respiratory transport chain, and such inhibition may represent a mechanism of metabolic suppression (Blackstone *et al.*, 2005, Kolluru *et al.*, 2013). In a similar fashion, the inhibition of cytochrome c by NO may also be a mechanism of metabolic suppression under hypoxic conditions (Fago *et al.*, 2012, Helbo *et al.*, 2013, Shiva *et al.*, 2007).

Inhalational administration of H_2S to mice can suppress metabolic rate and induce suspended animation (Blackstone *et al.*, 2005). Recently, changes in H_2S metabolism have been postulated to underlie the metabolic suppression found in hibernating bears (Revsbech *et al.*, 2014). H_2S production also increases during hypoxia (Kolluru *et al.*, 2013, Olson *et al.*, 2006). Consequently, although there is no evidence available in diving mammals and birds, H_2S could potentially contribute to a decrease in metabolic rate during hypoxia in these animals. Presumably, H_2S accumulation in the brain under such conditions would be minimal as suspended animation is not compatible with active diving.

The generation of NO from nitrite by the nitrite reductase activity of deoxyhemoglobin and deoxymyoglobin has been postulated to represent a mechanism whereby NO's inhibition of cytochrome oxidase may suppress metabolic rate during hypoxia as well as limit the generation of oxygen free radicals after ischemia-reperfusion events (Fago *et al.*, 2012, Flogel *et al.*, 2010, Jensen *et al.*, 2014, Shiva *et al.*, 2007, Soegaard *et al.*, 2012). Although the nitrite reductase activities of different whale myoglobins are not elevated, the high concentration of myoglobin in whale muscle creates the potential for significant NO formation by dexoymyoglobin from nitrite under hypoxic conditions (Helbo and Fago, 2012). Elevated blood nitrite levels in porpoises also support the potential for the use of this pathway to titrate metabolic rate during hypoxia (Soegaard *et al.*, 2012). The full significance of this potential pathway of metabolic regulation in diving animals remains to be investigated.

Metabolic suppression in diving marine mammals may also be partly secondary to ischemic preconditioning, a process in which a brief period of tissue ischemia can decrease ATP depletion rates and improve tissue survival during subsequent ischemic episodes (see Chapter 13). In addition, other mechanisms of suppression may involve the development of hypercarbia and intracellular acidosis during apnea (Elsner, 2015, Malan, 2014) and, at least in some marine mammals, elevated levels of endogenous carbon monoxide (Kajimura *et al.*, 2010, Pugh, 1959, Tift *et al.*, 2014). Hypercarbia has been associated with reduced metabolic rates during hibernation, and carbon monoxide binds to cytochrome oxidase, decreasing oxygen flux.

9.3 Basal and resting metabolic rates

Basal metabolic rate is the oxygen consumption of an awake, adult, post-absorptive animal at rest in a thermoneutral environment; it is typically calculated on the basis of classic allometric equations for birds ($L\ O_2\ min^{-1} = 0.0132$ (body mass$^{0.729}$) with mass in kilograms) or mammals ($L\ O_2\ min^{-1} = 0.0101$(body mass$^{0.75}$) with mass in kilograms) (Aschoff and Pohl, 1970, Kleiber, 1975). The term *standard metabolic rate* has been considered equivalent to the basal metabolic rate, but it has also been used to distinguish the metabolic rate measured under the above conditions as opposed to basal metabolic rate predicted by allometric equations (Hurley and Costa, 2001).

The conditions specified for basal metabolic rate determinations are difficult to achieve in experimental situations, and rarely occur in the wild. Indeed, elevations of oxygen consumption in animals "at rest" during experiments have been attributed to anticipatory increases in metabolic rate under these conditions (Taylor *et al.*, 1987, Williams *et al.*, 1993). Thus, resting metabolic rate of an animal in a study may be different from predicted basal metabolic rate. Furthermore, whether the standard conditions have been satisfied during measurements of metabolic rate of marine mammals at rest has been controversial (Hurley and Costa, 2001, Lavigne *et al.*,

1986, Williams *et al.*, 2004). Suffice it to say that metabolic rate measured at rest in marine mammals is often 1.4–3.0 times the allometrically predicted basal metabolic rate (see previous references). This elevation in metabolic rate at rest in marine mammals has also been attributed to thermoregulatory requirements, body composition, and high-protein diets, and has even been found to be associated with large alimentary tracts (specifically, the length of the small intestine) (Lavigne *et al.*, 1986, Williams *et al.*, 2001b).

9.4 Metabolic rate measurements

Oxygen consumption measurements in mammals and birds have most commonly been measured with respirometry techniques, doubly labeled water techniques, and heart rate measurements. Another approach, the Fick method, can also provide estimation of oxygen consumption, but it requires accurate measurements of cardiac output and the arterio-mixed venous oxygen content difference. Despite the value of this approach, the required arterial and pulmonary artery catheterizations have made it impractical in marine mammals and seabirds. In terrestrial mammals, the Fick method is more often used to calculate cardiac output on the basis of measured oxygen consumption and the arterio-venous oxygen difference (Taylor *et al.*, 1987). Consequently, respirometry, doubly labeled water, and heart rate have been the three basic approaches used to assess oxygen consumption in diving mammals and birds.

Open flow respirometry requires the animal to breathe into a chamber or mask that allows for accurate measurement of (a) the rate of airflow through the system; and (b) the fractional oxygen concentration exiting the system (Fedak *et al.*, 1981, Withers, 2001). While this approach has been applied to many marine mammals and seabirds, its disadvantage is that the animal must return to the breathing chamber in order for the analysis to be conducted. Potentially, this technique is applicable to animals in laboratory tanks and swim flumes, to animals trained to breathe beneath a dome at sea, and to wild animals freely diving under sea ice at an isolated dive hole to which they must return. However, it is not applicable to free-diving animals at sea. In addition, even at an isolated dive hole on the sea ice of McMurdo Sound, although this technique has been used with great success in Weddell seals (Castellini *et al.*, 1992b, Kooyman *et al.*, 1973a, Ponganis *et al.*, 1993a, Williams *et al.*, 2004), this approach is not necessarily feasible for other species. Emperor penguins *(Aptenodytes forsteri)*, for example, prefer to dive in groups and will not tolerate a dome over the dive hole as they usually exit the water after each dive.

Another limitation of respirometry in estimation of oxygen consumption during diving is that the measurements are only made during the surface interval. The oxygen consumed during the surface interval is a function of the metabolic rate and duration of the dive as well as the metabolic rate and duration of the surface interval. Thus, this measurement does not directly represent measurement of the actual oxygen consumption rate during the dive. Various approaches have been used to calculate "diving metabolic rate" with respirometry. In a method based on Scholander's original study in 1940, oxygen consumption during the submersion period was calculated from the volume of oxygen consumed above the baseline level during the surface interval until the post-submersion oxygen consumption returned to the baseline level. That oxygen volume was then divided by the duration of the submersion and was considered to represent the oxygen consumption rate during the submersion (Hurley and Costa, 2001, Williams *et al.*, 2004). The limitation to this method is that, on a theoretical basis, oxygen consumption during the hyperventilation and tachycardia of the surface interval is not necessarily equivalent to that consumed at rest when the heart rate is lower, organ perfusion is lower, and the cost of breathing is less.

In another approach used by Castellini and co-workers, the entire amount of oxygen consumed during a surface interval between dives (or until oxygen consumption returned to baseline levels during prolonged surface intervals) was divided by the time of the dive and surface interval (or the time of the dive and the surface interval time until baseline levels were reached) (Castellini *et al.*, 1992b). This provided a metabolic rate of the dive event (dive plus surface interval). The limitation of this method is that it does not represent the actual rate of oxygen consumption during the dive. In ducks and Humboldt penguins (*Spheniscus humbodti*), multiple linear regressions of volume of oxygen consumed against surface interval duration and dive duration have provided estimation of the oxygen consumed during the dive (Bevan *et al.*, 1992, Butler and Woakes, 1984, Stephenson, 1994, Woakes and Butler, 1983). The limitation here is that the regression coefficients only represent the mean oxygen consumption rates of dive and surface intervals at their mean durations.

Another approach to estimate metabolic rate of marine mammals and seabirds is the doubly labeled water (DLW) technique (Nagy, 1980, Speakman, 1998). It is based on the facts that labeled oxygen is removed from the body by both carbon dioxide and water, while labeled hydrogen is removed only via water. Hence, the rate of carbon dioxide production (and indirectly, oxygen consumption) can be calculated from the difference between the elimination of the two labels (Nagy, 1980, Speakman, 1998). These techniques have been applied extensively to many species of marine mammals and seabirds to measure field metabolic rates in the wild (Nagy *et al.*, 1999, Shaffer, 2011). While valuable in ecological modeling, application of DLW field metabolic rates to estimation of diving metabolic rate is limited because the DLW field metabolic rate is determined over a period of many days, and, thus, is not directly applicable to individual dives, groups of dives, or different types of dives. For example, use of either the field metabolic rate or the estimated foraging metabolic rate determined with DLW in emperor penguins significantly underestimated the duration of aerobic metabolism during dives of these birds (Nagy *et al.*, 2001). Furthermore, many validation studies of the DLW technique in comparison to respirometry have revealed a wide range of individual error (i.e., –40 to +80% differences), although differences between mean values of groups are usually much closer and acceptable (Bevan *et al.*, 1995a, 1995b, Boyd *et al.*, 1995, Butler *et al.*, 2004, Sparling *et al.*, 2008). Although field metabolic rates may be useful to begin to assess the potential range for diving metabolic rate (Costa *et al.*, 2001), these limitations in use of the DLW data should be remembered.

The third approach to estimate metabolic rate in marine mammals and seabirds is the heart rate method, in which the linear relationship between oxygen consumption and heart rate in an exercising animal (Bevan *et al.*, 1995b, Boyd *et al.*, 1995, Fedak, 1986, Kooyman and Ponganis, 1994, Williams *et al.*, 1991, 1992c) is used to predict metabolic rate (Butler, 1994, Butler *et al.*, 2004). This technique is based on the assumption that stroke volume and the arterio-venous difference in oxygen content is constant (or at least the average values over the exercise period are constant). However, stroke volume in seals is not necessarily constant between eupnea and apnea, or surface and submerged swimming (Ponganis *et al.*, 1990). In addition, the arterio-venous difference in oxygen content can also change throughout a breath hold (Kerem and Elsner, 1973). Therefore, use of a heart rate–oxygen consumption relationship determined in a swim flume may not be accurate to predict the oxygen consumption during or throughout a dive. However, given that values will average out over long periods of time, such an approach has been used to estimate a field metabolic rate (Bevan *et al.*, 1995a, 2002, Boyd *et al.*, 1999, Butler *et al.*, 2004, Froget *et al.*, 2004, Green *et al.*, 2003).

9.5 Diving metabolic rates: marine mammals

In consideration of diving metabolism and hypometabolism, it is important to define the resting value to which "hypometabolism" is referred. For example, mean oxygen consumption values of prolonged submergences of voluntarily submerged seals and sea lions, although less than values "at rest," were still not less than the allometrically predicted basal metabolic rates (Hurley and Costa, 2001, Webb *et al.*, 1998). The values "at rest" were 1.3 to 3.0 times the allometrically predicted rate. Therefore, it is not surprising that metabolic rates associated with longer submergences were less than values measured "at rest" (Hurley and Costa, 2001, Scholander, 1940, Webb *et al.*, 1998) or that metabolic rate declined as breath-hold time or percent of time submerged increased in other studies (Castellini *et al.*, 1992b, Fedak *et al.*, 1988, Gallivan, 1981, Thorson, 1993, Tift *et al.*, 2013, Webb *et al.*, 1998). As discussed in prior chapters, these changes in metabolic rate are probably secondary to perfusion-related declines in tissue oxygen consumption associated with slower heart rates during submersion as well as to possible changes in thermoregulatory requirements. However, in consideration of the need for further "downregulation" of metabolism at the cellular level, it is notable that the metabolic rates associated with submergences in almost all these studies were not below the allometrically predicted basal metabolic rate. Most of these studies were in laboratory situations. Field metabolic rates of young, translocated elephant seals at sea were, on average, about 1.5 times greater than allometrically predicted resting rates. (Maresh *et al.*, 2014). In contrast, estimations of the metabolic rate required to maintain aerobic metabolism during the repetitive, long-duration dives of elephant seals at sea predict significant "hypometabolism" with estimated metabolic rates as much as 60% below the allometrically predicted resting metabolic rate (Hindell *et al.*, 1992, Le Boeuf *et al.*, 1988).

Diving metabolic rate has been determined with respirometry in two phocid pinnipeds, the Weddell seal and the gray seal (*Halichoerus grypus*). In Weddell seals diving at an isolated dive hole beneath the sea ice of McMurdo Sound, diving metabolic rates were variable and decreased with increasing dive duration (Castellini *et al.*, 1992b, Kooyman *et al.*, 1973a, Ponganis *et al.*, 1993a). Average diving metabolic rate was 4.5 ml O_2 kg^{-1} min^{-1}, about 1.6 times the predicted basal metabolic rate, but not significantly different from values measured from seals at rest (Castellini *et al.*, 1992b). In the gray seal study, which involved captive animals "diving" in an underwater maze, diving metabolic rate was also variable and decreased with slower swim speeds and longer dive durations. Diving metabolic rate again averaged 1.7 times the predicted basal metabolic rate and was less than the measured resting metabolic rate at the surface (Sparling and Fedak, 2004).

A later study of Weddell seals equipped with a video camera and accelerometer further analyzed the contribution of feeding and stroke rate to the variability in diving metabolic rate (Williams *et al.*, 2004). In general, diving metabolic rates were similar to those of the previous Weddell seal studies. Feeding during dives resulted in a 45% increase in diving metabolic rate in comparison to that of fasting dives of similar duration and distance traveled. The elevation in diving metabolic rate could continue as long as five hours after fish ingestion. Locomotor activity increased diving metabolic rate 1.3–3.5 times above resting rates dependent on dive duration. Oxygen consumption increased linearly with the number of strokes during a dive, and the average cost per stroke was 0.044 ml O_2 kg^{-1} min^{-1}. Thus, diving metabolic rate was found to be a function of dive duration, number of strokes, and feeding.

However, in other species, prey ingestion may not always immediately result in digestion and an increase in metabolic rate. In gray seals, the elevation in metabolic rate associated with digestion (specific dynamic action) can be delayed for periods of many hours after foraging behavior (Sparling *et al.*, 2007). In summary, although average diving metabolic rates of seals are above allometrically predicted basal metabolic rates, they are usually near the metabolic rate of an animal at rest. Overall, the cost of diving is quite low.

The cost of diving also appears to be low in trained Steller sea lions (*Eumetopias jubatus*) in that diving metabolic rates determined by respirometry were near resting metabolic rates at the surface (Fahlman *et al.*, 2008, Hastie *et al.*, 2007). Resting rates were 55–85% above the allometrically predicted resting metabolic rate. Diving metabolic rate did decrease with dive duration; metabolic rate during short-duration dives could be as high as 25 ml O_2 kg^{-1} min^{-1}, while most longer-duration dives (4–7 min) had metabolic rates of 6–10 ml O_2 kg^{-1} min^{-1}.

In sea otters (*Enhydra lutris*), mean diving metabolic rate of single dives was 17.6 ml O_2 kg^{-1} min^{-1}, and that of foraging dives was 22 ml O_2 kg^{-1} min^{-1} (Yeates *et al.*, 2007). The single-dive value, equivalent to that of a sea otter swimming horizontally underwater (Williams, 1989), was 1.3 times measured resting metabolic rate and 3.7 times the allometrically predicted metabolic rate. The foraging dive value was 1.6 and 4.8 times those respective values. Once again, diving metabolic rate was relatively close to the measured resting metabolic rate. The relatively high diving and resting metabolic rates of sea otters were attributed to high maintenance costs (thermoregulation, digestion) and high locomotory costs.

9.6 Diving metabolic rates: birds

Oxygen consumption rates in voluntarily diving birds were above metabolic rates measured at rest. In tufted ducks and lesser scaups (*Aythya fuligula, A. affinis*), oxygen consumption measurements from surface intervals revealed metabolic rates approximately 3.5 times that measured at rest (Bevan and Butler, 1992b, Bevan *et al.*, 1992, Stephenson, 1994, Woakes, 1988, Woakes and Butler, 1983). Oxygen consumption in Humboldt penguins voluntarily diving in a tank was less elevated, about 26% above resting levels (Butler and Woakes, 1984). In all these species, heart rate during dives was also elevated above resting rate (see Chapter 5) in agreement with the oxygen consumption data. Thus, in these voluntarily diving birds in tanks, there was no evidence of a reduction in oxygen consumption relative to rates at rest.

However, as reviewed in Chapter 6, the longer dive durations of deep-diving species at sea have raised the question of potential metabolic rate reductions by other mechanisms such as hypothermia and severe bradycardia (Bevan *et al.*, 1997, 2002, Boyd and Croxall, 1996, Butler, 2000, 2004, Handrich *et al.*, 1997, Wright *et al.*, 2014). Although foraging metabolic rates of penguins range from four to nine times basal metabolic rate (for a review, see Nagy *et al.*, 2001), oxygen consumption measurements during dives of penguins at sea have yet to be determined.

Due to the difficulty and theoretical limitations of respirometry, DLW, and heart rate techniques in estimating diving metabolic rate, the actual oxygen store depletion rate during dives of emperor penguins at an isolated dive hole was assessed by measurement of changes in air sac P_{O2}, arterial and venous hemoglobin saturation, and muscle myoglobin saturation (Meir and Ponganis, 2009, Stockard *et al.*, 2005, Williams *et al.*, 2011a). Based on the changes in these oxygen stores for dives of about 6-min duration (such dives are assumed to be aerobic as they are not associated with increases in post-dive lactate concentrations (Ponganis *et al.*, 1997c)), the average actual oxygen store depletion rate during such dives was 6.8 ml O_2 kg^{-1} min^{-1}, slightly greater than the allometrically predicted basal metabolic rate (Williams *et al.*, 2011a). That value is also about one-third the field metabolic rate determined with DLW in penguins at the isolated dive hole, and equivalent to the oxygen consumption rates of emperor penguins resting in water in a flume, or standing in their thermoneutral zone (Williams *et al.*, 2011a). Thus, although oxygen store depletion rates are probably variable secondary to the physiological responses associated with the nature and circumstances of a given dive, this finding confirmed that the cost of diving was indeed quite low in emperor penguins, just as in seals.

9.7 Diving metabolic rates: summary

In conclusion, available diving metabolic rates in both seals and penguins indicate that the cost of diving is quite low. However, as yet, there is no evidence of unique molecular mechanisms of metabolic suppression in actively diving birds and mammals. In seals, mean diving metabolic rates are typically above the allometrically predicted basal metabolic rate, but often near or below metabolic rate measured in the animal at rest. In the emperor penguin, the actual oxygen store depletion rate is near both the rate measured at rest and also that predicted by allometric equations. In contrast, in other avian species, diving metabolic rates are typically above the metabolic rate measured at rest. The variability in diving metabolic rate is secondary to multiple factors, including dive duration, the degree of bradycardia and peripheral vasoconstriction, regional hypothermia, locomotory effort, and digestion. The role of other mechanisms such as hypoxic metabolic suppression in reduction of diving metabolic rate is not yet established, but has often been invoked to account for the routine long-duration dives of "surfacers" such as elephant seals.

10 The aerobic dive limit

The aerobic dive limit (ADL) has become the most fundamental concept in the interpretation of the diving physiology, diving behavior, and foraging ecology of marine mammals and diving birds. The ADL, as originally defined by Kooyman and co-workers, is the dive duration associated with the onset of post-dive blood lactate accumulation (Kooyman, 1985, Kooyman *et al.*, 1980, 1983). Studies of adult Weddell seals (*Leptonychotes weddellii*) diving at an isolated dive hole demonstrated that post-dive blood lactate concentration was distinctly elevated after dives of 26-min duration and that blood lactate began to rise after dives of about 20-min duration (Fig. 10.1). Furthermore, 97% of dives of Weddell seals were under 26 min. Thus, most dives were completely aerobic, and it was the efficiency of aerobic metabolism that allowed frequent, repetitive dives to depth.

Since these landmark publications in the early 1980s, the ADL hypothesis (dives are primarily aerobic) has become fundamental in the interpretation of diving behavior and foraging ecology, as well as diving physiology (Boyd and Croxall, 1996, Butler, 2004, 2006, Costa *et al.*, 2001, 2004, Davis and Kanatous, 1999). The number of species in which an ADL has been estimated are too numerous to cite. The concept of an aerobic limit re-emphasized the importance of hypoxic tolerance and the rate of oxygen store depletion that had been so well demonstrated by Scholander and Elsner (see prior chapters). In addition, the emphasis in biochemical investigations also shifted – from glycolysis and anaerobic capacity to aerobic metabolism and the magnitude of oxygen stores. Later studies of Weddell seals have also reaffirmed this finding, with post-dive increases in blood lactate concentration after dive durations of 17–23 minutes in adults and 4–5 minutes in pups (Burns and Castellini, 1996, Guppy *et al.*, 1986, Williams *et al.*, 2004).

Although the ADL may be one of the most frequently cited concepts in diving physiology and behavior, it has been rarely measured due to technical limitations. In this chapter, ADLs that have been determined by measurements of blood lactate levels in different species will first be reviewed. Then, various indirect methods to estimate the ADL will be assessed. Lastly, the physiological basis of the ADL will be examined.

10.1 Aerobic dive limits determined by blood lactate measurements

In addition to the Weddell seal, an ADL has been determined by post-dive blood lactate measurements in five other species under various conditions (Table 10.1). A 5.6-min ADL for emperor penguins (*Aptenodytes forsteri*) was measured in birds

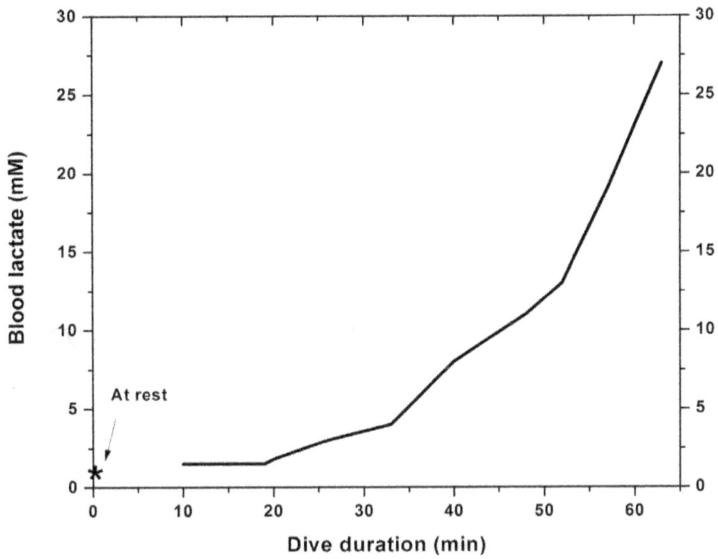

Figure 10.1 The lactate endurance curve for the Weddell seal (*Leptonychotes weddellii*) documented the rise in post-dive blood lactate concentration after dives of various duration, and established the basis for the aerobic dive limit concept. Adapted from Kooyman, 2006, Kooyman et al. (1980a).

diving at an isolated dive hole on the sea ice of McMurdo Sound (Ponganis *et al.*, 1997c). The experimental approach was basically similar to that used for Weddell seal. In Baikal seals (*Phoca sibirica*), a 15-min aerobic submersion limit was measured during spontaneous submersions of captive animals in shallow tanks (Ponganis *et al.*, 1997b). The term *submersion limit* was used because the seals were not actively swimming; they merely submerged and lay on the bottom of the tank. These seals had been captured in the wild and were untamed. If humans were in sight, they would stay underwater as long as 25 minutes, with heart rates as low as 5 bpm for the last 5 min.

The three other ADL measurements involved captive, trained animals. A 2.3-min ADL was measured in young California sea lions (*Zalophus californianus*), trained to swim submerged following a target in the ring tank that had been built as part of Scholander's Physiological Research Lab at Scripps Institution of Oceanography (Ponganis *et al.*, 1997d). This again was not actual diving as the animals were not descending to depth and undergoing pressure-related changes in lung volume and buoyancy. A 3.3-min ADL was measured in bottlenose dolphins (*Tursiops truncatus*) trained to dive to a target as deep as 210 m (Williams *et al.*, 1999). These animals were trained for fluke presentation to allow blood sampling from blood vessels in the fluke. In a similar experiment, two beluga whales (*Delphinapterus leucas*) were trained to dive and station at a target at up to 300-m depth for dive durations as long as 11 min (Shaffer *et al.*, 1997). An ADL as long as 9–10 min for the beluga whale was reported on the basis of elevated blood lactate concentrations after dives of 9–11 min. However, it is unknown if lactate might begin to increase after shorter dive durations of beluga whales

Table 10.1 Aerobic limits determined with post-dive blood lactate measurements in different divers.

Species	Aerobic limit (min)	Technique	Reference
Weddell seal *Leptonchotes weddellii*		Isolated dive hole	
Adult	17–23		A
Juvenile	10–13		B
Pup	4–5		C
Emperor penguin *Aptenodytes forsteri*	5.6	Isolated dive hole	D
Bottlenose dolphin *Tursiops truncatus*	3.3	Trained dives	E
Beluga whale *Delphinapterus leucas*	$\leq$9–11	Trained dives	F
Baikal seal *Phoca sibirica*	15	Stationary, voluntary dives in a tank	G
California sea lion *Zalophus californianus* (juvenile)	2.3	Submerged swimming	H
Sea otter *Enhydra lutris*	>1.6	Voluntary dives in a tank	I

References: A: Guppy *et al.*, 1986, Kooyman *et al.*, 1980, Williams *et al.*, 2004; B: Kooyman *et al.*, 1983; C: Burns and Castellini, 1996; D: Ponganis *et al.*, 1997c; E: Williams et al., 1999; F: Shaffer *et al.*, 1997; G: Ponganis *et al.*, 1997b; H: Ponganis *et al.*, 1997d; I: Yeates *et al.*, 2007.

Results for beluga whales and sea otters represent upper and lower limits, respectively, for an aerobic dive limit. In beluga whales, lactate was elevated after 9–11-min dives, but no data were obtained for dives of lesser duration. In sea otters, lactate was not elevated after dives up to 1.6 min in duration, but no data were obtained for dives of longer duration.

as there were no dives of shorter duration in the study. Lastly, although an ADL_M has not been determined in sea otters, it is at least 1.6 minutes as blood lactates did not rise after dives as long as 100 sec in captive otters (Yeates *et al.*, 2007).

These studies were remarkable technical achievements that have provided the best available estimates of the duration of predominant aerobic metabolism during diving. Nonetheless, potential limitations have been raised in application of these data to animals diving freely in the wild. First, for seals and penguins diving at an isolated dive hole, it could be questioned whether these animals are primarily searching for an escape hole and that physiological responses during such dives are different than in animals diving in a completely unrestricted situation. Although such search or "escape" dives undoubtedly occur, both seals and penguins appear to adapt to the situation and begin feeding regularly (Davis *et al.*, 1999, Kooyman, 1968, Ponganis *et al.*, 1993a, 2000, Williams *et al.*, 2004). It should also be noted that emperor penguins at the isolated dive hole predominantly dive to less than 50-m depth, so the ADL of 5.6 min may not be applicable to deep dives because of potential differences in buoyancy, stroke rate, diving air volume, and cardiovascular responses (Sato *et al.*, 2011, Williams *et al.*, 2012, Wright *et al.*, 2014). The submerged swimming sea lion study is also subject to the same limitation. Heart rate responses during submerged swimming were similar to those during trained dives to shallow depths at sea, but were higher than during deep (400 m) dives (McDonald and Ponganis, 2014, Ponganis *et al.*, 1997d).

Lastly, a potential criticism of techniques involving trained dives is that the cardiovascular responses and oxygen depletion patterns may be different between a natural dive and a dive in which the animal may not know the depth of the target or the duration it may have to station at the target. On the other hand, a foraging animal searching for prey may also not know the depth or duration of dive needed to find its prey. In conclusion, despite such potential limitations, these studies remain remarkable accomplishments and still provide the best available data on the duration of aerobic metabolism and onset of lactate accumulation during diving. In the future, development and refinement of an intravascular blood lactate sensor (Baker and Gough, 1995) may allow investigations in the wild.

10.2 Behavioral and calculated aerobic dive limits: indirect techniques

Due to the difficulty of actually measuring an ADL with blood lactate determinations (i.e., a measured ADL, ADL_M), many researchers have resorted to other approaches to examine the ADL. The first technique is the estimation of a behavioral ADL (ADL_B). This is based upon the original observations of Kooyman and co-workers that 92–96% of dives were less than the ADL_M, and that dives beyond the ADL had longer surface intervals (Kooyman *et al.*, 1980, 1983). Thus, the ADL_B has been estimated as the dive duration below which 95–97% of dives occur (Burns and Castellini, 1996, Hindle *et al.*, 2011). Alternatively, although many factors may contribute to the duration of a surface interval, the ADL_B can be considered the dive duration after which surface intervals begin to increase (Burns, 1999, Cook *et al.*, 2008, Kooyman and Kooyman, 1995). Analysis of minimum surface intervals in relation to dive duration provided an ADL_B of 8 min in emperor penguins at sea, about 2.4 min longer than the ADL_M at the isolated dive hole (Kooyman and Kooyman, 1995). Only about 4% of foraging dives of emperor penguins at sea were greater than that ADL_B. A similar type of analysis was used to estimate ADL_B in different age groups of Weddell seals (Burns, 1999). A recent elegant constraint line analysis of scattergrams of dive durations and surface intervals has been used to estimate an ADL_B of 3.2 minutes in adult, female Galapagos fur seals (*Arctocephalus galapagoensis*) (Horning, 2012).

The ADL has also been calculated (ADL_C) by dividing total body oxygen stores by the diving metabolic rate. This approach was again based on the findings of Kooyman and co-workers. They made such calculations from prior measurements of body oxygen stores and diving metabolic rate in Weddell seals and found that it was quite close to the inflection point in their lactate concentration–dive duration curve (Kooyman *et al.*, 1980, 1983). Although this formula has become the most common technique to estimate an ADL, there are several limitations to its use. First, as reviewed in Chapter 4, accurate assessment of oxygen stores is dependent not just on myoglobin concentration, hemoglobin concentration, and blood volume, but also on the less frequently documented respiratory volume and muscle mass. Second, and probably most important, what diving metabolic rate should be used in the formula? As originally calculated, as reaffirmed in a subsequent study of Weddell seals, and as demonstrated for submersions of Baikal seals, the dive event metabolic rate (total oxygen consumed during the surface interval divided by the sum of the dive duration plus surface interval) was used to correctly calculate the ADL (Kooyman *et al.*, 1980, Ponganis *et al.*, 1993a, 1997b,). In the Weddell seal studies, that diving metabolic rate was approximately two times the allometrically predicted resting metabolic rate. Note that neither the field metabolic rate, the estimated metabolic rate during a dive (based on accelerometry, heart rate, or excess oxygen consumption above baseline during the surface interval divided by dive duration), nor the actual oxygen store depletion rate during a dive were used. As reviewed in Chapter 9, dive event metabolic rates have been determined in few species. Consequently, the diving metabolic rate in the denominator of the ADL_C formula is the least documented variable in the equation. For mammalian divers, investigators will often use a value twice the allometrically predicted resting metabolic rate (as found in the Weddell seal data). Based on the most recent estimates of oxygen stores in emperor penguins, the 5.6-min ADL_M is also best predicted by a calculation with a diving metabolic rate twice the allometrically predicted resting metabolic rate (Ponganis *et al.*, 2010a).

It has been emphasized many times that the ADL_C formula should be considered to simply predict the dive duration with the onset of post-dive blood lactate accumulation (Kooyman and Ponganis, 1998, Kooyman *et al.*, 1999, Ponganis, 2011, Ponganis *et al.*, 1997d, 2003a, 2010a). The formula does not define the status of the oxygen stores at the ADL. Physiologically, the ADL has been a black box. Certainly, there is at least some localized depletion of oxygen and subsequent glycolysis at the ADL, but all oxygen stores are not necessarily depleted. After all, animals can dive much longer than the ADL. The formula, however, is convenient to at least begin to estimate the duration of aerobic dive time without resort to increased glycolysis and lactate accumulation. It should be remembered, however, that the accuracy of the formula has only been demonstrated in two species.

Because of the confusion over the physiological implications of the ADL_C formula, a new terminology was suggested in the late 1990s (Butler and Jones, 1997). The ADL_M, the ADL measured with lactate determinations, was labeled the diving lactate threshold (DLT). The dive duration calculated by the ADL_C formula, and at which all oxygen stores were consumed, was termed the calculated ADL or cADL. A diving lactate threshold is similar in concept to lactate thresholds in exercise studies of terrestrial mammals. The suggested cADL represents the point of complete oxygen depletion and is quite different from earlier use of the ADL_C to simply predict the onset of post-dive blood lactate accumulation. Although these new terms are now used in the literature, they have not gained universal use. Consequently, readers should be aware which terminology is being used. In order to avoid confusion and be consistent with the wording in earlier research, the suggested change in terminology is not used in this book.

10.3 Physiological basis of the aerobic dive limit

In order to examine the physiological basis of the aerobic dive limit and eliminate confusion caused by various interpretations of the ADL_C formula, oxygen store management was investigated in emperor penguins. In a series of projects, indwelling P_{O2}

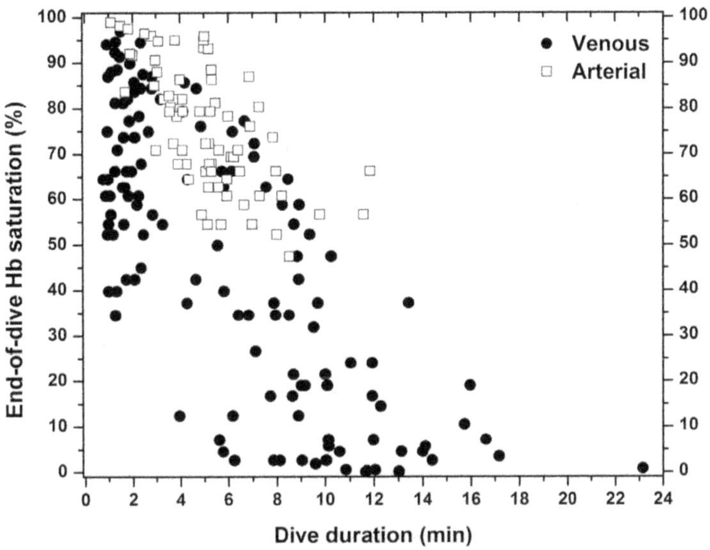

Figure 10.2 End-of-dive arterial and venous hemoglobin (Hb) saturations of the emperor penguin (*Aptenodytes forsteri*) were highly variable but demonstrated that blood oxygen was not completely depleted even in dives beyond the 5.6-min aerobic dive limit determined with post-dive blood lactate measurements. Adapted from Meir and Ponganis (2009).

electrodes and near-infrared myoglobin saturation probes were utilized to monitor O_2 store depletion in emperor penguins diving at an isolated dive hole (Meir and Ponganis, 2009, Ponganis *et al.*, 2007, 2009, Stockard *et al.*, 2005, Williams *et al.*, 2011a). This model was ideal because the 5.6-min ADL_M had been determined previously with post-dive blood lactate measurements (Ponganis *et al.*, 1997c).

Investigation of air sac, arterial, and venous P_{O2} revealed a wide range of end-of-dive values at the 5.6-min ADL_M. For example, end-of-dive respiratory O_2 fractions at the ADL ranged from 1% to 8% (Stockard *et al.*, 2005). Similarly, venous hemoglobin saturations ranged from 5% to 50% while arterial hemoglobin saturations were 55% to 95% (Fig. 10.2) (Meir and Ponganis, 2009). During two dives greater than 10 min, end-of-dive arterial hemoglobin saturations were still 60–65% (Fig. 10.2). Clearly, the respiratory and blood O_2 stores were not depleted at the ADL. However, myoglobin saturation profiles, although also variable, demonstrated complete depletion in some dives of 5–6-min duration (Fig. 10.3), close to the 5.6-min ADL_M (Williams *et al.*, 2011a). Given that muscle lactate accumulation begins at 10–20% saturation in seals (Scholander *et al.*, 1942a, Williams *et al.*, 2012), these investigations supported the hypothesis that the depletion of muscle oxygen is the physiological basis of the ADL (Kooyman and Ponganis, 1998). Near-complete myoglobin desaturation in the primary locomotory muscle results in increased glycolysis, accumulation of muscle lactate, and subsequent wash-out of lactate into the blood during the post-dive interval.

Further evidence for the lack of complete blood O_2 store depletion at the ADL was also found in hemoglobin saturation profiles of diving elephant seals and sea lions

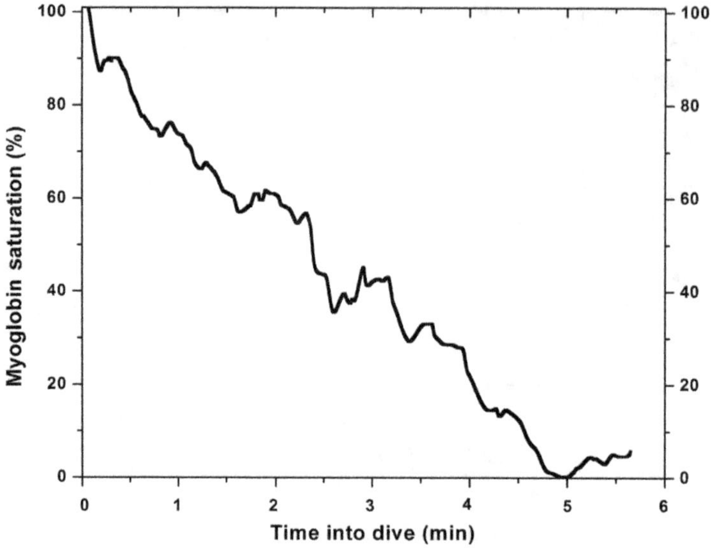

Figure 10.3 Myoglobin (Mb) saturation in the emperor penguin (*Aptenodytes forsteri*) demonstrated almost complete muscle oxygen depletion in dives near the 5.6-min aerobic dive limit consistent with the hypothesis that muscle oxygen depletion and subsequent lactate accumulation provided the physiological basis of the aerobic dive limit. Adapted from Williams et al. (2011a).

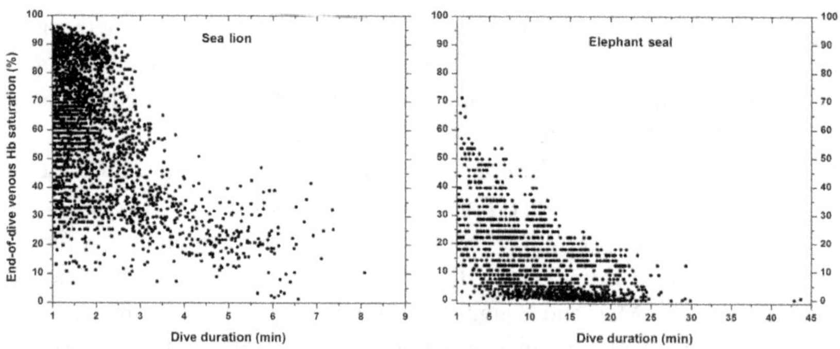

Figure 10.4 End-of-dive venous hemoglobin (Hb) saturations in lactating California sea lions (*Zalophus californianus*) and young northern elephant seals (*Mirounga angustirostris*) were highly variable, especially for dives of shorter duration. However, blood oxygen depletion was often incomplete even for longer dives. Sea lion data were from the posterior vena cava; elephant seal data from the hepatic sinus and extradural vein. Adapted from McDonald and Ponganis (2013), Meir et al. (2009).

(McDonald and Ponganis, 2012, 2013, Meir *et al.*, 2009). Although an at-sea ADL has not been determined with post-dive lactate measurements, end-of-dive venous hemoglobin saturations ranged as high as 40% in dives longer than 6 min in sea lions and 20 min in elephant seals (Fig. 10.4). End-of-dive arterial hemoglobin saturations were as high as 30% during dives of 20–22-min duration in elephant seals, and were greater than 80% during dives as long as 7 min in sea lions. For the sea lions, back calculation of alveolar P_{O2} from the measured arterial P_{O2} results in a lung oxygen fraction near 10%; approximately half of the lung O_2 store was not consumed in some of the deepest, longest dives of these sea lions. All these findings reinforce the fact that the calculation of an ADL based on O_2 stores and a diving metabolic rate does not represent the physiological basis of the onset of post-dive blood lactate accumulation. The calculated ADL may be useful in ecological models and prediction of the dive duration associated with the onset of lactate accumulation, but it does not address the physiological/biochemical processes underlying the ADL. Hence, the importance of recent research addressing myoglobin saturation, the degree of muscle perfusion and linkage of heart rate/stroke rate during diving (Chapter 5).

As already pointed out in many reviews and studies, the ADL probably has some variability, dependent on the physiological responses, locomotory effort, digestive processes, and ultimately metabolic rate during a dive (Kooyman and Ponganis, 1998, Meir and Ponganis, 2009, Ponganis, 2011, Ponganis *et al.*, 1993a, Sato *et al.*, 2011, Williams *et al.*, 2004, 2011a,). In surfacers such as elephant seals, even sleep may occur and potentially prolong the ADL by decreasing metabolic rate during a dive (Crocker *et al.*, 1997, Mitani *et al.*, 2010).

In a theoretical model, it has also been argued that the ADL can be optimized by coupling heart rate and muscle perfusion with muscle workload so as to simultaneously deplete the blood and muscle O_2 stores (Davis and Kanatous, 1999). Again, this emphasizes the significance of the linkage of heart rate (and perfusion) with flipper stroke rate during diving (see Chapter 5). However, the above findings in emperor penguins are not consistent with such a model of simultaneous depletion of the blood and muscle O_2 stores. Ideally, O_2 store management strategies and the onset of post-dive blood lactate accumulation can be further examined in other species with use of recorders and sensors for continuous P_{O2}, myoglobin saturation, and even lactate profiles. Translocated elephant seals, otariids on maternal foraging trips, trained cetaceans, and isolated dive hole studies of Weddell seals and emperor penguins are all feasible models.

11 Oxygen store depletion and hypoxemic tolerance

The principle determinants of the pattern and rate of depletion of O_2 stores during dives, as reviewed in prior chapters, are the cardiovascular responses, locomotory effort, and the thermoregulatory and digestive processes occurring during a given dive. In order to make full use of O_2 stores and maximize dive duration, divers should be tolerant of extremely low levels of O_2. Otherwise, they would not be able to take full advantage of their large O_2 storage capacities. This chapter expands upon a recent review and attempts to weave physiological responses with observed O_2 store depletion patterns (Ponganis *et al.*, 2011). The actual rates, patterns, and magnitude of O_2 store depletion during various breath-hold events (forced submersions, simulated dives, spontaneous breath holds (sleep apnea), trained dives and free dives) will be reviewed first. In the second part of the chapter, associated changes in blood gases, blood pH, and blood lactate concentrations will be examined. Lastly, potential mechanisms of hypoxemic tolerance at the tissue level will be reviewed.

11.1 Oxygen store depletion in marine mammals

11.1.1 O_2 depletion during forced submersions: seals

The rate, pattern, and magnitude of O_2 store depletion have been most completely studied in the forced submersion experiments of Scholander, Irving, and colleagues on young harbor seals (*Phoca vitulina*), gray seals (*Halichoerus grypus*), and hooded seals (*Cystophora cristata*) (Irving *et al.*, 1941b, Scholander, 1940, Scholander *et al.*, 1942a). The lung O_2 store declined to 3% O_2 (approximately 21 mm Hg or 2.8 kPa) (Scholander, 1940). The muscle O_2 store, isolated from the circulation by severe bradycardias and vasoconstriction (see Chapter 5), was almost completely depleted in about 10 min, while the blood (and respiratory) O_2 store lasted about 20–25 min (see Fig. 5.1). By comparison, arterial Hb saturations decreased to as low as 50–80% in less than 2 min during breath holds and dives of humans, and in less than 1 min during human sleep apnea (Dempsey *et al.*, 2010, Lindholm and Lundgren, 2009, Qvist *et al.*, 1993, Yumino and Bradley, 2008).

The muscle O_2 depletion rate during forced submersions, 10 ml O_2 kg^{-1} muscle min^{-1}, was about 2.5 times the resting rate due to struggling (Scholander *et al.*, 1942a). Under these conditions, muscle lactate began to accumulate at a myoglobin saturation of

10–20% (as in Fig. 5.1). Seventy-five years since their publication, these remain the only studies in marine mammals in which muscle lactate and O_2 contents have been measured simultaneously.

Elsner and colleagues later established that blood oxygen in seals could be depleted to arterial and venous P_{O2} values as low as 10 and 2 mm Hg, respectively (1.3 and 0.03 kPa) (Elsner *et al.*, 1970b, Kerem and Elsner, 1973). In forced submersions of Weddell seals (*Leptonychotes weddellii*), arterial P_{O2} averaged 32 mm Hg (4.5 kPa) at 8–12 min of submersion (Zapol *et al.*, 1979); arterial and mixed venous P_{O2} reached 20 mm Hg (<3 kPa) during a 46-min forced submersion (Hochachka *et al.*, 1977). In addition, as demonstrated by hepatic sinus O_2 contents that were greater than arterial values near the end of forced submersions, the lungs did not contribute O_2 to the blood during the latter portions of forced submersions of elephant seals (*Mirounga angustirostris*) (Elsner *et al.*, 1964b).

Comparison of the blood O_2 depletion data in the forcibly submerged harbor seal to that during asphyxia in the anesthetized, paralyzed dog revealed that the arterial blood O_2 content decreased at a rate 2–4 times slower in the seal (Kerem and Elsner, 1973). Declines in arterial and venous O_2 contents of seals during forced submersions were about 1–2 ml O_2 dl^{-1} min^{-1} (Elsner, 1969, Elsner *et al.*, 1964b, Irving *et al.*, 1941b, Kerem and Elsner, 1973, Scholander, 1940). Arterial P_{O2} at the asphyxial end point (hypoxemic EEG threshold) was 14 mm Hg (about 2 kPa) in the dog vs 10 mm Hg (1.3 kPa) in the seal. However, that asphyxial end point was reached much later in the seal (18.5 min) than in the dog (4.25 min). Arterial blood O_2 contents at the asphyxial end points were similar in both species.

11.1.2 O_2 depletion during sleep apnea: elephant seals

As reviewed in Chapter 5, during sleep apnea, heart rate was near 50 bpm, cardiac output was maintained at resting levels, and muscle blood flow declined but persisted at an average value near 50% of the eupneic (during a breath) level (Ponganis *et al.*, 2006b). Hence, the muscle O_2 store was not completely isolated from the circulation during these breath-holds. The lack of changes in blood lactate concentrations during and after the apneas was also consistent with adequate organ perfusion and O_2 delivery during the breath hold (Castellini *et al.*, 1986).

During sleep apneas in young elephant seals, serial blood gas analyses revealed that arterial and venous P_{O2}s quickly equilibrated within the first minute, indicative of minimum gas exchange in the lung after the first minute (Stockard *et al.*, 2007). The minimal role of the lung as an O_2 store was consistent with the exhalation of air observed at the start of the apnea, the low lung volumes determined in simulated dives (Kooyman *et al.*, 1973b), prior observations of the equilibration of arterial and venous O_2 contents during forced submersions (Elsner *et al.*, 1964b), and Elsner's proposal that the large hepatic sinus-vena cava of the seal acted as a venous blood O_2 reservoir (Elsner *et al.*, 1964b).

As demonstrated in Fig. 11.1 (Stockard *et al.*, 2007), the arterial P_{O2} and the corresponding Hb saturation during most of the breath hold (i.e., beyond 1 min) were

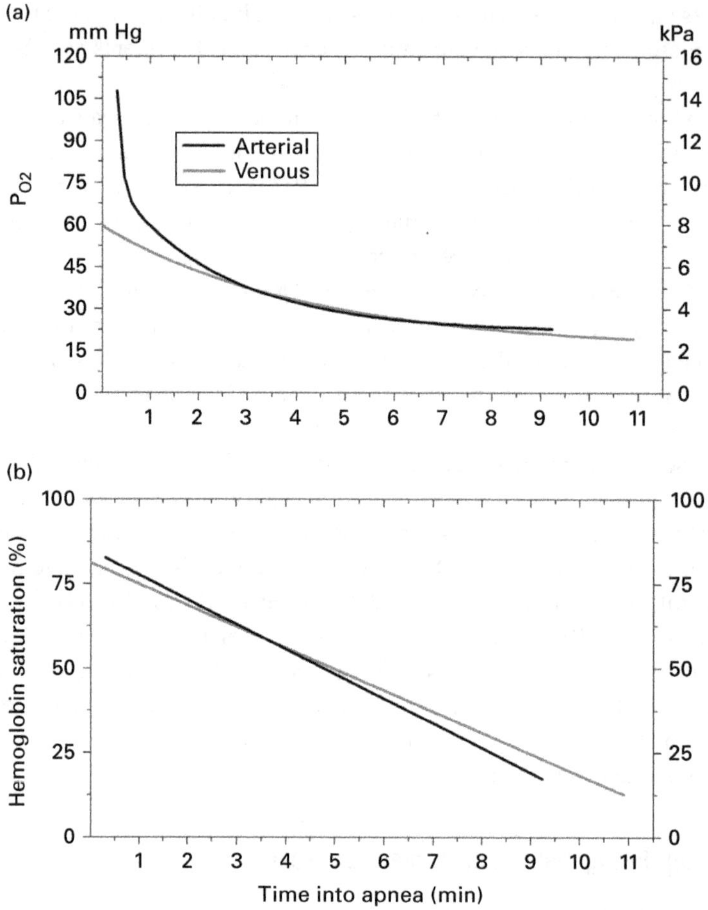

Figure 11.1 Changes in arterial and venous P_{O2} and hemoglobin saturation during sleep apnea of young elephant seals (*Mirounga angustirostris*). Adapted from Stockard et al. (2007).

in a range that would be considered critical (<60 mm Hg (7.8 kPa) and 90% saturation), and, indeed, severe (<50 mm Hg (6.5 kPa) and 80% saturation) in most human patients (Mason *et al.*, 2005, Nunn, 1977). The lowest arterial P_{O2}s during sleep apnea were less than the mean value (25 mm Hg, 3.3 kPa) of humans breathing ambient air near the summit of Mt. Everest (Grocott *et al.*, 2009).

During sleep apnea, blood O_2 content declined about 2 ml O_2 dl^{-1} min^{-1}. This was about twice the rate previously observed during forced submersions and was consistent with the maintenance of cardiac output during the breath hold. For a typical 7-min apnea in these young seals, about 56% of the blood O_2 store was consumed (Stockard *et al.*, 2007). That portion of the blood O_2 store alone was sufficient to provide enough O_2 for a metabolic rate of 4.2 ml O_2 kg^{-1} min^{-1} during the apnea (Stockard *et al.*, 2007).

Application of ^1H NMR spectroscopy techniques to analyze Mb saturation by Dr. Thomas Jue and associates allowed investigation of Mb desaturation in elephant seals

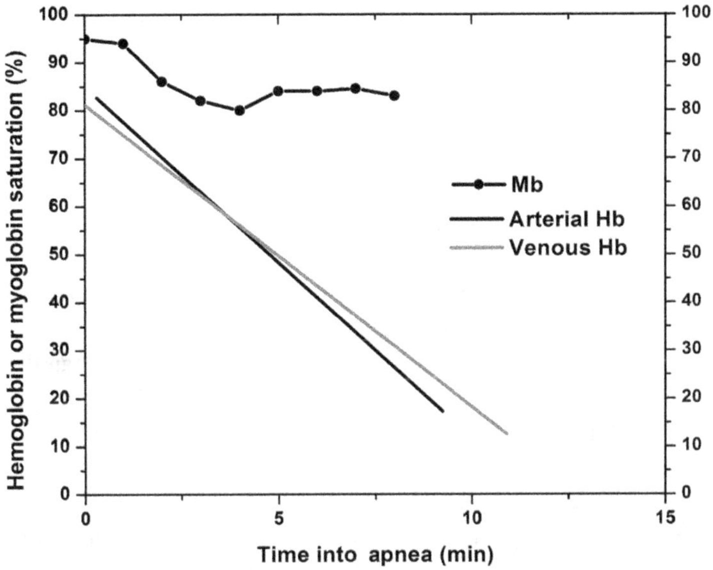

Figure 11.2 During sleep apnea, myoglobin (Mb) saturation remains near 80% while arterial and venous hemoglobin (Hb) saturations decline to as low as 12–15%. Adapted from Ponganis et al. (2008), Stockard et al. (2007).

sleeping inside a magnetic resonance imaging scanner (Ponganis *et al.*, 2008). Myoglobin desaturated rapidly with the onset of apnea (Fig. 11.2), and settled at a steady-state level near 80% saturation by 4 min into the apneic period. In these young seals, that initial desaturation rate corresponded to a muscle O_2 depletion rate of 1–2.3 ml O_2 kg^{-1} muscle min^{-1} during the first 4 min of the apnea. The Mb saturation remained at the 80–85% level until the end of apnea, after which it quickly re-saturated to >95% saturation. Maintenance of some muscle blood flow and blood-to-muscle O_2 transfer during the apnea allowed for a steady-state 80% saturation throughout most of the apnea. Assuming all muscle in the body desaturated to that level, the O_2 from muscle, in addition to that provided by the blood, would yield a metabolic rate of 4.7 ml O_2 kg^{-1} min^{-1} during a 7-min apnea in these young elephant seals. This O_2 store depletion rate of 4.7 O_2 kg^{-1} min^{-1} is 26% greater than the metabolic rate predicted at rest for a mammal of this size (Kleiber, 1975), and is consistent with the maintenance of heart rate and cardiac output during the breath-hold period.

During sleep apnea in elephant seals, it is the blood O_2 store that is primarily utilized. Blood O_2 depletion rates are greater, and muscle O_2 depletion rates slower during sleep apnea than during forced submersion (Fig. 11.2). Higher heart rates and tissue perfusion result in greater blood O_2 extraction. The lower rate of muscle O_2 depletion is secondary to both maintenance of some blood O_2 delivery as well as a lower muscle metabolic rate during sleep than during the stress of a forced submersion.

For typical 7-min apneas of these young elephant seals, the blood O_2 store is far from depleted; more than 40% of the initial blood O_2 still remains at the end of the breath hold. In addition, the muscle O_2 store is only 20% depleted during sleep apnea. Thus, although heart rate, cardiac output, and metabolic rate are maintained near resting levels, O_2 store depletion does not limit the breath-hold duration during sleep apneas.

11.1.3 O_2 depletion during free dives: elephant seals

Blood O_2 store depletion during diving was evaluated with the use of indwelling P_{O2} electrodes in translocated, juvenile elephant seals, and conversion of the P_{O2} profiles to Hb saturation profiles with use of the elephant seal O_2-Hb dissociation curve (Meir *et al.*, 2009). The translocation model, developed by Oliver and LeBoeuf (Oliver *et al.*, 1998), had also been used in heart rate studies, which had revealed that mean dive heart rates ranged from 31 to 48 bpm, values that were lower than during sleep apnea (Andrews *et al.*, 1997). The typical two- to three-day return trips from release sites to the rookery provided routine dives of 10–20-min duration and 100–200-m depth, and occasional dives as long as 44 min and as deep as 700 m (Meir *et al.*, 2009). During these trips, dives were spontaneous, voluntary, and, in contrast to Weddell seals diving under sea ice at an isolated dive hole (Kooyman, 1968), unrestricted in access to the surface. Blood P_{O2} was recorded at three sites – the aorta, the hepatic sinus, and the extradural vein.

Arterial P_{O2} peaked early during descent and then progressively decreased throughout the dive, consistent with a small lung O_2 store and lung "collapse" at depth (Fig. 11.3). Arterial P_{O2} values, as low as 12–23 mm Hg (1.6–3.07 kPa) at the end of dives, corresponded to routine Hb saturations of 8–26%, consistent with exceptional hypoxemic tolerance. By comparison, mean arterial P_{O2} and Hb saturation of climbers breathing ambient air at 8400 m altitude were 25 mm Hg (3.3 kPa) and 54%, respectively (Grocott *et al.*, 2009). The lowest arterial P_{O2} measured in this study (12 mm Hg; 1.60 kPa) was almost as low as the "critical P_{O2}" of harbor seals and Weddell seals (10 mm Hg; 1.33 kPa) (Elsner *et al.*, 1970b, Kerem and Elsner, 1973).

Routine end-of-dive venous P_{O2}s (Figs 10.4, 11.4, 11.5, 11.7) and Hb saturations were as low as 2–10 mm Hg (0.27–1.33 kPa, 0–4% Hb saturation), equivalent to or even lower than those measured in forced submersion studies, and even lower than the hypoxemic extremes of horses performing strenuous exercise (Bayly *et al.*, 1989, Manohar *et al.*, 2001). These low end-of-dive P_{O2} values resulted in near complete depletion of blood O_2 stores during routine dives, with net O_2 content depletion values up to 91% and 100% in the arterial and venous stores, respectively. In general, almost complete venous oxygen depletion occurred much more frequently in elephant seals than in sea lions (see Fig. 10.4).

It was notable that venous P_{O2} continued to increase during the early phase of the dive, sometimes reflecting arterial values, especially in the hepatic sinus (Fig. 11.4, 11.5). Such "arterialization" of venous blood (P_{O2} values greater than those typically found in venous blood) was not consistent with blood extraction by the tissues or blood flow to muscle in this period, and was suggestive of an arterio-venous shunt. There was also substantial overlap between end-of-dive arterial and hepatic sinus Hb saturation values. This was consistent with previous forced submersion and sleep apnea studies, and was expected in a breath-hold diver with collapsed lungs. It supports the concept of the large hepatic sinus and paired venae cavae as a significant O_2 reservoir.

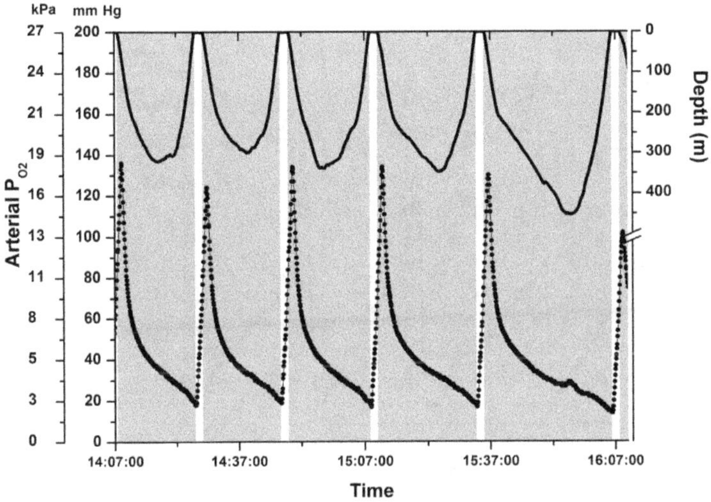

Figure 11.3. Arterial P_{O2} peaks during early descent and then progressively declines during dives of juvenile elephant seals (*Mirounga angustirostris*). Adapted from Meir et al. (2009).

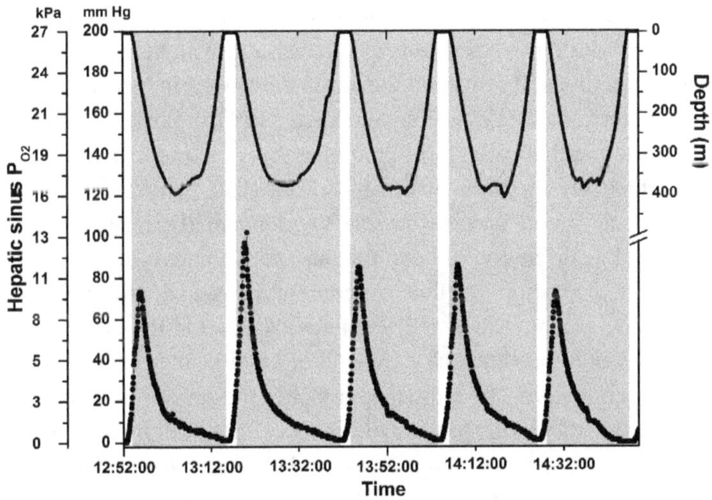

Figure 11.4 Hepatic sinus P_{O2} can peak at near arterial levels and then progressively decline to almost zero during some dives of young elephant seals (*Mirounga angustirostris*). Adapted from Meir et al. (2009).

The rate of decrease in venous P_{O2} and venous Hb saturation often became steeper concurrent with the ascent of the dive (Fig. 11.5), usually most pronounced in the final 15–45 sec. This period is coincident with both the ascent or "anticipatory" tachycardia (increased heart rate) (Andrews *et al.*, 1997) and the intense stroking that occur in this species (Williams *et al.*, 2000). These P_{O2} data support the hypotheses that (a) the ascent tachycardia serves to increase blood flow and O_2 delivery to depleted tissues at

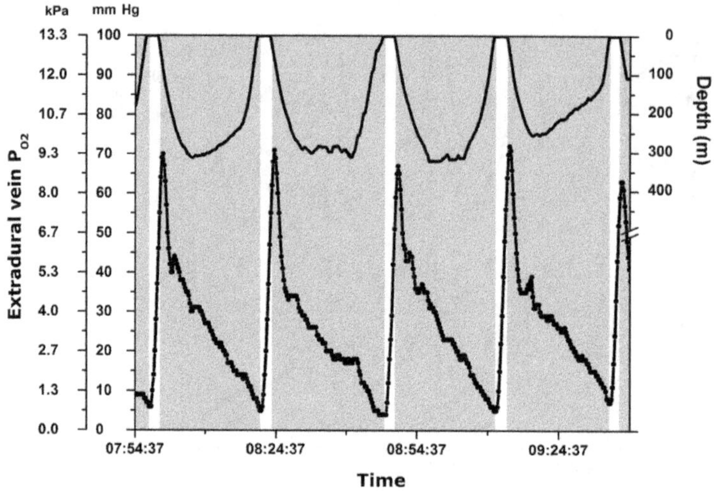

Figure 11.5 Extradural vein P_{O2} profiles have the same general pattern as those in the hepatic sinus during dives of young elephant seals (*Mirounga angustirostris*). Adapted from Meir *et al.* (2009).

the end of the dive, maximizing the gradient for O_2 uptake at the surface, and that (b) some blood flow to muscle occurs in this period (Thompson and Fedak, 1993). Some muscle blood flow during this period is also consistent with the lack of elevated muscle temperature during dives of seals and the slight elevations in blood lactate near the end of long dives (Guppy *et al.*, 1986, Ponganis *et al.*, 1993b). Increased muscle blood flow during increased exertion was also predicted by a numerical model of blood O_2 transport, in which the duration of aerobic metabolism during a dive was optimized by coupling muscle blood flow to muscle O_2 demand (Davis and Kanatous, 1999). Recent findings of a correlation of heart rate and stroke rate during short-duration dives of Weddell seals also supports this concept of increased muscle blood flow with exertion during the ascent (Davis and Williams, 2012). On the other hand, these more rapid declines in venous O_2 content at the end of the dive may also be reflective of ever lower arterial O_2 contents as the animal approaches the surface near the end of the dive.

The arterialization of venous blood early in a dive does not increase the magnitude of the total body O_2 stores as this increase in blood O_2 is due to O_2 transfer from the lung store (already included in the O_2 store calculation (Kooyman *et al.*, 1999, Meir *et al.*, 2009)). However, if depletion of arterial O_2 to below 20% Hb saturation (typical end-of-dive arterial Hb saturation value used for such calculations) occurs during dives (as documented in this study with values as low as 8% arterial oxygen saturation), the available O_2 store would increase by about 3 ml O_2 kg^{-1} to 88 ml O_2 kg^{-1} (see Chapter 4 for oxygen store calculations).

The results from this study also indicate that at the whole-animal level, juvenile elephant seals are not "hypometabolic" during diving, and that they do not require any significant anaerobic metabolism during routine dives. For example, the contribution of the venous O_2 store alone toward metabolic rate is >100% of the allometrically predicted basal metabolic rate, even for routine dives >10 min.

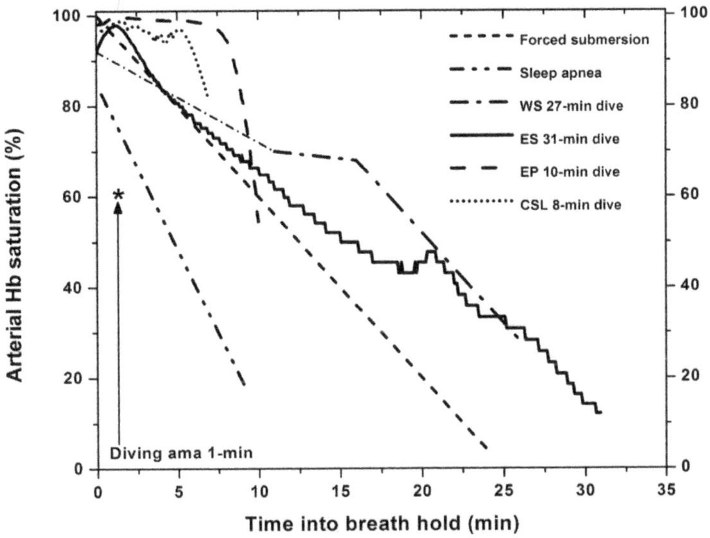

Figure 11.6 Arterial hemoglobin (Hb) saturation profiles during forced submersion of seals, sleep apnea of elephant seals (*Mirounga angustirostris*), and free dives of a Weddell seal (*Leptonychotes weddellii*), elephant seal, California sea lion (*Zalophus californianus*), and emperor penguin (*Aptenodytes forsteri*). Data were only sampled during the latter part of the Weddell seal dive. Arterial saturation after a 1-min dive of a diving ama is also provided. Adapted from McDonald and Ponganis (2012, 2013), Meir et al. (2009), Meir and Ponganis (2009), Qvist et al. (1986, 1993), Scholander (1940), Stockard et al. (2007).

11.1.4 O_2 depletion during free dives: Weddell seals

Investigations of diving physiology and O_2 store depletion in free-diving seals have been primarily and most extensively conducted with the isolated dive hole technique developed by Kooyman on the sea ice of McMurdo Sound (Kooyman, 1968, 1985, 2006, Kooyman and Kooyman, 2009). In a blood sampler study conducted by Zapol and colleagues, arterial blood O_2 depletion rates in free-diving Weddell seals (*Leptonychotes weddellii*) were about 0.8 ml O_2 dl^{-1} min^{-1} (Qvist *et al.*, 1986), in the same range as described above during forced submersions of much smaller seals (Fig. 11.6). In comparison, arterial O_2 content declined at 6 ml O_2 dl^{-1} min^{-1} in Korean diving ama (Qvist *et al.*, 1993).

The lowest arterial P_{O2} reported near the end of a dive of a Weddell seal was 18 mm Hg (2.4 kPa), corresponding to 28% Hb saturation (Qvist *et al.*, 1986). End tidal P_{O2} values were in a similar range (Kooyman *et al.*, 1973a, Ponganis *et al.*, 1993a). After dives of 26- and 34-min duration, end tidal P_{O2}s were 13–14 mm Hg (<2 kPa). These arterial and end tidal values, which were much greater than those at the end of extreme forced submersions, also raised the question as to how far the blood O_2 store was depleted during routine diving. This question was reinforced by findings during spontaneous sleep apneas of Weddell seals, in which, arterial P_{O2} declined at variable rates to only about 25 mm Hg (about 3.5 kPa) (Kooyman *et al.*, 1980).

In contrast to forced submersions of seals, muscle myoglobin desaturation was incomplete in free-diving Weddell seals (Guyton *et al.*, 1995) (see Fig. 11.8 for

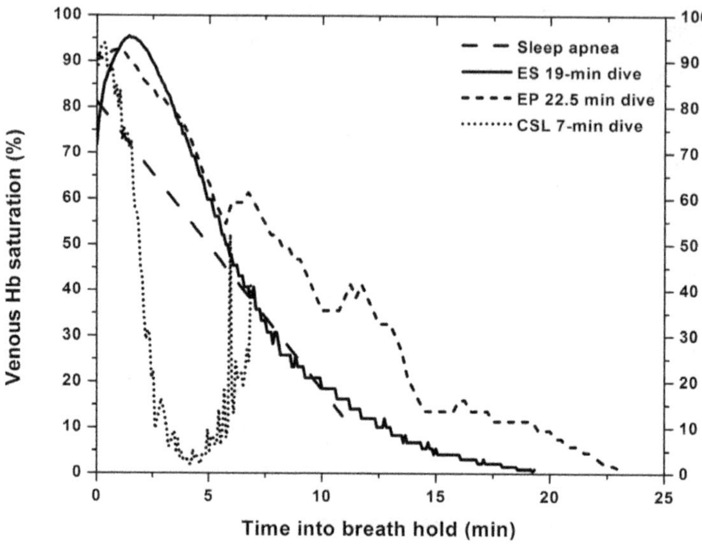

Figure 11.7 Venous hemoglobin (Hb) saturation profiles of elephant seals (*Mirounga angustirostris*) during sleep apnea and a dive; an emperor penguin (*Aptenodytes forsteri*) during one of the longest recorded dives in that species; and a California sea lion (*Zalophus californianus*) during a 310-m deep dive. Adapted from McDonald and Ponganis (2013), Meir et al. (2009), Meir and Ponganis (2009), Stockard et al. (2007).

comparison). Combined with the relatively mild bradycardias of free-diving Weddell seals (Hill *et al.*, 1987), this finding suggested that muscle blood flow persisted to some degree during free dives and that the blood O_2 store was not isolated from muscle. This contrasted with the severe bradycardia and peripheral vasoconstriction during forced submersions (Chapter 5). Muscle O_2 depletion rates in free-diving Weddell seals averaged about 7 ml O_2 kg^{-1} muscle min^{-1} (5.1% Mb saturation Δ min^{-1}) during dives less than 17 min (near the ADL), and 3.4 ml O_2 kg^{-1} muscle min^{-1} (2.5% Mb saturation Δ min^{-1}) in longer dives (Guyton *et al.*, 1995). These values were lower than those during Scholander's forced submersions experiments, but greater than the 2 ml O_2 kg^{-1} muscle min^{-1} of tourniqueted human muscle at rest (Blei *et al.*, 1993, Tran *et al.*, 1999). The low Mb desaturation rates in swimming Weddell seals could be consistent with blood O_2 supplementation of muscle metabolism, and/or a lower muscle metabolic rate due to hydrodynamics, prolonged gliding, and a low cost of swimming (Williams *et al.*, 2000, 2004). In support of the concept of such blood O_2 supplementation of muscle, it is notable that maintenance of some muscle blood flow during restrained submersions of trained vs. naïve seals was accompanied by a slower rate of Mb desaturation in the trained seals (Jobsis *et al.*, 2001).

11.1.5 O_2 depletion during free dives: sea lions

Blood O_2 store depletion in sea lions during maternal foraging trips to sea has been evaluated with a similar approach used in elephant seals – with use of backpack P_{O2}

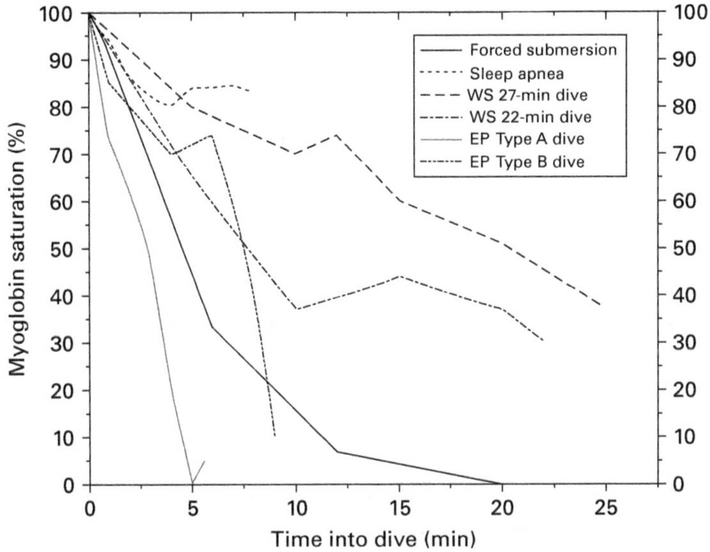

Figure 11.8 Myoglobin saturation profiles during forced submersion of seals; sleep apnea of an elephant seal (*Mirounga angustirostris*); two dives of a Weddell seal (WS) (*Leptonychotes weddellii*); and two dives of an emperor penguin (EP) (*Aptenodytes forsteri*). Muscle sampling sites included the longissimus dorsi during forced submersion and sleep apnea, the latissimus dorsi in the Weddell seal, and the pectoral muscle in the emperor penguin. There were two distinct patterns of desaturation in the emperor penguin, consistent with a lack of muscle blood flow in Type A dives, and intermittent muscle blood flow in Type B dives. Adapted from Guyton et al. (1995), Ponganis et al. (2008), Scholander (1940), Williams et al. (2011).

recorders to provide a P_{O2} profile and application of the sea lion O_2–Hb dissociation curve to provide Hb saturation profiles (McDonald and Ponganis, 2012, 2013). Arterial P_{O2} data during deep dives (200–360 m) revealed a double-peaked P_{O2} profile, consistent with lung collapse and re-expansion near 225-m depth (see Chapter 12). Remarkably, arterial Hb saturations were maintained greater than 90% during most of the dive, with end-of-dive saturations greater than 80% even for dives as long as 7 min (Figs 5.7, 11.6). Thus, lung collapse not only limited N_2 absorption at depth, but also preserved the respiratory O_2 store for use later during the dive, during ascent.

Another mechanism that has been proposed to preserve arterial oxygenation in otariids is air exhalation during ascent, with maintenance of a pulmonary shunt ("lung collapse") and prevention of possible reverse transfer of O_2 from blood into the lungs (Hooker *et al.*, 2005). Air exhalation was documented during ascent from deep dives of Antarctic fur seals (*Arctocephalus gazella*) in the study. The authors postulated that lung P_{O2} during ascent might be lower than blood P_{O2} due to both the depletion of lung O_2 during the dive and the decrease in ambient pressure during ascent. Gas exchange in this situation could result in reverse transfer of O_2 into the lung, with a net lowering of blood O_2 content (Lanphier and Rahn, 1963). Exhalation of air during ascent was postulated to maintain collapse of alveoli, prevent such reverse transfer of O_2, and decrease the risk of shallow water blackout.

Venous P_{O2} and saturation profiles were complex and highly variable in California sea lions. Start-of-dive Hb saturations were greater than 78% in 40% of dives, and were sometimes as high as 95%. Such arterialization of venous blood was consistent with arterio-venous shunting prior to the dive. End-of-dive Hb saturations ranged from 10% to 80% for dives of three-minutes duration, the dive duration often calculated to be the aerobic dive limit of California sea lions. These end-of-dive saturations were not necessarily the minimum values; minimum values often occurred earlier in the dive and ranged between 7% and 70% for 3-min dives. In 15% of all dives, end-of-dive saturations were greater than initial dive saturations, consistent with continued gas exchange during dives and, again, with possible arterio-venous shunting during dives, especially in dives with end-of-dive saturations greater than 90%. In deep dives greater than four minutes in duration, there was usually near complete venous O_2 depletion mid-dive, but venous Hb saturation typically increased during ascent, with end-of-dive venous saturations as high as 40% for 7-min dives (Figs, 5.7, 10.4, 11.7). The late dive increase in venous Hb saturation was consistent with resumption of gas exchange and peripheral blood flow during ascent.

The mean blood O_2 depletion rate in these sea lions was 4.3 ml O_2 dl^{-1} min^{-1}. Provided the venous O_2 profiles in the distal vena cava are representative of the entire venous O_2 store, the venous O_2 contribution alone to metabolic rate for a 3-min dive was estimated to be about 4.4 ml O_2 dl^{-1} min^{-1}, about 33% greater than the allometrically predicted resting metabolic rate. Thus, sea lions, during their most common dives under 3-min duration, do not appear to be hypometabolic.

11.1.6 O_2 depletion in cetaceans: trained dives and stationary breath holds

Limited data were also available for cetaceans during trained dives and stationary breath holds. After dives of a trained bottlenose dolphin (*Tursiops truncatus*), the O_2 concentration of exhaled air was as low as 3% or 22 mm Hg (3 kPa) (Ridgway *et al.*, 1969). During stationary breath holds of dolphins, venous P_{O2} in blood vessels of the tail fluke reached values as low as 17–22 mm Hg (about 2–3 kPa) in bottlenose dolphins (3.3- to 4.5-min apneas), 30–37 mm Hg (about 4–5 kPa) in Pacific white-sided dolphins (2.4- to 3.5-min apneas), and 31–40 mm Hg (about 4–5 kPa) in killer whales (9.6- to 13.3-min apneas) (Noren *et al.*, 2012b, Williams *et al.*, 1999). In beluga whales (*Delphinapterus leucas*), fluke venous P_{O2}s were near 20 mm Hg (about 3 kPa) after stationary breath holds between 10- and 17-min duration (Shaffer *et al.*, 1997).

11.2 Oxygen store depletion in seabirds

11.2.1 O_2 depletion during forced submersions: penguins and ducks

In his 1940 monograph, Scholander examined blood and muscle O_2 depletion in macaroni (*Eudyptes chrysolophus*) and gentoo (*Pygoscelis papua*) penguins (Scholander, 1940). Arterial blood O_2 content declined quickly from 21 ml O_2 dl^{-1} to 3 ml O_2 dl^{-1}

within 5 min, yielding an overall blood O_2 depletion rate of 3.6 ml O_2 dl^{-1} min^{-1} (Fig. 5.9). Muscle O_2 content declined from 40 to 0 ml O_2 kg^{-1} muscle by 5 min, yielding a minimum muscle O_2 depletion rate of 8 ml O_2 kg^{-1} muscle min^{-1}. This muscle O_2 depletion rate during forced submersion of the penguin was in the same range as resting muscle O_2 consumption (11 O_2 kg^{-1} muscle min^{-1}) reported in the pekin duck (*Anas platyrhynchos*) (Grubb, 1981).

During forced submersion studies of pekin ducks by Jones and colleagues, air-sac P_{O2} was near 30 mm Hg (4 kPa, 25% of the initial value), and arterial and venous P_{O2} values were 30 and 23 mm Hg, respectively (4 and 3 kPa), at the point of "imminent cardiovascular collapse" (Hudson and Jones, 1986). Due to the P_{50} (P_{O2} at 50% Hb saturation) of duck Hb and the Bohr effect, blood O_2 contents were nil at those blood P_{O2}s. In comparison to Scholander's penguins, both blood and muscle O_2 depletion rates were lower in the smaller pekin duck. Arterial and venous O_2 content depletion rates during forced submersions in ducks were 2.2 and 1.4 ml O_2 dl^{-1} min^{-1}, respectively (Stephenson and Jones, 1992b). The muscle O_2 store was estimated to be depleted within 45 sec; this would result in a minimum O_2 depletion rate of about 5 ml O_2 kg^{-1} muscle min^{-1} (Stephenson and Jones, 1992b). Since muscle O_2 depletion was not directly measured in this study, the actual depletion rate may have been faster, i.e., near the previously cited 11 ml O_2 kg^{-1} muscle min^{-1} resting value (Grubb, 1981).

11.2.2 Blood oxygen depletion during simulated dives: penguins

During simulated dives of 5-min duration in Adélie and gentoo (*Pygoscelis adeliae*, *P. papua*) penguins in a pressure chamber, air-sac O_2 concentration decreased at a rate of 2.2% min^{-1} to minimum values near 2% (about 15 mm Hg or 2 kPa) (Kooyman *et al.*, 1973c). Arterial P_{O2} declined from 80 mm Hg (10.2 kPa) to 20–30 mm Hg (2.6–4.3 kPa). These end-of-dive values were similar to those above in pekin ducks at the point of "imminent cardiovascular collapse." However, the lowest tolerable level of hypoxemia in penguins was not determined as that was not part of the study. Blood O_2 content and muscle O_2 depletion were also not examined in this study.

11.2.3 O_2 depletion during free dives: emperor penguins

Oxygen store depletion has also been investigated in emperor penguins (*Aptenodytes forsteri*) diving freely at an isolated dive hole in McMurdo Sound. This approach, again pioneered by Kooyman (Kooyman *et al.*, 1971a), has allowed examination of heart rate, swim speed, stroke frequency, and temperature during diving (Kooyman *et al.*, 1992b, Meir *et al.*, 2008, Ponganis *et al.*, 2001, 2003b, 2004, van Dam *et al.*, 2002). Most importantly, an ADL of 5.6 min has been documented by blood lactate determinations in birds diving at the isolated dive hole (Ponganis *et al.*, 1997c). During these dives, the penguins travel as far as 1.2 km from the dive hole, and primarily feed on the sub-ice fish, *Pagothenia borchgrevinki* (Ponganis *et al.*, 2000, Shiomi *et al.*, 2008). Because of the availability of this prey item, most dives are less than 100 m, relatively shallow for emperor penguins.

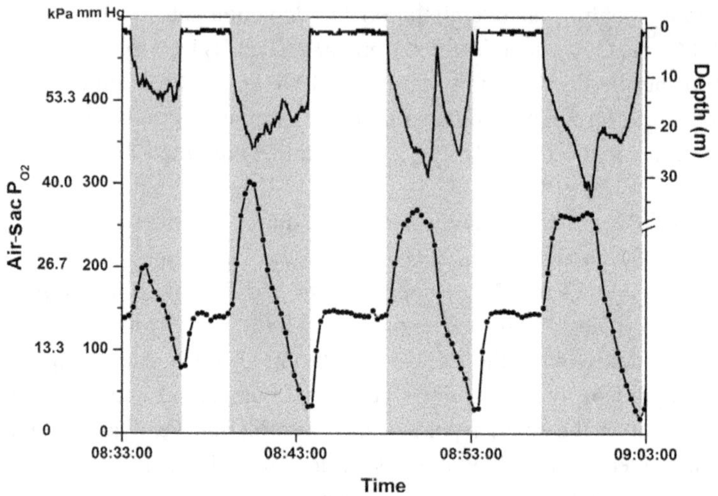

Figure 11.9 Air-sac P_{O2} profiles during shallow dives of emperor penguins (*Aptenodytes forsteri*) illustrate classic compression hyperoxia during descent followed by a decrease in P_{O2} later in the dive due to both oxygen consumption and to the decrease in ambient pressure during ascent. Adapted from Stockard et al. (2005).

Air-sac and blood O_2 depletion during dives have been examined with use of a P_{O2} electrode and custom-built recorder (Ponganis *et al.*, 2007, 2009, Stockard *et al.*, 2005). As in research with the elephant seal, P_{O2} profiles have been converted to Hb saturation profiles with determination of O_2–Hb dissociation curves of emperor penguins, and application of the curve to the P_{O2} data (Meir and Ponganis, 2009). Investigation of pectoral muscle O_2 depletion has been conducted with the development of a backpack near-infrared recorder and probe (Williams *et al.*, 2011a).

11.2.3.1 Air-sac O_2 depletion during free dives: emperor penguins

Air-sac P_{O2} profiles in 5- to 75-m deep dives of up to 11-min duration revealed a compression hyperoxia followed by a decrease in P_{O2} as the respiratory O_2 fraction declined secondary to O_2 consumption and as the ambient pressure decreased during ascent (Fig. 11.9) (Stockard *et al.*, 2005). Final P_{O2} declined exponentially and reached values as low as 0 mm Hg; it was less than 20 mm Hg (2.7 kPa) in 42% of dives. By comparison, the inspiratory P_{O2} for a bar-headed goose (*Anser indicus*) at a simulated altitude of 11 580 m was 23 mm Hg (3.1 kPa), and air-sac P_{O2} of an Adélie penguin at the end of a simulated dive was 15 mm Hg (2.0 kPa) (Black and Tenney, 1980, Kooyman *et al.*, 1973c). End tidal P_{O2} of a climber breathing ambient air on Mt. Everest (35 mm Hg, 4.7 kPa) and human shallow-water black-out thresholds (25 mm Hg, 3.3 kPa) were also greater than many of the end-of-dive P_{O2}s in the air sacs of emperor penguins (Ferretti *et al.*, 1991, Ferrigno and Lundgren, 2003, West *et al.*, 1983).

The low end-of-dive air-sac P_{O2}s in emperor penguins contrast with the air sac P_{O2}s near 30 mm Hg (4 kPa) of ducks at the point of "imminent cardiovascular collapse"

(Hudson and Jones, 1986). Such hypoxic tolerance is afforded, in part, by a left shift of the O_2–Hb dissociation curve of the emperor penguin (and other penguins as well as high-altitude flying birds) in comparison to that of the duck (Meir and Ponganis, 2009). At a P_{O2} of 20 mm Hg (2.7 kPa), the Hb of the duck would be devoid of O_2 while the Hb of the emperor penguin would still be 27% saturated.

Initial air-sac P_{O2}s during dives indicated that initial O_2 fractions could be as high as 19%, which were greater than the 15% value observed in simulated dives in a pressure chamber (Kooyman *et al.*, 1973c). Complete consumption of the 19% O_2 fraction resulted in a greater respiratory O_2 store than that typically calculated with the pressure chamber results (Kooyman, 1989, Ponganis *et al.*, 2010a). The overall rate of change in the respiratory O_2 fraction during dives of up to 11-min duration ranged between 5% and 2% min^{-1}. By comparison, the respiratory O_2 fraction changed about 2% min^{-1} during simulated dives of penguins (Kooyman *et al.*, 1973c).

11.2.3.2 Blood O_2 depletion during free dives: emperor penguins

Arterial P_{O2} profiles of diving emperor penguins were characterized by a compression hyperoxia followed by a decline to values as low as 26–30 mm Hg during dives of up to 12-min duration (Ponganis *et al.*, 2007, 2009). As an example, see Fig. 11.11. The pattern was similar to that observed in the air sac, but it was not clear why these end-of-dive arterial values were generally greater than air-sac values for dives of similar duration. The magnitude of air-sac P_{O2} depletion was quite variable, and it is possible that differences in physiological responses and depth profiles of individual dives may have at least partially contributed to differences between the air-sac and arterial studies. Regardless of the mechanisms, it is notable that arterial Hb saturation of emperor penguins can be maintained near 100% during almost the entire dive (Fig. 11.6), even during dives as long as 12 min (Meir and Ponganis, 2009).

Arterial Hb saturation declined primarily during the ascent, and reached a lowest value of only 47%. These final arterial P_{O2}s and Hb saturations were greater than those of the bar-headed goose at a simulated altitude of 11 580 m (22 mm Hg, 28% saturation) and similar to those of humans breathing ambient air near the summit of Mt. Everest (25 mm Hg, 54% saturation) (Black and Tenney, 1980, Grocott *et al.*, 2009). Whether lower arterial P_{O2}s and Hb saturation occur in other dives of emperor penguins is unknown and awaits further investigation. During the ascent, the arterial Hb desaturation rate was about 25% min^{-1}, although the average value over a dive of 9-min duration was much lower, about 5–6% min^{-1} (Meir and Ponganis, 2009).

The ability to maintain arterial Hb saturation near 100% during most of the duration of a dive in emperor penguins is similar to the saturation profile of deep-diving sea lions, but contrasts with that in elephant seals, where arterial Hb saturation is usually between 70% and 30% during the majority of the dive (Fig. 11.6). This highlights the size and role of the respiratory O_2 store in penguins (and sea lions) and emphasizes the significance of lung-to-blood O_2 transfer during dives of penguins (and sea lions).

Venous P_{O2} and Hb saturation profiles during dives of emperor penguins were remarkable for the variability in the rates and patterns of decline throughout the dive (Figs 11.7, 11.10). Venous P_{O2} could transiently rise during the early portion of some

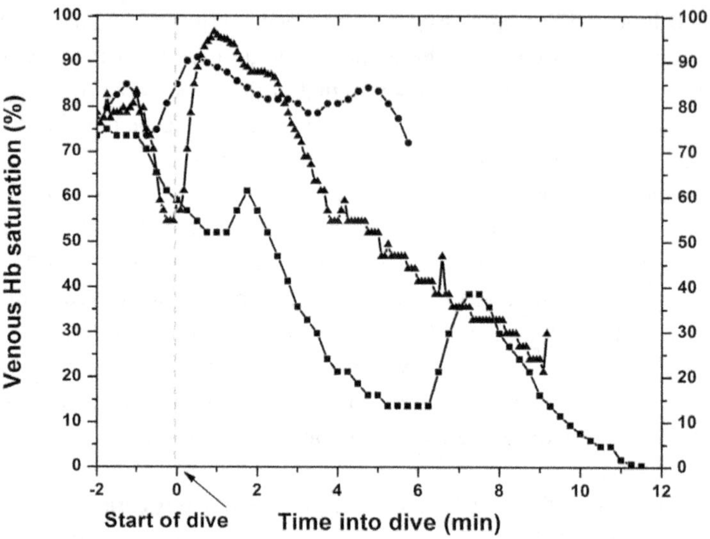

Figure 11.10 Variable venous hemoglobin (Hb) desaturation profiles during dives of 6- to 12-min duration in emperor penguins (*Aptenodytes forsteri*). Adapted from Meir and Ponganis (2009).

dives while it declined in others. Initial or peak P_{O2} values were often consistent with early, even pre-dive arterialization of venous blood (i.e., >90% Hb saturation).

Venous P_{O2} and Hb saturation declined at variable rates and, at times, with fluctuations during the later portions of dives (Figs 11.7 11.10). The transient elevations in P_{O2} and Hb saturations during dives were not consistent with muscle blood flow and blood O_2 extraction by muscle during such dives. Rather, these increases in P_{O2} emphasized (a) the potential role of arterio-venous (A-V) shunts in arterialization of the venous O_2 store, and, again, (b) the significance of net lung-to-blood O_2 transfer during dives of penguins.

On the other hand, muscle blood flow may occur in those dives in which venous P_{O2} and Hb saturation decline significantly during the early portion of the dive. This plasticity in the peripheral blood flow is also suggested by the range of venous P_{O2}s and Hb saturations for a dive of a given duration. For example, near 6-min dive duration, final venous Hb saturations ranged from 3% to 75% (Fig. 10.2). End-of-dive venous Hb saturations indicated that the entire venous O_2 store could be consumed; 15% of dives had final venous Hb saturation $\leq$ 5%.

11.2.3.3 Muscle O_2 depletion during free dives: emperor penguins

As reviewed in Chapter 5, Mb desaturation profiles of diving emperor penguins revealed two distinct patterns (Fig. 11.8), a monotonic decline (Type A) and a mid-dive plateau pattern (Type B) (Williams *et al.*, 2011a). The monotonic decline in Mb saturation throughout Type A dives allowed calculation of a mean desaturation rate of 14.4 ± 3.8% min^{-1}. As discussed in Chapter 5, this desaturation rate was similar to rapid Mb desaturation observed in forced submersion studies (Scholander *et al.*, 1942a), and was consistent with a lack of muscle blood flow. It was also higher than desaturation rates observed in Weddell seals, where muscle perfusion was suspected during dives (Guyton *et al.*, 1995; see Fig. 11.8). Mean muscle O_2 consumption (12.4 ± 3.3 ml O_2 kg^{-1} muscle min^{-1}) based on the Mb desaturation rate in diving emperor penguins was low, less than one-tenth the pectoralis-supracoracoideus muscle O_2 consumption calculated from emperor penguins swimming maximally in a flume (160 ml O_2 kg^{-1} muscle min^{-1}) (Kooyman and Ponganis, 1994, Ponganis *et al.*, 1997a), demonstrating the efficiency of underwater locomotion in diving emperor penguins.

In Type B dives, after an initial decline, Mb saturation plateaued, consistent with some muscle perfusion, and then declined again during the ascent, presumably due to a decrease in muscle blood flow (see Chapter 5). Such intermittent muscle blood flow is consistent with the variable venous P_{O2} profiles discussed above (Fig. 11.10). The Type B plateau period of Mb desaturation in diving emperor penguins is similar to that during sleep apnea in elephant seals, during some dives of Weddell seals, and during exercise in humans (Guyton *et al.*, 1995, Ponganis *et al.*, 2008, Richardson *et al.*, 1995) Saturation values in Type B dives of 8- to 10-min durations often eventually declined to below 10%; mean desaturation rate in Type B dives was 9.8 ± 2.4% min^{-1}. Despite the potential supplementation of the Mb–O_2 stores during dives, the near complete depletion of Mb in longer Type B dives (Fig. 11.8) is supportive of the concept that the onset of post-dive lactate accumulation is secondary to muscle O_2 depletion.

The results of these studies of O_2 store depletion in emperor penguins emphasize the significance of the relatively large respiratory O_2 store in penguins, as well as the apparent plasticity in peripheral vascular responses. In contrast to seals, but similar to sea lions, this allows for maintenance of the penguin's arterial Hb saturation during much of the dive (Fig. 11.6). It has been hypothesized that the transient elevations in heart rate during the early segments of dives of penguins are not only important for lung-to-blood O_2 transfer, but that the increased cardiac output can be utilized either to enhance the venous O_2 store through the use of A-V shunts (Fig. 11.11), or to supplement the muscle O_2 store through maintenance of muscle blood flow (Fig. 11.8, Type B profile).

The average depletion rates of all the body O_2 stores of the emperor penguin contributed 6.8 ml O_2 kg^{-1} min^{-1} to whole-body metabolic rate for dives of about 10-min duration. The muscle O_2 store contributed most (53%), while the respiratory and blood compartments contributed 31% and 16%, respectively (Williams *et al.*, 2011a). This average total O_2 store depletion rate during diving demonstrates the low metabolic cost of diving in emperor penguins as it is similar to measured and predicted resting metabolic rates. It should be emphasized that O_2 depletion rates are highly variable, and that there is probably a wide range of diving metabolic rates dependent on the activity and duration of an individual dive. In addition, the actual metabolic rate during dives, especially dives longer than the ADL, may be greater due to phosphocreatine breakdown and glycolysis.

11.3 Blood gases, blood pH, and lactate

Depletion of O_2 stores during breath holds results in a concomitant increase in the partial pressure of carbon dioxide (P_{CO2}) and a decline in pH. Although P_{O2} can become

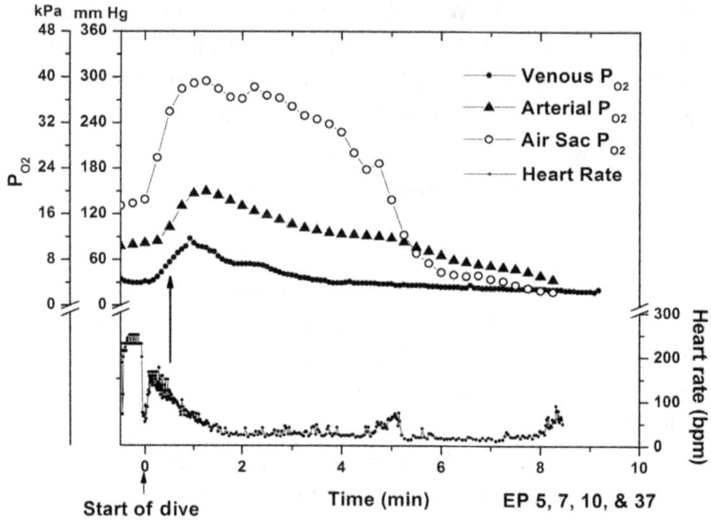

Figure 11.11 Air-sac, arterial and venous P_{O2} profiles, and a heart rate profile during shallow dives of about 9-min duration from four different emperor penguins (EP, *Aptenodytes forsteri*). It is hypothesized that the increased cardiac output due to the relatively high heart rate early in the dive not only serves to enhance lung-to-arterial blood O_2 transfer, but also can transfer O_2 directly into the venous system via arterio-venous shunts. Venous blood is thus arterialized to increase the size of the blood oxygen store during the remainder of the dive. Adapted from Ponganis et al. (2009).

remarkably low during free dives of marine mammals, the blood P_{CO2} and pH are usually within ranges tolerated by human patients with severe lung disease and by patients during permissive hypercapneic ventilation protocols (Baker *et al.*, 2005, Nunn, 1977, West, 2007). The P_{CO2}s measured during dives are also less than that associated with CO_2 narcosis in humans (Nunn, 1977). The changes in blood P_{CO2} and pH during the dive are limited by (a) the finite size of the body O_2 store, (b) the continued decline in heart rate and metabolic rate as a dive progresses ever longer, (c) the isolation of CO_2 and lactate in tissues that are not perfused at such low heart rates, and (d) the buffering capacity of blood. The largest changes in blood pH in diving mammals, as demonstrated by the lactate wash-out in Scholander's original work, occur during the post-dive interval when tissue perfusion is restored (Scholander, 1940).

11.3.1 P_{O2}, P_{CO2}, and pH: forced submersion

During forced submersions of harbor seals, arterial P_{O2} decreased and P_{CO2} increased to 42 and 70 mm Hg (5.6 and 9.3 kPa) at 10 min, respectively, and were 10 and 98 mm Hg (1.3 and 13.0 kPa) at 22 min (Kerem and Elsner, 1973). The venous P_{O2} and P_{CO2} were 30 and 84 mmHg (4.0 and 11.2 kPa) at 10 min, and 3 and 105 mm Hg (0.4 and 14.0 kPa) at 22 min. Arterial and venous pH were near 7.3 at 10 min and 7.2 at 22 min. With end-of-submersion arterial and venous Hb saturations near 10% and 0%, almost all of the blood O_2 store was consumed during a 22-min submersion. As also reviewed in

Chapter 4, these data and those of Scholander (Scholander, 1940) provide the criteria with which available O_2 stores are calculated (Kooyman, 1989, Lenfant *et al.*, 1970). Weddell seals displayed similar results. During 8–12-min forced submersions of Weddell seals, mean arterial P_{O2}, P_{CO2}, and pH were 32 mm Hg (4.3 kPa), 59 mm Hg (7.9 kPa), and 7.25 pH units, respectively (Zapol *et al.*, 1979). At 55 min of submersion of Weddell seals, corresponding arterial values were 10 mm Hg (1.3 kPa), 84 mm Hg (11.2 kPa), and 7.11 pH units (Elsner *et al.*, 1970b).

During forced submersions of ducks taken to the point of "imminent cardiovascular collapse" (Hudson and Jones, 1986), final arterial and venous P_{O2}s averaged 30 and 23 mm Hg (4.0 and 3.0 kPa), arterial and venous P_{CO2}s, 63 and 66 mm Hg (8.4 and 8.8 kPa), and arterial and venous pH, 7.11 and 7.04 pH units. The end arterial and venous P_{O2} values in these ducks were considerably higher than in the maximally submerged seal.

11.3.2 P_{O2}, P_{CO2}, and pH: sleep apnea and stationary breath holds

Changes in blood gases during sleep apneas of phocid seals were not as extreme as during forced submersions. In Weddell seals and young elephant seals, end-of-apnea arterial and venous P_{O2}s were 18–26 and 15–31 mm Hg (2.4–3.5 and 2.0–4.1 kPa), respectively, for breath holds up to 11 min in duration (Kooyman *et al.*, 1980, Stockard *et al.*, 2007, Tift *et al.*, 2013). The corresponding arterial and venous P_{CO2}s were both 55–72 mm Hg (7.3–9.6 kPa) in range and resulted in pH values near 7.3.

In cetaceans, trained stationary submersions provided the opportunity to obtain fluke blood samples (potentially a mix of arterial and venous blood) at the end of the submersion. In 2.4–4.5-min submersions of Pacific white-sided dolphins (*Lagenorhynchus obliquidens*) and bottlenose dolphins, P_{O2} ranged from 17 to 37 mm Hg (2.3–4.9 kPa), with higher values in the former; P_{CO2} from 57 to 70 mm Hg (7.6–9.3 kPa), again with higher values in the former; and pH from 7.31 to 7.40 pH units (Noren *et al.*, 2012b). In the killer whale (*Orcinus orca*), for breath holds of 8.6 to 13.3 min, P_{O2} ranged from 31 to 40 mm Hg (4.1–5.3 kPa); corresponding P_{CO2} and pH values were 63–75 mm Hg (8.4–10.0 kPa), and 7.21–7.27 pH units (Noren *et al.*, 2012b). In the beluga whale (*Delphinapterus leucas*), after 18-min apneas, P_{O2} was 20–23 mm Hg (2.7–3.1 kPa), P_{CO2}, 83 mm Hg (11.1 kPa), and pH, 7.17 (Shaffer *et al.*, 1997).

In comparison to the above data after trained submersions, resting ventilatory rates in young gray whale calves (*Eschrichtius robustus*) resulted in arterial pH, P_{CO2}, and P_{O2} ranges of 7.23–7.35 pH units, 41–69 mm Hg (5.5–9.2 kPa), and 55–62 mm Hg (7.3–8.3 kPa), respectively (Sumich, 2001, Wahrenbrock *et al.*, 1974). During medical exams of bottlenose dolphins, arterial pH and blood gas values were in the same range (Houser *et al.*, 2010).

11.3.3 P_{O2}, P_{CO2}, and pH: diving

There have been relatively few blood gas determinations during dives of marine mammals. Near the end of a 27-min dive of a Weddell seal (Qvist *et al.*, 1986), arterial P_{O2}, P_{CO2}, and pH were 18 mm Hg (2.4 kPa), 55 mm Hg (7.3 kPa), and 7.3 pH units, respectively. P_{O2} rapidly increased to above 60 mm Hg (8 kPa) within the first few min of the post-dive interval. End tidal P_{O2} profiles of Weddell seals after dives as long as 32 min revealed a similar range of P_{O2} values and also a rapid recovery of P_{O2} and presumed Hb saturation (Ponganis *et al.*, 1993a). The minimum P_{O2}s during these dives in seals were less than the minimum end-of-dive arterial P_{O2} values (near 30 mm Hg, 4 kPa) found in human breath-hold divers (Ferrigno and Lundgren, 2003, Qvist *et al.*, 1993). In translocated juvenile elephant seals with mean dive durations near 10 min, and occasional dives as long as 40 min (Meir *et al.*, 2009), minimum P_{O2}s recorded with an indwelling P_{O2} electrode were $\leq$10 mm Hg (1.3 kPa) in the extradural vein in 51% of dives (lowest value, 2 mm Hg or $<$0.3 kPa), $\leq$10 mm Hg in the hepatic sinus in 21% of dives (lowest value, 2 mm Hg), and $\leq$30 mm Hg (4 kPa) in the aorta in 46% of dives ($\leq$20 mm Hg (2.7 kPa) in 10% of dives; lowest value, 12 mm Hg (1.6 kPa)). During the surface interval, mean times of recovery of arterial P_{O2}, hepatic sinus P_{O2}, and extradural vein P_{O2} were 82 sec, 149 sec, and 123 sec, respectively. These data from both Weddell seals and elephant seals indicated that these animals are capable of almost complete blood O_2 depletion during some dives.

End tidal P_{O2}s for dives of other species were not always as low as in Weddell seals and elephant seals at sea. After dives up to 12 min in duration in captive gray seals (Reed *et al.*, 1994b), end tidal P_{O2}s were near 60 mm Hg (8 kPa). Post-dive end tidal gas analyses from bottlenose dolphins after dives to 200-m depth also revealed a similar ranges of values with mean P_{O2}s as low as 39 mm Hg (5.2 kPa), and P_{CO2} as high as 57 mm Hg (7.6 kPa) (Ridgway *et al.*, 1969). After 25-sec breath holds of harbor porpoises (*Phocoena phocoena*), end tidal P_{O2} and P_{CO2} were 53–60 mm Hg (about 7–8 kPa), and near 58 mm Hg (7.7 kPa), respectively (Reed *et al.*, 2000). In the manatee (*Trichechus manutus*), end tidal values after 10-min dives were 27 mm Hg (3.6 kPa) (P_{O2}) and 87 mm Hg (11.6 kPa) (P_{CO2}) (Gallivan *et al.*, 1986). In these examples, all P_{O2}s are consistent with arterial Hb saturations $\geq$50% and less than complete blood O_2 depletion under such conditions.

There are even fewer blood gas data available for diving birds. In free-diving emperor penguins, arterial P_{CO2} was 44–52 mm Hg (5.9–6.9 kPa) with corresponding pHs of 7.45 and 7.32 at 2.1–2.8-min into a dive (Ponganis *et al.*, 2009). At 1–3-min into dives, venous P_{CO2}s and pHs ranged between 48 and 62 mm Hg (6.4–8.3 kPa), and 7.47 to 7.37 pH units, respectively (pH at rest was 7.5). As already discussed, the lowest arterial and venous P_{O2}s observed in free-diving emperor penguins were near 25 mm Hg (3.3 kPa) and 0 mm Hg (0 kPa), respectively (Meir and Ponganis, 2009, Ponganis *et al.*, 2009).

11.3.4 Hypoxic and hypercapneic respiratory drives during diving

One final comment concerning blood gases is the question of what stimulates a diver to end a dive and breathe. It has been postulated that P_{CO2} may be the most significant factor determining routine dive durations and dive behavior (Stephenson, 2005). This is consistent with research in human divers, where a P_{CO2} threshold appears to be the break point for the urge of human divers to end a breath hold and breathe (Chapter 2). Hypoxic and hypercapneic respiratory drives to breathe appear to be just as sensitive in diving mammals and birds as in terrestrial animals, although the threshold may be higher in divers (Butler, 1982, Butler and Jones, 1997, Butler and Taylor, 1973, 1983, Butler and Woakes, 1982, Craig Jr and Påsche, 1980, Elsner *et al.*, 1977, Furilla and Jones, 1986, Gallivan, 1980, 1981, Kohin *et al.*, 1999, Kooyman *et al.*, 1971b, Milsom *et al.*, 1981, 1996, Parkos and Wahrenbrock, 1987, Påsche, 1976a, 1976b, Skinner and Milsom, 2004). In addition, as described in these references, exposure to hypercapneic gas mixtures typically decreases spontaneous apneas and diving activity, whereas hypoxic mixtures usually have less of an effect, although that may be dependent on the level of hypoxia. Although these responses are mediated through the carotid bodies (P_{O2} and P_{CO2}) and central chemoreceptors (P_{CO2}), input from higher cortical centers undoubtedly can override these reflexes to allow for longer dives. Furthermore, for a seal, such as a Weddell seal making a 70-min round trip from a dive hole, that CO_2 threshold would presumably have to occur at the turn-around point of the dive. Clearly, our understanding of the neuroregulation of dive duration is far from complete. Interestingly, in a three-dimensional dive profile study, the turn-around point for emperor penguins diving under ice was associated with a limit for the number of wing strokes during the dive (Shiomi *et al.*, 2012).

11.3.5 Blood lactate concentrations during and after dives

As demonstrated in Scholander's early forced submersion experiments, blood lactate concentration increased during the submersion and then spiked even higher during the recovery period (Scholander, 1940). During the forced submersion, blood lactate increased in seals, penguins, and ducks to concentrations of 3–5 mM. During the recovery period, lactate concentrations were as high as 20 mM, 24 mM, and 13 mM in seals, penguins, and ducks, respectively.

In contrast to forced submersions, blood lactate levels do not change significantly during sleep apneas, short dives, or even long dives. Although blood lactate concentrations began to increase slightly toward the end of long dives of Weddell seals and Baikal seals (*Phoca sibirica*), those concentrations were less than 2 mM (Guppy *et al.*, 1986, Ponganis *et al.*, 1997b, Qvist *et al.*, 1986). As observed in the Weddell seals, these minimal increases in lactate may be secondary to onset of the ascent tachycardia and potential reperfusion of previously ischemic tissues (Guppy *et al.*, 1986). However, in the Baikal seal study, these small elevations in blood lactate occurred when heart rate was only 5–10 bpm (Ponganis *et al.*, 1997b). This suggests that these small changes in blood lactate concentrations during long dives may also be secondary to increased glycolysis in the heart and brain, which are still perfused at such low heart rates, but with blood with low O_2 content and P_{O2}. In emperor penguins, blood lactate concentrations also remained less than 2 mM, even as far as 10.5 min into a dive (Ponganis *et al.*, 2009).

In contrast to during the dive, blood lactate concentrations can increase markedly after long dives. In Weddell seals and emperor penguins, blood lactate can increase to 10–13 mM, and in Baikal seals, bottlenose dolphins, and beluga whales to 4–6 mM (Kooyman *et al.*, 1980, Ponganis *et al.*, 1997b, 1997c, Shaffer *et al.*, 1997, Williams *et al.*, 1999).

The wash-out of lactate after long dives or forced submersions is associated with a further decrease in blood pH. For example, after dives of 27 and 37 min, arterial pH values of Weddell seals were 7.29, and 7.16, respectively (Qvist *et al.*, 1986). These approached values measured after forced submersions (7.0 to 7.2) (Elsner *et al.*, 1970b, Liggins *et al.*, 1980). After a 61-min dive of a Weddell seal, arterial pH was 6.8 and P_{CO2} was 55 mm Hg (7.3 kPa) (Kooyman *et al.*, 1980). Post-dive P_{CO2} values after long dives of Weddell seals were high, but less than 60 mm Hg (8.0 kPa) (Qvist *et al.*, 1986). These observations reinforce the significance and dynamic nature of the surface interval. Not only are blood and muscle O_2 stores restored, but CO_2 and lactate are also washed out from under-perfused tissues. The lowest blood pH values and greatest potential effects on the O_2 affinity of Hb are not during the dive, but during the surface interval.

Lastly, although prolonged dives are usually associated with longer surface recovery periods in Weddell seals, it is notable that both Weddell seals and Baikal seals will dive before blood lactate levels return to baseline levels (Castellini *et al.*, 1988, Kooyman, 2006, Kooyman *et al.*, 1980, Ponganis *et al.*, 1997b). In addition, blood lactate concentrations can continue to decrease during diving activity (Castellini *et al.*, 1988, Kooyman, 2006, Ponganis *et al.*, 1997b).

11.4 Mechanisms of hypoxemic tolerance

To fully deplete and take advantage of their large O_2 stores, diving mammals and birds should have extreme hypoxemic tolerance. Although such tolerance had been demonstrated during forced submersions of seals, extreme respiratory and blood O_2 depletion have now been recorded during free dives of emperor penguins, elephant seals, and sea lions. As already reviewed, some of these extremes are below inspiratory values found in bar-headed geese at simulated altitude, as well as below end tidal and blood levels in humans breathing ambient air near the summit of Mt. Everest. This tolerance is best exemplified in the elephant seal. During most of a dive, this seal functions under what would be considered hypoxic conditions in human patients. Indeed, human breath-hold divers experience cardiac arrhythmias and neurologic complications with arterial saturations that, while low for humans, are far higher than those routinely observed in seals (Andersson *et al.*, 2009a, 2009b, Lindholm and Lundgren, 2009, Liner and Andersson, 2009). Sleep apnea patients develop pulmonary hypertension and right ventricular failure with chronic episodic exposures to hypoxia (Dempsey *et al.*, 2010).

The mechanisms underlying such tolerance as well as avoidance of reperfusion injury remain to be fully determined. Perhaps the best example of tissue tolerance to complete ischemia and anoxia is the survival of seal kidneys after one hour of complete, warm ischemia (Elsner, 1988, Halasz *et al.*, 1974). Potential adaptations for survival under such conditions involve shorter O_2 diffusion distances, greater anaerobic metabolic capacity, hypothermia, other mechanisms of tissue hypoxemic tolerance, and anti-oxidant function.

11.4.1 Hypoxemic tolerance: shorter O_2 diffusion distances

Shorter O_2 diffusion distances have been suggested by higher brain capillary densities in seals and dolphins, as well as by tissue mitochondrial distribution and elevated mitochondrial volume densities in seals (Davis, 2014, Fuson *et al.*, 2003, Glezer *et al.*, 1987, Kanatous *et al.*, 1999, 2002, Kerem and Elsner, 1973). As for diving birds, tissue capillary densities are also generally higher in birds than in mammals (Faraci, 1991). In terms of O_2 delivery at low P_{O2}, the shift in the O_2-hemoglobin dissociation curve of penguins is also an advantage in comparison to other birds (Lenfant *et al.*, 1969b, Meir and Ponganis, 2009, Milsom *et al.*, 1973).

11.4.2 Hypoxemic tolerance: neuroglobin

The precise function of neuroglobin and its role in hypoxemic tolerance are still unknown (see Chapter 4). Nonetheless, the potential role of neuroglobin in cerebral hypoxemic tolerance of marine mammals has been addressed in several papers. In the first study which used a spectrophotometric technique (potentially non-specific for neuroglobin, cytoglobin, and hemoglobin), neuroglobin was found to be slightly elevated in marine mammals, with an inverse relationship of measured neuroglobin concentrations to maximum dive durations of different species (shorter-duration divers had the highest concentrations) (Williams *et al.*, 2008). It was suggested that elevated neuroglobin concentrations in active, shorter dive duration "swimmers" such as sea lions serve to facilitate O_2 diffusion because these animals have lower arterial P_{O2}, Hb content, and consequently blood O_2 content during diving (Williams *et al.*, 2008). Although an attractive hypothesis, it appears that sea lions can maintain arterial Hb saturations and blood O_2 content even during their longest, deepest dives, and therefore may not need extra neuroglobin to facilitate diffusion (McDonald and Ponganis, 2012, 2013). Conversely, elephant seals routinely experience much more significant arterial Hb desaturation (Meir *et al.*, 2009), and yet have lower measured neuroglobin content than sea lions (Williams *et al.*, 2008).

In the second marine mammal neuroglobin paper, an anti-neuroglobin antibody-based Western blotting technique was used to quantify neuroglobin concentration; no differences were found in brain neuroglobin concentrations between mouse, rat, and hooded seal (*Cystophora cristata*) (Mitz *et al.*, 2009). These researchers considered the assays of Williams *et al.* to be non-specific for neuroglobin. Although Mitz and co-workers did not find any difference in neuroglobin concentration between seals and rodents, they did find that neuroglobin in the seal was found predominantly in the glial cells (astrocytes) of the brain and not in the neurons. This contrasted with terrestrial mammals. Cytochrome c was also primarily localized in the glial cells of the seals, whereas, in terrestrial mammals, it was again located in the neurons. Thus, there appears to be a basic difference in oxidative capacity/neuroglobin distribution between seals and terrestrial mammals. In the seals, neurons appear to be predominantly anaerobic, and the glial cells aerobic, while in terrestrial mammals the opposite occurs. Mitz *et al.* have suggested that it is the anaerobic metabolism/capacity of seal neurons that are the basis of the seal's cerebral hypoxemic tolerance. The oxidative glial cells were proposed to utilize lactate during recovery periods, thus decreasing brain lactate concentrations. These remarkable findings clearly hold promise as to a better understanding of the hypoxemic tolerance of seals.

Further investigations of neuroglobin mRNA expression and immuno-histochemical localization of neuroglobin confirmed similar findings in harp seals (*Pagophilus groenlandicus*) (Schneuer *et al.*, 2012). However, in contrast to the seals, but similar to terrestrial controls, neuroglobin and cytochrome c were localized in the neurons of cetaceans (harbor porpoise and minke whale (*Balaenoptera acutorostrata*)). In addition, neuroglobin mRNA expression was enhanced 4- to 15-fold in the cetaceans relative to terrestrial controls. Thus, in contrast to the phocid seals, neuroglobin was elevated in concentration and was located in the neurons of these two cetacean species. Presumably, elevated neuroglobin concentrations in the neurons may play a role in cerebral protection in these cetaceans. It is unknown whether the same pattern occurs in long-duration, deep-diving cetaceans such as beaked whales. Clearly, more studies with further refinement of functional and quantitative assays are needed to further define the potential roles of this relatively newly discovered globin in different species of diving animals.

11.4.3 Hypoxemic tolerance: glycolytic and buffering capacities

Although glycolytic enzymes activities are not elevated or only slightly elevated in marine mammal tissues, elevated glycogen concentrations, at least in the brain and heart of Weddell seals, presumably provide greater anaerobic metabolic capacity (Castellini *et al.*, 1981, Davis and Guderley, 1990, Fuson *et al.*, 2003, Kerem *et al.*, 1973, Ponganis and Pierce, 1978, Ponganis *et al.*, 1997a). However, skeletal muscle glycogen and phosphocreatine concentrations are not elevated in marine mammals and diving birds (Blix, 1971, Goforth, 1986, Groscolas and Rodriquez, 1982, Kerem *et al.*, 1973, Stephenson and Jones, 1992b, Stephenson *et al.*, 1997, Williams *et al.*, 2012). Recent modeling of muscle glycogen and phosphocreatine depletion in dives of emperor penguins suggests that, even for an exceptionally long 23-min dive, muscle glycogen content would be decreased less than 30%, and muscle lactate content would be near 31 mmole kg^{-1} muscle – very high, but less than that in the muscle of forcibly submerged seals or in thoroughbred horses at full gallop (Scholander, 1940, Snow *et al.*, 1985, Williams *et al.*, 2012).

Because of the potential for such high lactate concentrations, it is notable that muscle buffering capacity is elevated in marine mammals and penguins, most probably secondary to imidazole-containing compounds such as carnosine, anserine, and balenine (Baldwin, 1988, Baldwin *et al.*, 1984, Castellini and Somero, 1981, Lestyk *et al.*, 2009, Lukton and Olcott, 1958, Okuma and Abe, 1992). In contrast, buffering capacity of muscle is not elevated in flighted diving birds such as murres and puffins (Davis and Guderley, 1987, 1990). In marine mammals and penguins (except for the little blue penguin (*Eudyptula minor*), the buffering capacity of blood is much higher than in terrestrial mammals and other birds (Lenfant *et al.*, 1968, 1969a, 1969b, 1970, Murrish, 1982, Nicol, 1991). All in all, although glycolytic flux rate may not be especially enhanced in marine mammals and penguins, the capacity for prolonged anaerobic metabolism is suggested by elevated tissue buffering capacities, and by elevated glycogen concentrations in critical organs such as the brain and heart, which must maintain function even during extreme hypoxemia and ischemia. In other tissues, such as muscle and splanchnic organs, metabolic demand and the rate of glycolytic flux presumably decline secondary to decreased perfusion of organs, potential hypothermia and, in muscle, decreased workload due to the economy of locomotion.

11.4.4 Hypoxemic tolerance: hypothermia

Hypothermia reduces metabolic rate through the well-known Q_{10} effect, in which a two- to three-fold reduction in metabolic rate is associated with a 10 °C decline in temperature (Guppy and Withers, 1999, Quinones *et al.*, 2014). This temperature-associated decline in metabolic rate has formed the basis for cold preservation of organs for transplantation, hypothermic protection during cardiac and vascular surgery, and hypothermic treatment regimens for victims of cardiac arrest and ischemic insult (Gunn and Thoresen, 2006, McAnulty, 2010, Quinones *et al.*, 2014, Shankaran *et al.*, 2005). In addition, therapeutic hypothermia has potential beneficial effects during the post-ischemic period, including decreased accumulation of excitatory amino acids in the brain, inhibition of apoptotic pathways, decreased free radical production, and decreased inflammatory processes (Yenari and Han, 2012).

However, although temperature fluctuations and regional heterothermy occur during dives (Chapter 8), it should be noted that prolonged core hypothermia has not been observed in emperor penguins and elephant seals (Meir and Ponganis, 2010, Ponganis *et al.*, 2004). Nonetheless, it is notable that (a) brain temperatures do decline during the extreme bradycardias of forced submersions, (b) occasional large, transient fluctuations in blood temperatures can occur in free-diving elephant seals, and (c) arterial temperatures are 2–3 °C lower after long dives of Weddell seals (Hill *et al.*, 1987, Kooyman *et al.*, 1980, Meir and Ponganis, 2010, Odden *et al.*, 1999, Scholander *et al.*, 1942b).

11.4.5 Hypoxemic tolerance: hypoxia-linked mechanisms

Hypoxia-linked reduction in energy production and energy requirements represent another potential route of protection against the effects of hypoxia. Downregulation of metabolism (reversed Pasteur effect and ion channel arrest) in marine mammals and diving birds has often been cited as a mechanism that provides hypoxic tolerance as well as reduces the metabolic cost of diving (see Chapter 9). Although such protective mechanisms occur during periods of chronic hypoxia and torpor in lower vertebrates and during hibernation in mammals (Bickler and Buck, 2007, Guppy and Withers, 1999, Hochachka 1986b, Hochachka and Lutz, 2001, Quinones *et al.*, 2014, Ramirez *et al.*, 2007), evidence for such processes in active marine mammals and diving birds is lacking. Furthermore, as reviewed in Chapter 9, the exposure time and degree of hypoxia required for induction of hypoxia-induced protective mechanisms (i.e., decreased protein synthesis and activation of hypoxia inducible factor (HIF)) are much longer and more severe than the hypoxia during most dives of marine mammals and seabirds. The increased tissue buffering capacity of marine mammals and penguins previously reviewed would also minimize decreases in pH during hypoxia and argue against metabolic suppression linked to acidosis (Guppy and Withers, 1999, Malan, 2014).

Lastly, although decreased protein synthesis and HIF activation may occur in terrestrial mammalian models, they do not afford the degree of protection against hypoxia which the seal tissues (kidney and brain) have (Folkow *et al.*, 2008, Halasz *et al.*, 1974). Indeed, the intrinsic hypoxic tolerance of seal neurons in comparison to rat neurons has been demonstrated by much longer maintenance of membrane potentials and of the ability to discharge (Folkow *et al.*, 2008). Therefore, it appears that either further modification of these responses or other mechanisms are necessary to provide hypoxic protection in marine mammals and diving birds. Other possibilities, such as ischemic pre-conditioning, are reviewed in Chapter 13. It has also been proposed that protection of the central nervous system in diving seals against hypoxia may be afforded through a combination of hypometabolism, neuroprotection, and maintenance of "functional integrity" (through reconfiguration of neural networks) (Ramirez *et al.*, 2007). Readers are referred to this very thorough discussion of hypoxic tolerance for review of the potential roles of (a) glycolysis, (b) avoidance of N-methyl-D-aspartate (NMDA) receptor activation and an increase in intracellular calcium concentrations (which can lead to cell necrosis and apoptosis), (c) ion channel modulation, and (d) network reconfiguration.

11.4.6 Hypoxemic tolerance: avoidance of reperfusion injury

Finally, much of the tissue damage associated with hypoxic exposure occurs through the generation of oxygen free radicals when the tissue is reperfused with oxygen-rich blood (Powers and Jackson, 2008). Such reperfusion injury may be avoided with enhanced O_2 free radical scavenging, especially through elevated glutathione levels and increased activities of enzymes involved in glutathione recycling (Elsner *et al.*, 1998, Vázquez-Medina *et al.*, 2006, 2007, Zenteno-Savin *et al.*, 2010). The peroxidase activity of myoglobin in muscle is also elevated in divers secondary to their high myoglobin concentrations; thus, myoglobin probably also plays a significant role in hydrogen peroxide breakdown (Helbo and Fago, 2012, Helbo *et al.*, 2013). In addition, the nitrite reductase of myoglobin, and the potential generation of nitric oxide, could result in decreased reactive oxygen species formation through nitric oxide's inhibition of cytochrome c oxidase (Helbo and Fago, 2012, Helbo *et al.*, 2013). The potential roles of anti-oxidants and O_2 free radical scavenging are discussed in detail in Chapter 13.

Lastly, carbon monoxide, which is elevated in Weddell seal and elephant seal blood (see Chapter 4), may also afford protection against reperfusion injury (Kajimura *et al.*, 2010, Lancel *et al.*, 2009, Nakaoa *et al.*, 2005, Piantadosi, 2008, Pugh, 1959, Rhodes *et al.*, 2009, Tift *et al.*, 2014, Wu and Wang, 2005). The carbon monoxide elevations in the seals are potentially beneficial in decreasing mitochondrial oxygen flux, oxygen free radical formation, inflammation, and apoptosis.

12 Pressure tolerance

Pressure tolerance in diving mammals and birds remains perhaps the least investigated aspect of diving physiology. Anatomical and physiological adaptations are necessary in deep divers to avoid four major complications of increased ambient pressure around the body. These problems include physical distortion of tissue and gas spaces (barotrauma), excess nitrogen absorption and the formation of nitrogen bubbles on ascent or at the surface (decompression sickness), the narcotic effect of high nitrogen pressures on the nervous system (nitrogen narcosis), and the onset of neurological symptoms (tremors, myoclonic jerks, somnolence, nausea, decreased mental performance) associated with the pressure at depth and/or with rapid changes in depth (high-pressure nervous syndrome). Detailed reviews of the symptoms and mechanisms of these pathologies are available in diving physiology texts (Brubakk and Neuman, 2003, Sebert, 2010).

In this chapter, general aspects of decompression sickness, nitrogen narcosis, and high-pressure nervous syndrome will first be reviewed. Then, the anatomical and physiological adaptations responsible for the pressure tolerance of marine mammals and diving birds will be examined.

12.1 Decompression sickness, nitrogen narcosis, and high-pressure nervous syndrome

These three syndromes are the consequence of the effects of pressure. Typically, these complications occur in human divers during compressed-air diving. However, decompression sickness (DCS) has been reported in human breath-hold divers (Cross, 1965, Ferrigno and Lundgren, 2003, Paulev, 1965). DCS results from excess N_2 absorption at depth and inadequate N_2 wash-out during ascent. The resulting elevations in P_{N2} are associated with the onset of symptoms and bubble formation. Symptomatic bubble formation in cats requires a minimum P_{N2} of 2476 mm Hg (330 kPa or 3.3 ATA) (Harvey *et al.*, 1944), although asymptomatic venous gas emboli have been detected in humans on decompression from steady-state pressure exposure of 1023 mm Hg (135 kPa, 1.35 ATA or <5 m depth) (Eckenhoff *et al.*, 1990).

12.1.1 Decompression sickness

Decompression sickness is considered secondary to an increase in the P_{N2} of tissue or blood relative to ambient pressure. The dissolved gas excess (supersaturation) leads to a phase disequilibrium that can result in bubble formation (Tikuisis and Gerth, 2003). Gas bubbles can form through tribonucleation, a process wherein the separation of two surfaces within a liquid (or shear within a soft material, i.e., tissue) leads to the formation of gas nuclei (Goldman, 2010, Ikels, 1970). Another mechanism of bubble formation and expansion, which can occur on exposure of gas micronuclei to ultrasound, is a process known as rectified diffusion, whereby the bubble size oscillates and increases in size secondary to acoustic pressure fluctuations (Crum and Mao, 1996, Houser *et al.*, 2001). As will be reviewed in Section 12.7, rectified diffusion has been postulated to contribute to development of DCS in whales after exposure to sonar.

Once formed, venous bubbles may be associated with (a) circulatory obstruction and vascular congestion, (b) platelet aggregation, (c) activation of the coagulation system and onset of disseminated intravascular coagulation, and (d) activation of complement and the immune system (Francis and Mitchell, 2003). Accumulation of N_2 in tissues with high N_2 solubility and low perfusion may also lead to formation of tissue bubbles (autochthonous formation), with localized tissue damage and hemorrhage (Francis and Mitchell, 2003). Poorly perfused tissues include spinal cord white matter, periarticular tissues, fat, and the inner ear. It is notable that the spinal cord is a much more common site than the brain for CNS symptoms and pathological lesions of DCS (Caruso, 2003, Dutka, 2003). However, it is not completely clear why the spinal cord is more susceptible to DCS than the brain. As already stated, cerebral perfusion (and presumably N_2 wash-out) is often described as more luxuriant than spinal cord blood flow (Francis and Mitchell, 2003), but simultaneous measurements of spinal cord and cerebral perfusion rates are quite similar (Hickey *et al.*, 1986, Sakamoto and Monafo, 1989).

12.1.2 Nitrogen narcosis and high-pressure nervous syndrome

Nitrogen narcosis begins in humans at a depth of 30 m; this would correspond approximately to a P_{N2} of 2400 mm Hg (320 kPa or 3.2 ATA) (Halsey, 1982). The onset of high-pressure nervous syndrome (HPNS) occurs at 190 m (20 ATA) in human divers (Halsey, 1982). Many cetaceans and pinnipeds not only dive deeper than this HPNS threshold, but they also descend faster than safe compression rates for humans. In addition, many marine mammals remain below that symptomatic threshold depth for time periods well beyond that required for HPNS to occur in humans (Bennett and Rostain, 2003, Ponganis *et al.*, 2003a). It is unknown whether unique anatomical or physiological adaptations protect against HPNS in deep-diving mammals. However, since N_2 is known to suppress the onset of HPNS in human divers, an elevated tissue P_{N2} in marine mammals that is less than that associated with N_2 narcosis may be an important factor in the prevention of HPNS in these animals (Ponganis *et al.*, 2003a). However, in order to avoid N_2 narcosis and DCS in deep-diving mammals, excess

N_2 absorption at depth must still be minimized. Hence, the significance of Scholander's hypothesis of alveolar collapse and lack of gas exchange at depth.

12.2 Avoidance of barotrauma in marine mammals

Tissue trauma induced by descent to depth is due to mechanical distortion and tissue compression. In humans, this can occur in both breath-hold divers and compressed air divers (Ferrigno and Lundgren, 2003). Air-filled cavities are especially susceptible to such injury. Air-filled cranial sinuses are absent in pinnipeds (King, 1983), and middle ear cavities are lined with venous plexuses, which have been postulated to become engorged with blood and obliterate the air space at depth (Odend'hal and Poulter, 1966, Welsch and Riedel-Sheimer, 1997). In cetaceans, the large pterygoid sinus which communicates with the middle ear via the Eustachian tube is also lined with prominent blood vessels (Ponganis *et al.*, 2003a). The extensive venous vasculature of the cranial air sinuses of the bottlenose dolphin (*Tursiops truncatus*) has been recently described and illustrated in exquisite detail (Costidis and Rommel, 2012).

Several adaptations appear to protect against barotrauma in the airways of marine mammals. First, actual narrowing of the flexible trachea, but not the more reinforced bronchi of seals occurs during compression and has been demonstrated with air tracheograms and bronchograms obtained in a pressure chamber (Kooyman *et al.*, 1970). Second, as found in the striped dolphin (*Stenella coeruleoalba*), the trachea, although stiffer than those in goats and pigs, has a higher breaking point and readily returns to its original shape after compression (Cozzi *et al.*, 2005). In addition, large venous sinuses observed in the tracheal walls of striped dolphins may become engorged at depth to obliterate space and again prevent barotrauma (Cozzi *et al.*, 2005). Such tracheal wall engorgement had been previously suggested in sperm whales (*Physeter macrocephalus*) (Leith, 1989). Engorgement of the extensive vasculature observed in the terminal air spaces of beaked whales has also been postulated to prevent "lung squeeze" (Cozzi *et al.*, 2005, Ninomiya *et al.*, 2005). Such central pooling of blood within venous vessels of the thorax and lung has been demonstrated in human divers (Craig, 1968, Ferrigno and Lundgren, 2003, Schaeffer *et al.*, 1968), and has been postulated to also occur in the thoracic retia mirabilia of some cetaceans (McFarland *et al.*, 1979, Slijper, 1962, Vogl and Fisher, 1982). A recent modeling study of the fin whale thoracic rete suggested that the volume of blood in the rete was adequate for pressure equilibration in the thorax (Lillie *et al.*, 2013). It is notable that intrathoracic vascular structures of deep-diving kogiids occupy 9% of thoracic volume while those in the bottlenose dolphin occupy about 5% (Fig. 12.1) (Piscitelli *et al.*, 2010). The retia mirabilia are reviewed in more detail in Chapter 6.

In addition to the above mechanisms that protect the reinforced airways of diving mammals against lung squeeze, volume pressure curves of both the chest wall and lung of the ribbon seal (*Histriophoca fasciata*) demonstrate nearly limitless compression collapse (Leith, 1976). Such compressibility has also been demonstrated in the lungs of sea lions and cetaceans (Denison *et al.*, 1971, Kooyman and Sinnett, 1979, Leith, 1976,

(a)

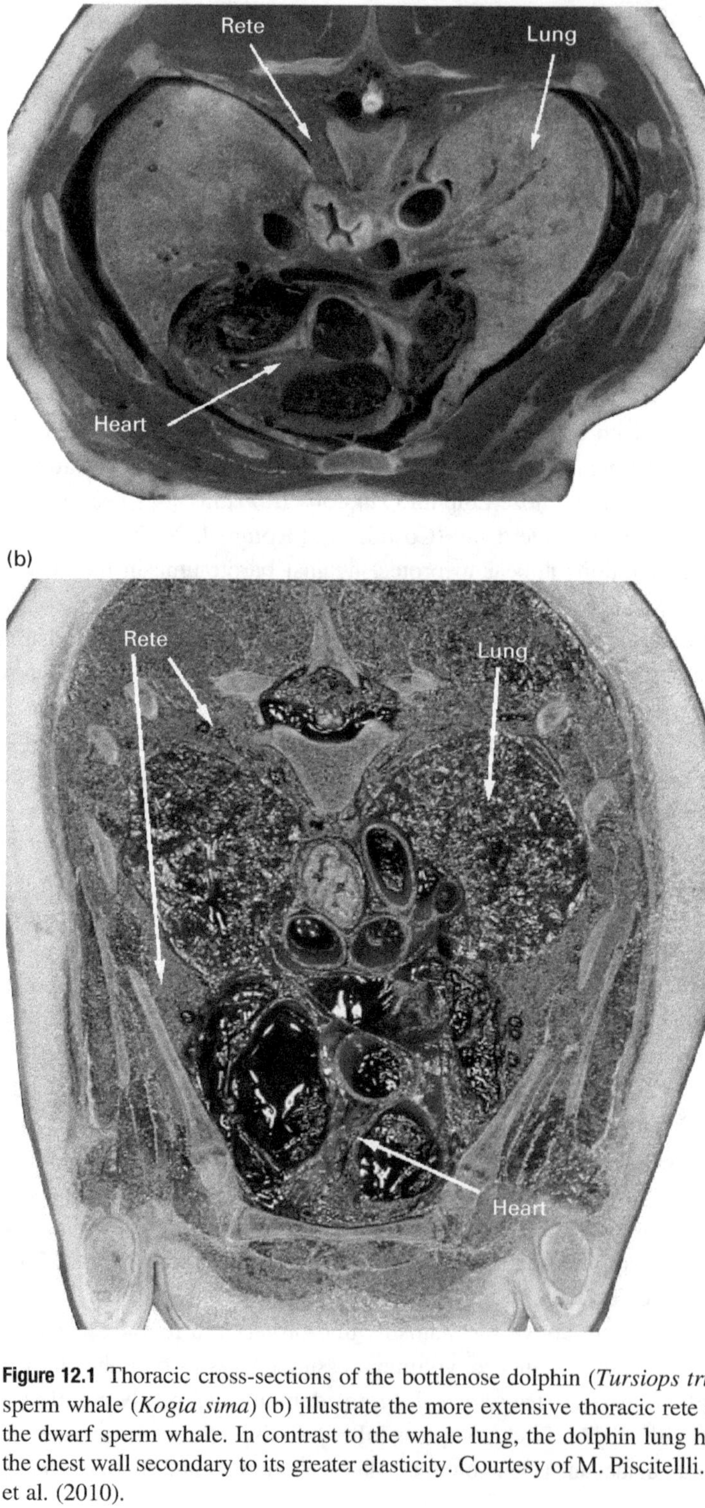

Figure 12.1 Thoracic cross-sections of the bottlenose dolphin (*Tursiops truncatus*) (a), and dwarf sperm whale (*Kogia sima*) (b) illustrate the more extensive thoracic rete and flaccid lungs in the dwarf sperm whale. In contrast to the whale lung, the dolphin lung has retracted away from the chest wall secondary to its greater elasticity. Courtesy of M. Piscitellli. Adapted from Piscitelli et al. (2010).

Ridgway *et al.*, 1969). In these studies, relaxation volumes of isolated lungs were minimal and considered evidence of the collapsibility of alveoli. Actual compression of the chest wall has been photographically documented in the bottlenose dolphin at depth (Ridgway *et al.*, 1969). In Weddell seals free-diving to 80 m, however, chest circumference measurements during dives did not reveal chest compression (Falke *et al.*, 2008). It was postulated that elevation of the diaphragm contributed to lung compression in this case.

Biomechanical studies of the excised thorax in cetaceans have revealed that thoracic compression may be limited anatomically, and that the maximum relative decrease in thoracic volume of deep divers (*Kogia* sp.) was 21%, similar to that of the shallow-diving bottlenose dolphin (Piscitelli *et al.*, 2010). However, in *Kogia* sp., the lungs occupied only 15% of thoracic volume while in bottlenose dolphins the lungs comprised 37% of that volume. In addition, as described above, the retia are twice as large in the kogiids as in the dolphins. These observations emphasize the potential importance of engorgement of the retia as a mechanism to prevent lung barotrauma in deep-diving whales. Despite these potential biomechanical disadvantages of the bottlenose dolphin for diving to depth, it should be remembered (see Chapter 1) that recent studies have documented deep-dive behavior (500 m) in bottlenose dolphins (Klatsky *et al.*, 2007).

In addition to the role of the retia and thoracic compression in prevention of lung barotrauma, the intrathoracic abdominal cavity occupies a significant portion of the muculoskeletal thorax of the kogiid whales (Piscitelli *et al.*, 2010, 2013). Elevation of the diaphragm due to cranial movement of abdominal contents may also compress the thoracic cavity during deep dives. Numerical modeling of the thoracic rete of the fin whale suggests that blood volume of the rete is sufficient to achieve pressure equilibration in the thorax (Lillie *et al.*, 2013). Lastly, a rigid, non-compliant thoracic aorta has also been postulated to resist collapse due to changes in the transmural pressure gradients that might be secondary to a lag in equilibration of intrathoracic pressure during a dive (Lillie *et al.*, 2013).

12.3 Lung collapse and minimization of N_2 absorption in marine mammals

Further support for Scholander's model of alveolar collapse and the cessation of gas exchange at depth are found in blood N_2 investigations of seals, both during simulated dives and free dives (Falke *et al.*, 1985, Kooyman and Sinnett, 1982, Kooyman *et al.*, 1973a). During dives, arterial P_{N2} rises, plateaus and, then, gradually decreases despite further increases in ambient pressure (Fig. 12.2). That plateau has been considered indicative of the cessation of gas exchange during a dive (see Section 12.6 for a detailed description). The level at which P_{N2} plateaued corresponded to a depth of 20–50 m in elephant seals (*Mirounga angustirostris*) and Weddell seals (*Leponychotes weddellii*) (Falke *et al.*, 1985, Kooyman *et al.*, 1973a).

More support for a depth threshold for cessation of gas exchange was found in arterial P_{O2} profiles of free-diving elephant seals (Meir *et al.*, 2009). Arterial P_{O2} transiently rises early in a dive, which is indicative of continued gas exchange at least

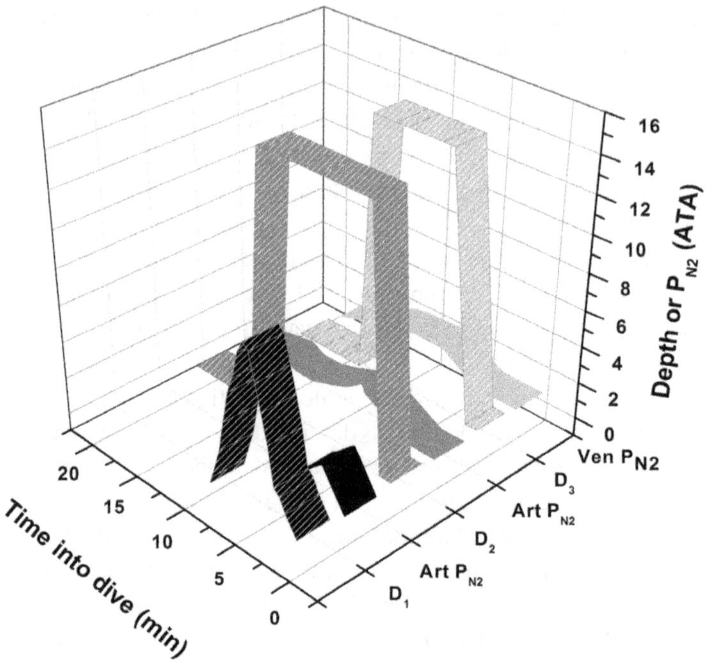

Figure 12.2 Blood P_{O2} and depth profiles during a free dive (D_1) of a Weddell seal (*Leptonychotes weddellii*), and simulated dives (D_2, D_3) of elephant seals (*Mirounga angusitrostris*) and a Weddell seal in a pressure chamber provided evidence for lung collapse and lack of gas exchange at depth. Pressure is in atmospheres absolute (ATA). Abbreviations: Art: arterial, Ven: venous. Adapted from Falke et al. (1985), Kooyman et al. (1973a).

until the depth at which P_{O2} peaks. Mean depth of the maximum arterial P_{O2} in these free-diving seals was 20 m, similar to the depth of "lung collapse" in pressure chamber studies. However, in individual dives, the depth of the maximum arterial P_{O2} ranged as deep at 82 m. Although still supportive of the Scholander hypothesis, this greater depth for the peak arterial P_{O2} suggests that the diving lung volume at the start of some free dives may be greater than that determined in pressure chamber dives.

Depth of lung collapse has also been estimated in bottlenose dolphins, based on measurements of muscle N_2 wash-out (Ridgway and Howard, 1979). In this classic study, mass spectrometry was used to measure muscle N_2 levels in dolphins trained to beach themselves after a one-hour session of serial dives to 100-m depth. Muscle N_2 measurements, made between 8 and 22 min after the dives, were back extrapolated on a semilogarithmic linear plot to estimate the end-of-dive muscle N_2 level (2.2 to 2.8 ATA) and tissue half-time for N_2 washout. A 70-meter depth of lung collapse was based on the dive depth patterns of the dolphin prior to sampling, and the assumptions of a constant tissue half-time, and a symmetrical tissue wash-in and wash-out pattern during dives. A limitation of this study is that such assumptions may not be valid. In a recent heart rate study of dolphins trained to perform similar serial dives to depth, heart rates were slower during the post dive-session period than during surface intervals between individual dives, suggesting that tissue half-times might be shorter during the surface intervals between dives than during the recovery period after the dive session (Houser *et al.*, 2010). In addition, heart rate profiles were not symmetrical during ascent and descent. Consequently, tissue wash-in and wash-out may also not be symmetrical, as assumed in the model.

More recently, hyperbaric computerized tomographic (CT) scans of marine mammal carcasses have allowed three-dimensional reconstruction of lung volumes at different pressures and subsequent estimation of depth of lung alveolar collapse (Moore *et al.*, 2011). This novel approach was limited by potential post mortem changes in lung and airway compliance of carcasses and also by an only indirect index of alveolar collapse. In addition, pulmonary hemorrhage and edema were noted in some specimens, and prior reports have documented pulmonary edema, hemorrhage, alveolar rupture, and rupture/contraction of bronchiolar sphincters in by-catch cetacean carcasses (Cowan and Curry, 2008, Knieriem and Hartmann, 2001). The results of the CT analysis predicted depths of lung collapse of 58 m in a gray seal (*Halichoerus grypus*), and 133 m in a harbor porpoise (*Phocoena phocoena*). Although deeper than estimated by physiological investigations, differences in compliance of airway structures may be especially important in determination of the depth of alveolar collapse (Bostrom *et al.*, 2008).

12.4 Pulmonary shunts in marine mammals

The degree of pulmonary shunt has been determined in one study of simulated dives in harbor seals (*Phoca vitulina*) and California sea lions (*Zalophus californianus*) (Kooyman and Sinnett, 1982). Pulmonary shunt is the relative blood flow through the lung that does not participate in gas exchange. The lung collapse proposed by Scholander is essentially a 100% pulmonary shunt. Pulmonary shunt flow was 8% and 13% at rest in harbor seals and sea lions, respectively. The shunt progressively increased with depth of compression to 70% in harbor seals at 10 ATA (90 m depth), and to 57% in sea lions at 7.8 ATA (68 m depth). From these data, it was estimated that total lung collapse (100% shunt) would occur at depths of 170 m in harbor seals and 160 m in sea lions. These are much deeper threshold depths for lung collapse than those determined in Weddell seals and elephant seals, and indicate that species differences probably exist. It was also notable that these depth thresholds for lung collapse were similar in harbor seals and sea lions, two species with very different terminal airways. As reviewed previously in Chapter 3, sea lions have cartilaginous rings in the distal airway all the way to the alveolus, whereas harbor seals lack such cartilaginous reinforcement. Although the harbor seal airway is more flexible, it was concluded that the muscle and fibrous tissue in the walls of the distal airway provided sufficient support to allow emptying and collapse of the alveoli at depth.

Further evidence for lung collapse or pulmonary shunting in sea lions has been demonstrated in arterial P_{O2} profiles of free-diving animals during maternal foraging trips to sea (McDonald and Ponganis, 2012). In deep dives to 300–400 m, arterial P_{O2} abruptly declined near 200-m depth during descent, and again abruptly rose at the same

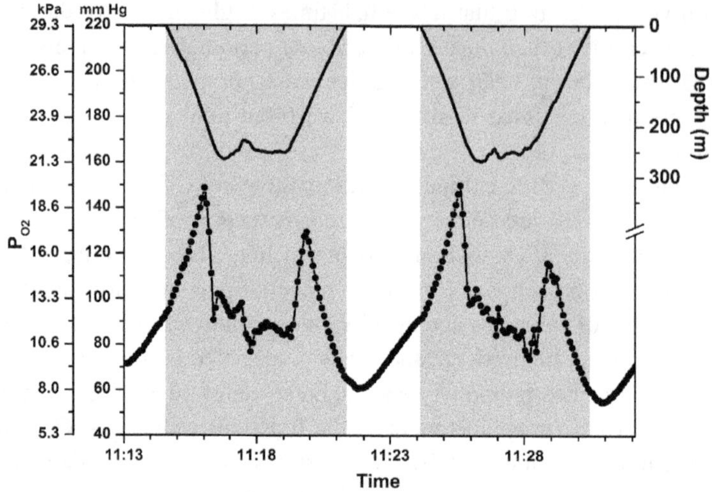

Figure 12.3 Double-peaked arterial P_{O2} profile of free-diving California sea lion (*Zalophus californianus*) was consistent with lung collapse at depth. Adapted from McDonald and Ponganis (2012).

depth range during ascent (Fig. 12.3). This profile was consistent with lung collapse and decreased gas exchange at depth; the depth of collapse was only slightly greater than those estimated by Kooyman and Sinnett in their pressure chamber studies. The transient rise in P_{O2} during ascent contributed to the maintenance of arterial saturation during the final portion of the dive. Thus, decreased gas exchange at depth not only limits N_2 absorption at depth, but may also preserve respiratory O_2 for use during the ascent.

Finally, it should be emphasized that lung collapse is most probably not an all-or-none phenomenon. As demonstrated in the experiments of Kooyman and Sinnett, and as hypothesized in modeling studies (Bostrom *et al.*, 2008, Fahlman *et al.*, 2006, 2009), the degree of pulmonary shunt probably increases with depth and only eventually reaches 100% shunt or complete alveolar collapse at a critical depth. Additionally, the start-of-dive lung volume will also affect the depth at which complete collapse occurs. The depths of lung collapse in the arterial P_{O2} study of sea lions increased with maximum depth of dive, suggesting that sea lions inhaled a large air volume prior to deeper dives (McDonald and Ponganis, 2012).

The routine collapse and re-expansion of alveoli during dives to depth raise the question of surfactant function and its role in the maintenance of gas exchange in these animals. A detailed review of surfactant in marine mammals is provided in Chapter 13.

12.5 Avoidance of decompression sickness in marine mammals

Despite the potential protective effect of lung collapse and decreased gas exchange at depth against increased N_2 absorption, evidence of possible DCS has been found in stranded cetaceans. In particular, necropsies of beaked whales stranded in association with naval sonar exercises revealed intravascular/tissue bubble lesions and fat emboli consistent with, although not diagnostic of, DCS (Fernandez *et al.*, 2005, Jepson *et al.*, 2003). Pathologic findings consistent with dysbaric osteonecrosis in sperm whales also raised the same question as to possible bubble formation and the magnitude of N_2 absorption in deep-diving mammals (Moore and Early, 2004). Over the last decade, these initial pathological findings have led to many searches for evidence of bubble formation as well as theoretical modeling studies of N_2 absorption and distribution (Hooker *et al.*, 2012).

The roles of N_2 supersaturation, sound effects, and changes in dive behavior in the possible development of DCS after exposure of beaked whales to sonar will be reviewed in Section 12.7. In this section, anatomical/physiological investigations and pathology findings will be reviewed.

12.5.1 Avoidance of decompression sickness in marine mammals: thoracic rete hypothesis

In addition to lung collapse, the thoraco-spinal retia of cetaceans, which supply cerebral perfusion via their connection to the spinal meningeal artery (see Chapter 6), have also been postulated to provide protection against DCS. In the harbor porpoise (*Phocoena phocoena*), these arterial vessels branch into smaller arteries and eventually into thin-walled sinusoids prior to entry into the spinal meningeal artery (Blix *et al.*, 2013). The small arteries and sinusoids are embedded in fat. Based on the high solubility of N_2 in fat and the increased surface area of the thin-walled sinusoids, Blix and associates have suggested that N_2 from blood in the sinusoids diffuses into the surrounding fat, thereby decreasing the P_{N2} of blood going to the brain, and minimizing the risk of N_2 bubble formation in the cerebral arterial supply during ascent and at the surface. This is certainly a possibility.

Although spontaneous arterial N_2 bubble formation has been considered rare in DCS because of the rapid equilibration of gases between inspired air and pulmonary blood (Tikuisis and Gerth, 2003), transfer of venous gas emboli into the arterial system has been observed in human divers (Ljubkovic *et al.*, 2011). Brain DCS symptoms in human divers have usually been considered secondary to gas embolism due to pulmonary barotrauma or to transpulmonary or transatrial (patent foramen ovale) passage of nitrogen bubbles into the arterial system (Dutka, 2003, Francis and Mitchell, 2003). The transfer of bubbles to the arterial system in the human divers only occurred during heavy venous gas emboli loads, and it was postulated that this transfer was secondary to the presence of intrapulmonary shunts or to exceeding the lung's bubble filtration threshold (Ljubkovic *et al.*, 2011). There was no evidence of intracardiac shunts in the divers. Interestingly, none of the human divers with arterialized bubbles developed DCS symptoms. Nonetheless, the retial fat reservoir proposed by Blix and associates would presumably minimize the effect of any such N_2 bubbles in cetaceans.

In addition, the proposed retial fat reservoir might function to decrease arterial P_{N2} and N_2 delivery to the brain and spinal cord during a dive. The combination of a diving bradycardia and retial fat N_2 depot could lead to lower elevations in brain and spinal cord P_{N2}, potentially minimizing the risks of both DCS and nitrogen narcosis.

12.5.2 Avoidance of decompression sickness in marine mammals: N_2 supersaturation

Much recent research on the possibility of DCS in marine mammals has focused on the detection of gas bubbles. In seals and dolphins that had drowned at depth in gill nets, gas bubbles were detected in 65% of carcasses and considered as evidence of blood N_2 supersaturation (Moore *et al.*, 2009). Such bubble formation in carcasses after they have been brought up from depth confirmed the high blood P_{N2}s and bubble formation in blood sampling syringes that were reported by Kooyman in his pressure chamber studies of seals 36 years earlier (Kooyman *et al.*, 1973a). In contrast to the gill net carcass study, bubbles were not detectable by ultrasound in the portal vein of dolphins trained to make serial dives to 100-m depth (Houser *et al.*, 2010), nor in any of 20 dolphins live caught in shallow water for health exams (Dennison *et al.*, 2012b). However, bubbles were detected by ultrasound in 21 of 22 live, stranded bottlenose dolphins, primarily in the kidney region (Dennison *et al.*, 2012b). In necropsied animals, these bubbles were located primarily in subcapsular loose connective tissue, leading to the suggestion that the kidneys might act as a bubble filter (Dennison *et al.*, 2012a). Gas lesions in early necropsy reports were also predominantly in the liver and kidneys (Fernandez *et al.*, 2005, Jepson *et al.*, 2003, 2005).

More detailed examinations of bubble distribution and bubble gas content have also been conducted on cetaceans stranded by miscellaneous causes (Bernaldo de Quirós *et al.*, 2011, 2012,). Animals were classified as deep and non-deep divers on the basis of a 500-m threshold and known dive depths of the species. Intravascular bubbles were found in 58% of animals, and were more prevalent in deep divers (76%) than non-deep divers (52%). Subcapsular gas (primarily peri-renal) was found in 65% of all specimens, and in 66% and 71% of non-deep and deep divers, respectively. Analysis of gas content in deep divers revealed predominantly N_2 ($\geq$70%) and carbon dioxide. The higher prevalence of intravascular bubbles in deep divers and the high N_2 content of those bubbles led to the suggestion that these were most probably decompression related. It has also been suggested that elevated carbon dioxide levels may seed N_2 bubble formation (Fahlman *et al.*, 2014).

12.5.3 Avoidance of decompression sickness in marine mammals: N_2 solubility

Tissue N_2 solubility is also an important factor in determining the quantity of N_2 taken up by a tissue and its potential role in N_2 distribution, as well as in the potential risk of symptomatic bubble formation in the tissue. In the first ever study of N_2 solubility in cetacean tissue, Koopman and Westgate found almost two-fold higher N_2 solubilities in the blubber of deep-diving beaked whales than the lowest values in other cetaceans and control references (Koopman and Westgate, 2012). N_2 solubility increased with wax ester content in the blubber; wax esters accounted for 52% of the variation in N_2 solubility. Beaked whale blubber contained almost 100% wax esters, whereas porpoise blubber had no wax esters. N_2 solubility in the acoustic fat bodies of a Risso's dolphin (*Grampus griseus*) was 16% greater than in its blubber, demonstrating that N_2 solubilities in fat even vary within the body. These findings have significant implications for the assumptions and accuracy of mathematical modeling studies, as well as for the interpretation of the distribution of pathological lesions within an animal.

12.5.4 Avoidance of decompression sickness in marine mammals: blood N_2 levels

Despite the great interest in possible DCS in marine mammals, blood N_2 levels during dives have only been measured in two papers; the first involved simulated dives in a pressure chamber, and the second, free dives of Weddell seals at an isolated dive hole (Fig. 11.2). These valuable studies were both technological accomplishments. As such, these findings deserve further description.

In the pressure chamber study (Kooyman *et al.*, 1973a), peak arterial P_{N2} occurred early in the dive and was about 3 ATA in elephant seals and a Weddell seal during dives as deep as 14.6 ATA (136 m). Arterial P_{N2} gradually declined after the peak. Venous P_{N2} gradually increased, with peak values near 2 ATA in the latter half of dives. In harbor seals, peak arterial P_{N2} reached 5.3 ATA, and venous P_{N2}, 3 ATA, during at 14.6 ATA dives. The peak P_{N2} of 3 ATA in elephant seals would correspond to a depth of about 20 m, at which lung collapse, or at least a significant decrease in gas exchange, occurred. In the Weddell seal study (Falke *et al.*, 1985), arterial P_{N2} was never greater than 4.2 ATA during dives as deep as 230 m; serial samples revealed a gradual decline as the dive progressed. It was estimated that lung collapse occurred at 25–50-m depth.

In cetaceans, blood N_2 levels have only been examined during the post-dive period in a bottlenose dolphin trained to make serial dives to 100-m depth for a one-hour period (Houser *et al.*, 2010). In contrast to the elevated intramuscular P_{N2} found in dolphins trained to dive to similar depths (Ridgway and Howard, 1979), P_{N2} in fluke blood samples was not elevated. Interpretation of these findings was limited, however, because of the site of blood sampling. These limitations included: (a) blood samples from the fluke can be a mix of arterial and venous blood because of the fluke's vascular rete (see Fig. 6.2); and (b) the amount of N_2 absorbed in fluke tissue during the dive (and therefore washed out during the surface interval) is probably limited by the dive response and decreased fluke blood flow during the dive.

12.5.5 Avoidance of decompression sickness in marine mammals: summary

Our understanding of all these findings is still incomplete. Although lung collapse, cessation of gas exchange, and a diving bradycardia may limit N_2 absorption at depth, lung collapse may occur at deeper depths than those estimated by pressure chamber studies of pinnipeds and by muscle N_2 sampling after dive sessions of dolphins. Diving air volumes and differences in airway/lung compliance may contribute to these differences. High blood N_2 levels at depth certainly occur as documented by Kooyman, Falke, and co-workers. Asymptomatic (silent) bubbles would be not unexpected and would be consistent with the findings of dysbaric osteonecrosis in sperm whales. A key question regarding necropsies and exams of stranded animals is whether the prevalence and distribution of bubbles in stranded animals reflect bubble formation and distribution during normal dive behavior. Post-mortem or, even, post-moribund off-gassing of N_2 from saturated tissues has been considered and debated by all authors, and represents a potential limitation to the interpretation of stranding data.

12.6 Cellular and biochemical effects of pressure in marine mammals

Adaptations to pressure at the cellular and biochemical level also appear to occur in some marine mammals, although the number of studies and tissues examined are limited (Castellini *et al.*, 2002). For example, in contrast to terrestrial mammals, platelet activation on rapid decompression did not occur in elephant seals; this would decrease the risk of thrombosis in these deep divers during ascent (Field and Tablin, 2012). It was suggested that the lipid composition, and, in particular, the three-fold elevated cholesterol concentration of elephant seal platelets, may contribute to this protective response. In regard to membrane function, the ordering and fluidity of red blood cell membranes were higher in harbor seals and elephant seals than in northern fur seals and terrestrial mammals (Williams *et al.*, 2001a). Differences in the composition and function of the membranes of both red blood cells and platelets in deep divers suggest the possibility that alterations in neural membrane composition and function may contribute to avoidance of HPNS in these animals. Such research could provide valuable insights into pressure tolerance. Potential effects of pressure on cardiac tissue and heart rate have been reviewed in Chapter 2 (see Section 2.1.8). In addition to studies of membrane function, it has also been found that red blood cell glycolytic flux was either not altered or increased in marine mammals, but was depressed in terrestrial mammals (Castellini *et al.*, 2001). Lastly, in contrast to fish, but similar to the rabbit, the skeletal muscle lactate dehydrogenase Michaelis–Menten constant K_M for NADH was not affected by changes in pressure to 204 atmospheres in muscle samples from elephant seals, sperm whales, and emperor penguins (*Aptenodytes forsteri*) (Croll *et al.*, 1992b).

12.7 Sonar-associated whale strandings

A major conservation topic to which diving physiology and mechanisms of pressure tolerance are relevant is the stranding of beaked whales in association with sonar exposure. Physiological research potentially contributes to the mitigation of this problem by providing a knowledge base from which to develop possible solutions. This issue is of concern not only to the navies of the world, but also to the oil and gas industry as whale strandings have been reported in temporal association both with naval sonar exercises and with underwater seismic surveys (Frantzis, 1998, Simmonds and Lopez-Jurado, 1991, Taylor *et al.*, 2004, US Department of Commerce & US Navy, 2001).

Although evidence of hemorrhage in acoustic fat tissue on necropsy of such stranded whales had been previously reported (US Department of Commerce & US Navy, 2001), the presence of intravascular gas bubbles, diffuse severe vascular congestion, and microvascular hemorrhage associated with fat emboli in vital organs was first reported in 2003 from whales stranded on the Canary Islands (Jepson *et al.*, 2003). Further detailed description of bubble lesions and fat emboli were also soon provided for the Canary Island strandings as well as strandings in England not associated with sonar exposure (Fernandez *et al.*, 2005, Jepson *et al.*, 2005). In addition, gas analyses eventually revealed that many of the bubbles in these stranded whales were composed primarily of N_2 (Bernaldo de Quirós *et al.*, 2012). As previously described in this chapter, these pathological findings, in addition to evidence of dysbaric osteonecrosis in sperm whale bones (Moore and Early, 2004), led to the consideration of nitrogen supersaturation, "silent" bubble formation, and the increased risk of DCS in cetaceans secondary either to the effects of sound or changes in dive behavior (Cox, 2006).

From a workshop held in 2004 on the strandings (Cox, 2006), change in dive behavior was considered the most likely possible cause of the pathology underlying the strandings. As summarized in that paper, gas bubble formation, hypoxia, acidosis, and arrhythmias could result from (a) prolongation of dive duration and time at depth; (b) changes in ascent and descent rates; or (c) increased duration of surface intervals. The stranding itself could also give rise to some of the pathological findings. Basic knowledge of beaked whale anatomy/physiology, dive behavior, and nitrogen uptake/distribution were lacking. An excellent review of cetacean anatomy/physiology with special reference to available data on beaked whales is highly recommended to readers (Rommel *et al.*, 2006). Tissue damage to air-filled structures (lungs) due to sound resonance was considered unlikely due to minimal tissue displacement (NOAA, 2002). And although expansion of gas micronuclei due to rectified diffusion was a possibility (Crum *et al.*, 2005), that, of course, required the presence of gas micronuclei, and further research was recommended. However, the primary research recommendations of the workshop were (a) the conduct of controlled sonar exposure experiments on whales; (b) study of the anatomy, physiology, and pathology of beaked whales, especially post-stranding with a standardized necropsy protocol; (c) documentation of baseline diving behavior and physiology; (d) assessment of population effects; and (e) review of past and new strandings to examine common pathologies, sound reception levels, sound sources, and habitat (oceanographic/topographic) features that might affect the intensity and characteristics of transmitted sound.

This section will highlight some of the behavioral, anatomical, and physiological research that stemmed from that conference. As will be seen, the etiology of the strandings has not been resolved. Behavioral investigations have resulted in remarkable initial findings. Most physiological work has focused on the potential for N_2 supersaturation and how a change in dive behavior could lead to bubble formation and increased risk of symptomatic decompression sickness. Further documentation of dive behavior, physiological responses, and the kinetics of gas uptake and distribution are still needed.

12.7.1 Dive behavior

Controlled exposure studies have been expensive and difficult to conduct. In the first published controlled exposure study (Tyack *et al.*, 2011), findings were inconclusive. Blainville's beaked whales (*Mesoplodon densirostris*) stopped echolocating during foraging deep dives and moved away during playback of simulated sonar. Dive profiles also changed and were characterized by unusually long, slow ascents.

In a study of Cuvier's beaked whales (*Ziphius cavirostris*), exposure to mid-frequency sonar was initially associated with cessation of echolocation and stroking that was then followed by prolongation of normal dive duration with slow ascent rates but with high stroke rates and at fast swim speeds (2.6–3.1 m s^{-1}) away from the sound source (DeRuiter *et al.*, 2013). The inter-deep-dive interval was also prolonged to 6.6–7.6 hours. The sound reception levels of the sonar signal for the dramatic responses in these whales were 89–120 decibels, a range at which it has been assumed there would be no behavioral disturbances (DeRuiter *et al.*, 2013).

Prolonged dives at high swim speeds away from a sound source without echolocation may well be associated with increased risks of stranding due to disruption of normal dive metabolism/physiology, i.e., oxygen depletion, lactate formation and metabolic acidosis, hyperthermia, abnormal nitrogen uptake and distribution, loss of navigation. In a more recent observation, exposure of a Baird's beaked whale (*Berardius bairdii*) to sonar also resulted in apparent avoidance behavior with elevated stroke rates and swim speeds that lasted for about 1.5 hours after exposure (Stimpert *et al.*, 2014). Further controlled exposure experiments should provide more insight into the potential role of changes in dive behavior in the strandings of these whales.

Although baleen whales have not been associated with strandings after naval sonar exercises, it is notable that mid-frequency sonar may also affect their dive behavior. Controlled exposure experiments with blue whales (*Balaenoptera musculus*) resulted in disruption of some deep dives with cessation of feeding and high-speed swimming away from the sound source (Goldbogen *et al.*, 2013). Behavioral responses were complex, variable, and probably dependent on the level of received sound, the type of sound, and the behavioral state of the individual whale. Although the sonar exposure has not been associated with pathological events in baleen whales, disturbance of normal feeding behaviors (as in the beaked whales) has potential impact on energetics and foraging efficiency.

12.7.2 Anatomy, pathology, and physiology

As previously reviewed in this chapter, anatomical studies revealed that lungs in deep-diving kogiid whales were smaller, more compliant, occupied a smaller percentage of thoracic volume, and were associated with more vascular retia than the lungs of shallow-diving bottlenose dolphins (*Tursiops truncatus*) (Piscitelli *et al.*, 2010). Such differences are important in evaluation of the mathematical modeling of N_2 absorption and distribution. Similarly, the potential role of fat associated with the intrathoracic rete of cetaceans as an N_2 reservoir should also be considered (Blix *et al.*, 2013). Speaking of N_2 fat reservoirs, N_2 solubilities of whale blubbers vary and are different from typically assumed values (Koopman and Westgate, 2012, McClelland *et al.*, 2012).

In a novel approach to evaluate lung compression of seal and dolphin carcasses, a hyperbaric, computer tomographic scanning system was used to estimate depth of full lung compression (lung collapse) (Moore *et al.*, 2011). For seals, the estimated depth of collapse was similar to that predicted by Kooyman's earlier shunt studies, while, for dolphins, the estimated depth was deeper than that estimated previously in Ridgway and Howard's diving dolphin study (see Section 12.3). A deeper depth of lung collapse raised the question of whether there might be more gas exchange and N_2 absorption during dives than previously thought. These estimations were limited, however, by the use of carcasses (changes in tissue compliance) and by use of an indirect pixel density criterion to define "lung collapse" (atelectasis).

Ultrasound, computerized tomography, and magnetic resonance imaging have been used to search for intravascular bubbles and establish N_2 supersaturation as the underlying mechanism of bubble formation in cetaceans (Dennison *et al.*, 2012a, 2012b, Moore *et al.*, 2009). Blood N_2 supersaturation has been supported by detection of intravascular bubbles in stranded carcasses, carcasses of animals that died at depth in fishing nets, and live-stranded dolphins. Bubbles were especially prominent in the peri-renal region of dolphins. Interestingly, in live-stranded animals, these bubbles appeared to have no effect as most were successfully released. It was also notable that no bubbles were detectable in dolphins live-caught for health exams. This raised the question of whether bubbles in stranded animals were secondary to the stranding and to off-gassing of N_2 from tissues in the stranded animals since they could not return to depth as they normally would.

Among all these studies of possible bubble formation in cetaceans, there has been only one diving study with a living cetacean (Houser *et al.*, 2010). In a bottlenose dolphin trained to perform serial deep dives for a one-hour period, bubbles were not detected by ultrasound in both the portal and brachiocepalic vein. Post-dive N_2 levels were also not elevated in blood samples from the fluke, and heart rate profiles were characterized by rapid onset of severe bradycardias to 20–30 bpm (as reviewed previously in this chapter).

12.7.3 Models of N_2 uptake and distribution

In contrast to the lone study of a diving animal, most research related to diving physiology and the potential role of N_2 supersaturation in the pathology of the sonar-associated strandings has focused on mathematical modeling of N_2 absorption and distribution (Bostrom *et al.*, 2008, Fahlman *et al.*, 2006, 2007, 2009, Hooker *et al.*, 2009, Houser *et al.*, 2001, Kvadsheim *et al.*, 2012, Zimmer and Tyack, 2007). As emphasized by those authors, such models are of value because of the difficulty of conducting diving studies on cetaceans, let alone on beaked whales. The models can also provide hypotheses than might be tested eventually. Limitations in such models include lack of basic knowledge about the anatomy and physiology of the whales and the multiple assumptions involved. Indeed, N_2 uptake models for beaked whales have resulted in opposite conclusions – that shallow dives are the problem (Tyack *et al.*, 2011, Zimmer and Tyack, 2007) or that deep dives are the problem (Fahlman *et al.*, 2009, Hooker *et al.*, 2009). A most recent modeling study, based on actual dive profiles of whales exposed to sonar, concluded that beaked whales did not appear at increased risk of decompression sickness, but that sperm whales (*Physeter macrocephalus*) might be (Kvadsheim *et al.*, 2012). However, sperm whale strandings have not been reported in association with naval sonar exercises.

All mathematical models, however, do predict that N_2 supersaturation will occur, and this has been supported by the ultrasound bubble investigations of dead and stranded animals (Hooker *et al.*, 2012). A recent excellent review summarizes these findings, acknowledges that our understanding is still incomplete, and concludes that adaptations in marine mammals should be viewed as important not only in minimizing N_2 loading, but also in management of the N_2 load (Hooker *et al.*, 2012).

There are differences in the results and conclusions of some of these mathematical models for various reasons. Importantly, Fahlman and co-workers (Fahlman *et al.*, 2009) have pointed out that earlier models assumed instantaneous lung collapse and cessation of gas exchange at a threshold depth (Fahlman *et al.*, 2006, 2007, Houser *et al.*, 2001, Ridgway and Howard, 1979, Zimmer and Tyack, 2007). In other words, gas diffusion was assumed to linearly increase with depth until the threshold depth was reached, at which point it instantly stopped. Fahlman and colleagues have emphasized that it is much more probable that a pulmonary shunt gradually develops with depth due to progressive alveolar compression as Kooyman and Sinnett demonstrated in harbor seals and sea lions (Kooyman and Sinnett, 1982). A 100% shunt corresponds to complete alveolar compression, i.e., lung collapse. Although evident to physiologists for years, progressive shunting had not been incorporated into modeling efforts. It was with application of a mathematical model incorporating the development of a pulmonary shunt during dives (Bostrom *et al.*, 2008, Fahlman *et al.*, 2009) that it was predicted that the deep dives and not the shallow dives of beaked whales predisposed them to higher blood N_2 levels and increased risk of DCS (Hooker *et al.*, 2009). There are still limitations to these interpretations. The pulmonary shunt equations, for example, are based on a theoretical model of tracheal and alveolar compression, but not on actual data from a cetacean or even a marine mammal (Bostrom *et al.*, 2008).

In addition to the argument about pulmonary shunts, there are more limitations to these mathematical models. Many modeling studies assumed that gas exchange ceased in cetaceans at 70 meters, based on Ridgway and Howard's estimate from their diving dolphins (Ridgway and Howard, 1979). In part, that depth for lung collapse was based on the assumptions of (a) instantaneous cessation of gas exchange at the threshold depth; (b) symmetrical N_2 wash-in and wash-out profiles during descent and ascent; and (c) an N_2 tissue half-time calculated from back extrapolations of data collected from the beached dolphins 8–20 minutes after diving. Ridgway and Howard recognized these limitations, and more recent physiological data bring the 70-m depth for lung collapse into question. First, as pointed out by Fahlman *et al.*, and as demonstrated by Kooyman and Sinnett, it is much more probable that a pulmonary shunt gradually develops (Fahlman *et al.*, 2009, Kooyman and Sinnett, 1982). Second, tissue wash-in and wash-out rates as well as gas exchange in the lung are partly dependent on heart rate and perfusion, but heart rate profiles in diving dolphins are not symmetrical during descent and ascent as assumed in models (Houser *et al.*, 2010, Williams *et al.*, 1999). Third, heart rates of dolphins beached as in the original study are much lower than surface interval heart rates between dives (Houser *et al.*, 2010). Therefore, the N_2 tissue half-time may be less during surface intervals than during beaching. Consequently, the 70-m threshold, although it may be the only estimate available from a live animal, should be viewed with reservation. Indeed, the excised lung study by Moore previously reviewed in this chapter suggests the threshold depth for "lung collapse" is probably deeper. In addition, application of such data to models for deep-diving whales is also suspect, given the differences in lung compliance and relative lung volume between the bottlenose dolphin and deep-diving kogiid whales (Piscitelli *et al.*, 2010).

More recent anatomical and physiological research will undoubtedly make modeling more complex and lead to more model revisions. As already mentioned, blubber histology and N_2 solubility coefficients are not uniform among cetacean species, suggesting both that blubber blood flow may vary among species, and that N_2 solubilities are different from the commonly assumed olive oil coefficient (Koopman and Westgate, 2012, McClelland *et al.*, 2012). In addition, the actual diving lung volume is critical to modeling but remains a mystery. Arterial blood O_2 profiles suggest that gas exchange during free dives continues to deeper depths than during simulated dives in pressure chambers in both seals and sea lions (McDonald and Ponganis, 2012, Meir *et al.*, 2009). This would be consistent with larger inspired air volumes during free dives. In penguins, estimated diving air volumes increase with depth and are also greater than in simulated dives in pressure chambers (Sato *et al.*, 2002, 2011). Lastly, none of the models have considered the possibility of arterio-venous shunting during dives and the potential effect that would have on N_2 uptake and distribution. And yet arterializaton of venous blood during dives of elephant seals, sea lions, and emperor penguins suggests that such shunting may occur during dives (McDonald and Ponganis, 2013, Meir and Ponganis, 2009, Meir *et al.*, 2009).

Clearly, future modeling efforts would benefit from more anatomical and physiological data from cetaceans. Plans by several investigators to gather heart rate and other physiological data (Gutierrez-Herrera *et al.*, 2013) from free-diving whales would be especially valuable. Detailed investigation of physiological responses and N_2 kinetics in an actual diving animal are also needed to allow evaluation and refinement of mathematical models.

12.8 Avoidance of barotrauma in diving birds

Avoidance of tissue barotrauma in penguins is partly achieved by the absence of pneumatized bones. Not only does the lack of air space in the bone decrease buoyancy, but its absence also eliminates the potential for tissue barotrauma in an air-filled cavity. Although less studied in other diving birds, bone pneumaticity is absent or reduced in loons, cormorants, auks, and diving ducks (O'Connor, 2004, 2009, Casler, 1973, Smith, 2012). As in diving mammals, large venous sinuses in the middle ear and in the mucosa of the Eustachian tube in king penguins are considered to become engorged at depth and decrease the volume of the middle ear cavity to prevent barotrauma (Sadé *et al.*, 2008).

12.9 Avian lung structure and effects of pressure

When thick-billed murres (*Uria lomvia*) were found to dive as deep as 200 m, Croll and co-workers raised the question as to how these birds avoided collapse of a rigid lung at such depth (Croll *et al.*, 1992a). This was based on earlier findings that the air volume in the avian lung was about 10–15% of the combined air-sac and lung volume (Duncker, 1972, Scheid *et al.*, 1974). Based on this ratio, Croll *et al.* estimated that lung collapse should begin at about 70 m, much shallower than the deepest dives of the murres. Otherwise, pulmonary barotrauma or lung squeeze should occur. However, due to the small diameter and radius of curvature of the air capillaries in the avian lung, it has been argued that, once collapsed, the resulting high surface tension, despite the presence of surfactant, would prevent re-expansion of the air capillary (Duncker, 1972, 1974, Scheid *et al.*, 1974).

Croll and co-workers proposed several possible mechanisms to prevent pulmonary barotrauma in diving birds: (a) the lungs of deep-diving birds may be somehow resistant to collapse; (b) structural adaptations and/or perhaps different surfactant in diving birds might allow compression and re-expansion; or (3) the lungs in divers may comprise a smaller portion of the respiratory system. In 1999, Kooyman *et al.* proposed that engorgement of pulmonary blood capillaries during dives might obliterate the air space of the air capillary at depth, and, thus, both prevent pulmonary barotrauma and stop gas exchange at depth. This process would be analogous to the pooling of blood in the lungs and closed air spaces of human and marine mammals, as discussed earlier in both this chapter and Chapter 2.

The pulmonary blood capillary engorgement hypothesis of Kooyman *et al.* (1999) could prevent barotrauma as well as prevent gas exchange and the absorption of excess nitrogen. However, more recent data from chickens does not support this hypothesis. In comparison to mammals, avian pulmonary blood capillaries are essentially indistensible (Watson *et al.*, 2008). When pressure inside the capillary was increased from 0 to 25 cm H_2O, capillary diameter of the chicken increased only 13%, whereas that in the dog increased 125%. If pulmonary blood capillaries of diving birds are similar to those of chickens, the capillary engorgement hypothesis is unlikely. The relatively thickened blood gas barriers reported in Adélie and emperor penguins also suggest that the pulmonary blood capillary in penguins is not distensible. In order to further consider potential mechanisms which might prevent lung barotrauma or control gas exchange and nitrogen absorption at depth in diving birds, it is necessary to further review the structure and function of the avian lung.

In actuality, it is the air capillary that is most resistant to collapse. Compression studies of the duck lung revealed that the parabronchi and secondary bronchi have a finite compliance, but that lung compression did not alter arterial blood gases (Macklem *et al.*, 1979). This lack of change in arterial blood gases indicated that compression did not result in significant collapse of the air capillary or impairment of gas exchange. Therefore, what is critical for prevention of barotrauma in the avian diver's lung is the ratio of air-sac volume to the air capillary volume of the lungs. Theoretically, the numerator in this ratio would be more accurate if it were air-sac volume plus that portion of the air volume of the lung not in the air capillaries. In addition, dependent on the compliance of the trachea, air-sac air may also be distributed to the trachea in addition to the lung.

In the only quantitative morphometric study of the lung of a penguin in the literature (Maina and King, 1987), lung volume was 30 ml kg^{-1} in a 4.5 kg Humboldt penguin

(*Spheniscus humboldti*), and air capillary volume was only 18% of lung volume (Maina and Nathaniel, 2001). Although the Humboldt penguin is not a deep diver (Luna-Jorquera and Culik, 1999), it is notable that mass-specific air capillary volume in this penguin was at the low end of the range of avian values (Maina and Nathaniel, 2001). If the Adélie penguin (*Pygoscelis adeliae*) lung is similar to that of the Humboldt penguin lung, the maximum diving air volume of 200 ml kg^{-1} measured by Sato and co-workers (Sato *et al.*, 2002) would result in a 195:5 ratio of air-sac volume to air-capillary volume, based upon the morphometric data in the Humboldt penguin. This is sufficient to allow for a 39-fold compression of pulmonary air without risk of barotrauma to the air capillary. This is equivalent to 380 m depth, far greater than routine dive depths of this species. Although these calculations did not consider the volume of a potentially rigid trachea, such morphometric measurements are an essential first step to evaluating mechanisms for the avoidance of pulmonary barotrauma. The only other study of penguin lung morphometry was non-quantitative, but reported a thickened blood gas barrier in the emperor penguin just as was found in the Humboldt penguin (Welsch and Aschauer, 1986). Presumably, this thickened barrier lends support to the air capillary wall.

To further evaluate this hypothesis of a protective ratio of air-sac volume to lung air-capillary volume, lung and maximum air-sac volumes have been measured in three penguin species with use of computerized tomography and three-dimensional anatomic reconstructions (Ponganis *et al.*, 2015). Lung volumes scaled to body mass according to allometric predictions, but maximum air-sac volumes were two to three times predicted values. Based on the penguin lung morphometry data of Maina and King, the resulting air-sac to air-capillary ratios could easily account for pulmonary baroprotection in even the deepest dives of Adélie, king (*Aptenodytes patagonicus*), and emperor penguins. However, if the parabronchi and trachea are also incompressible, the air sac to combined air capillary + parabronchi + tracheal volumetric ratio was inadequate to account for baroprotection in the deeper dives, especially in king and emperor penguins. A decrease in the volume of the parabronchi and trachea may occur due to physical compression, vagally induced bronchoconstriction (Barnas *et al.*, 1978, King and Cowie, 1969), and possibly venous engorgement of the airway wall. Before final conclusions can be reached, species-specific lung morphometrics, tracheal volumes and compliances, and further verification of actual start-of-dive air-sac volumes are needed. In addition, although a large air sac to lung volume ratio may provide protection against barotrauma, this mechanism does not resolve the question of how penguins avoid excess nitrogen absorption at depth.

12.10 Gas exchange in penguins at depth: pulmonary shunts and heart rate

One mechanism which could limit gas exchange and prevent excess nitrogen absorption at depth would be the development of ventilation–perfusion mismatch, i.e., a pulmonary shunt, at depth. In this regard, the lung of the Humboldt penguin does contain a significant amount of blood; pulmonary capillary blood volume composes 51% of the exchange tissue, about one-third more than that in other birds (Maina and Nathaniel, 2001). It is also notable that a 28% intrapulmonary shunt has been estimated in emperor penguins at rest on the basis of blood and air-sac gas analyses (Meir and Ponganis, 2009). That value is 4–8 times greater than values reported in other birds. During the tachycardia and hyperventilation prior to a dive, the calculated shunt decreased to 14%. Therefore, it is not inconceivable that a significant shunt gradually develops later during the prolonged apnea and progressive bradycardia of a deep dive.

Changes in gas exchange due to changes in heart rate and pulmonary blood flow are supported by heart rate and blood P_{O2} profiles in emperor penguins (Meir *et al.*, 2008, Ponganis *et al.*, 2010a, Wright *et al.*, 2014). During the first few minutes of dives, there was always a transient relatively elevated heart rate, sometimes to as high as 140 bpm (see Figs 5.13d, 5.15, 5.16). During this same time period of dives, arterial and often venous P_{O2} rise, consistent with enhanced gas exchange, and transfer of lung O_2 into the blood O_2 store early in the dive (see Fig. 11.11). However, after this period of relatively high heart rates and enhanced gas exchange early in a dive at shallow depth, heart rate progressively declined later in the dive. In deep dives of emperor penguins at sea, heart rate could be as low at 10 bpm at maximum depth (Wright *et al.*, 2014). Such severe bradycardia most probably contributes to decreased gas exchange at depth in the emperor penguin as even moderate decreases in heart rate has been associated with decreased pulmonary gas exchange in human divers (Andersson *et al.*, 2002). This hypothesis remains to be investigated further in the future.

In summary, prevention of barotrauma and regulation of gas exchange at depth in diving birds remain a mystery. However, the hypotheses advanced here, a large air sac to air capillary volume ratio and development of an intrapulmonary shunt, are possible solutions and may account for the depths achieved by many diving birds.

Whether changes in surfactant might play a role in pressure tolerance of the lung of the avian diver is unknown. There have been no studies. Surfactant function in birds is reviewed in Chapter 13.

12.11 Blood N_2 levels in penguins

Similar to marine mammals, there have been few studies, only three, in which blood N_2 levels have been determined in penguins. During simulated dives of Adélie and gentoo penguins in a pressure chamber (Kooyman *et al.*, 1973b), arterial P_{N2} quickly reached values of 2.5 ATA to almost 4 ATA during dives to 4 ATA depth. In dives to 7.8 ATA, arterial P_{N2} was 2.5–5.0 ATA. Unlike seals, arterial P_{N2} did not decline during the dive (Fig. 12.4). There was no evidence in the 4 ATA (30 meters) deep dives that pulmonary gas exchange had significantly declined. In the 7.8 ATA (68 meters) dives, the relatively low arterial P_{N2} values suggested a pulmonary shunt, which was estimated to be 50–60% on the basis of a pulmonary shunt study reported in that paper.

In a pressure chamber study of simulated dives in king penguins (Ponganis *et al.*, 1999a), venous P_{N2} was less than 2.8 ATA during and after dives as deep as 11.2 ATA (102 m). In free-diving emperor penguins at an isolated dive hole (Ponganis *et al.*, 2009), arterial P_{N2} averaged 2.1 ATA early in dives in samples collected at 25–32 m (3.5–4.2 ATA); venous samples provided a similar value.

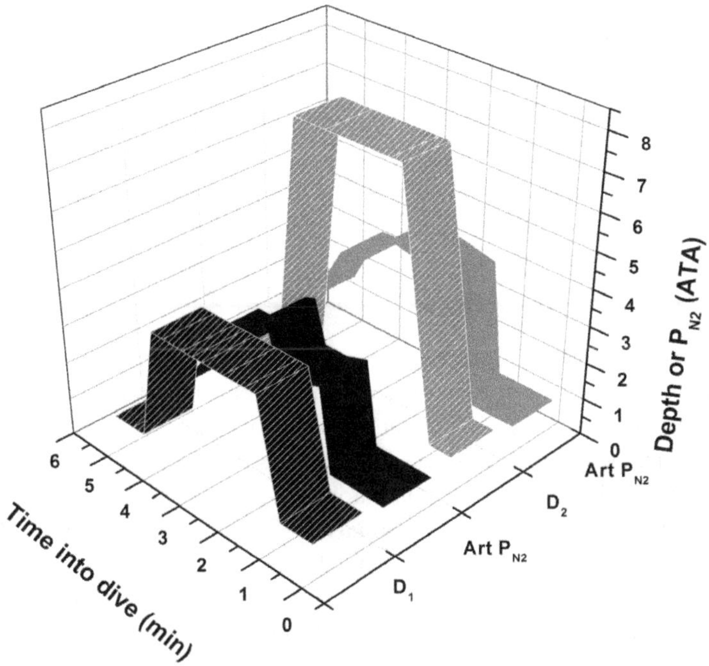

Figure 12.4 Arterial (Art) P_{N2} and depth profiles (D_1, D_2) during simulated dives of Adélie and gentoo penguins (*Pygoscelis adeliae, P. papua*) in a pressure chamber demonstrated significant increases in blood N_2, consistent with maintenance of gas exchange at depth. Pressure is in atmospheres absolute (ATA). Adapted from Kooyman et al. (1973b).

Although the results of these last two studies are indicative of continued gas exchange during dives, interpretation is limited due to lack of arterial data in the king penguin paper, and both a small sample size and depth range in the emperor penguin study. Further blood N_2 investigations in diving penguins require technological development of smaller blood sampling devices or an intravascular N_2 sensor.

13 Bio-medical relevance

The cellular and molecular mechanisms underlying the diving capacities of marine mammals and seabirds are relevant to both the understanding and treatment of human pathologies. For example, facial immersion in ice water, an old treatment to suppress supraventricular tachycardia, was based in part on the seal's diving bradycardia which is intensified by wetting of the face (Hamilton *et al.*, 1979, Kawakami *et al.*, 1967). In Chapter 6, the basic principles of myocardial oxygen supply and demand in the treatment of coronary artery disease were illustrated by the cardiovascular adaptations of the diving seal. In this chapter, the significance and potential application of diving adaptations to five topics in clinical medicine and basic cell biology will be examined. The subjects include hypoxemic/ischemic tolerance and avoidance of reperfusion injury; lung surfactant function; hypoxic pulmonary vasoconstriction; the regulation of myoglobin (Mb) gene expression; and the biophysics of Mb (and oxygen flux) within the muscle cell.

13.1 Avoidance of reperfusion injury and hypoxemic/ischemic tolerance

Reperfusion injury occurs upon the restoration of blood flow and O_2 delivery to tissues that had been previously ischemic (no blood flow). In the clinical situation, this most commonly occurs with the thrombolytic (clot lysis) and stent treatments of stroke and heart attack victims. It can also occur in tissues in which blood flow is temporarily occluded for surgical procedures, as well as in transplanted organs after implantation. The primary mechanisms responsible for such damage after the return of blood flow are considered to be the generation of reactive oxygen species (ROS), including superoxide radicals, hydrogen peroxide, hydroxyl radicals, peroxynitrite, and hyperchlorite. These oxidizing agents can result in DNA damage, protein nitration and fragmentation, and lipid peroxidation (Powers and Jackson, 2008). One such mechanism of superoxide production results from accumulation of hypoxanthine due to degradation of adenosine triphosphate (ATP) during the ischemic period. During reperfusion, excess hypoxanthine and O_2 react via the xanthine oxidase pathway to yield superoxide radicals.

Protection against reperfusion injury is afforded by ROS scavenging enzymes and intracellular antioxidant compounds (Powers and Jackson, 2008). Several significant enzymes include superoxide dismutase (SOD), which scavenges superoxide radicals; catalase (CAT), which breaks down hydrogen peroxide; glutathione peroxidase (GPX), which consumes hydrogen peroxide with conversion of glutathione to glutathione disulfide; glutathione-S-transferase (GST), which conjugates reactive intermediates with glutathione; and glutathione reductase (GR), which recycles glutathione from glutathione disulfide. Important antioxidant compounds include glutathione, α-lipoic acid, ascorbate, α-tocopherol, and carotenoids.

The tolerance of seal tissues to ischemic as well as hypoxemic (low O_2) insult was demonstrated in the 1960s by Halasz *et al.*, who found that seal kidneys, in contrast to dog kidneys, recovered function after an hour of complete ischemia in the absence of any protective measures, including cooling (Halasz *et al.*, 1974). Mitochondrial function of hepatocytes also appears to be better preserved in seals than rats during complete ischemia (Hochachka *et al.*, 1988). Seal brain cell preparations continue to function under severe hypoxic conditions (Folkow *et al.*, 2008). Tolerance to such low blood flow and O_2 tensions were also demonstrated in intact animals with forced submersion studies primarily conducted by Elsner and co-workers (Blix *et al.*, 1983, Elsner, 1965, Elsner and Gooden, 1983, Elsner *et al.*, 1964b , 1966a, 1970b, 1970c, 1971, 1985, Kerem and Elsner, 1973, Zapol *et al.*, 1979). More recently, evidence for such extreme hypoperfusion and tolerance of hypoxemia has also been demonstrated in free-diving animals (see Chapter 5). Heart rates less than 10 bpm can occur during free dives of seals, sea lions, and penguins (McDonald and Ponganis, 2014, Meir *et al.*, 2008, Thompson and Fedak, 1993, Wright *et al.*, 2014). One seal even performed a 14-min dive with an average dive heart rate near 7 bpm. Similarly, arterial P_{O2} and hemoglobin saturations of seals and penguins can reach levels considered dangerous and potentially catastrophic in humans (Meir and Ponganis, 2009, Meir *et al.*, 2009, Qvist *et al.*, 1986) (see Chapter 10).

The potential role of antioxidants in the ischemic tolerance of organs and tissues of diving animals was first investigated by Elsner and colleagues. They found that heart and kidney hypoxanthine concentrations were significantly less in seals than in pigs after 30 min of warm ischemia (Elsner *et al.*, 1998). The lower hyoxanthine concentrations in the seal could account for lower superoxide radical production during reperfusion and contribute to the ischemic tolerance of the seal tissues. It was postulated that less hypoxanthine accumulation might be secondary to enhanced glycolytic capacity in seal tissue as demonstrated by elevated glycogen concentrations in Weddell seals (Kerem *et al.*, 1973). The increased buffering capacity of seal tissues (Castellini and Somero, 1981) would also be consistent with this hypothesis. Notably, in this as well as later studies, antioxidant enzyme activity patterns were not as predictable. Although SOD activity was higher in seal than pig hearts as expected, the reverse occurred in the kidneys. This difference in tissue SOD patterns was attributed to the fact that the seal heart continued to work during ischemia, whereas kidney function ceased.

Investigation of superoxide radical production and lipid peroxidation (formation of thiobarbituric acid reactive substances (TBARS)) in seal and pig tissues (heart, kidney, and liver) also demonstrated unexpected findings; neither process was reduced in seals (Zenteno-Savín *et al.*, 2002). Thus, the capacity to produce superoxide radicals and the potential for lipid peroxidation were not decreased in the seal. However, in this study, tissue antioxidant capacity in each tissue was greater in seals than in pigs. This presumably afforded protection against any deleterious effects of superoxide radicals during the reperfusion period.

Further studies of antioxidant enzyme activities in seals versus pigs revealed different profiles in different tissues (Vázquez-Medina *et al.*, 2006). In the seal heart, SOD, GPX, and GST were elevated; in the seal liver, CAT was elevated; and in seal muscle, GPX was elevated. The lack of a consistent profile may well be due to differences in perfusion, metabolic rates, and oxygen fluxes in individual tissues. In addition, antioxidant enzymes may not be the sole protection against reperfusion injury. Glutathione levels were elevated in seal tissues (Vázquez-Medina *et al.*, 2007), and these authors have suggested that elevated glutathione concentrations and high recycling rates of glutathione are an important component of protection against reperfusion injury in seals.

In Adélie (*Pygoscelis adeliae*) and emperor penguins (*Aptenodytes forsteri*), blood antioxidant capacity was elevated in comparison to those in flying Antarctic birds (Corsolini *et al.*, 2001). Investigations in emperor penguins have also revealed that the capacity for superoxide production and the potential for lipid peroxidation are not reduced in liver and muscle in comparison to a variety of flying marine birds (Zenteno-Savin *et al.*, 2010). Although these findings were similar to those in seals, the antioxidant enzymes profiles of emperor penguins differed from those in the seals. CAT and GST were elevated in both the muscle and liver of the penguins. In addition, GPX was elevated in the liver. It would appear that the scavenging of hydrogen peroxide by CAT, the conjugation of reactive intermediates with glutathione by GST, and the conversion of peroxides by GPX are particularly important in protection against reperfusion injury in penguins. These differences in antioxidant mechanisms between seals and penguins may be secondary to their phylogenies and muscle fiber types, as well as to differences in the degree of hypoxemia, ischemia, and O_2 flux during routine dives. A thorough description of antioxidant mechanisms in diving mammals and birds is recommended to readers (Zenteno-Savín *et al.*, 2011).

The molecular mechanisms underlying the avoidance of reperfusion injury in diving mammals and birds remain incompletely understood. Other mechanisms which could protect against ischemic and hypoxemic damage were reviewed in Chapter 11, and include the elevated glycogen concentrations and enhanced glycolytic and buffering capacities already mentioned, hypothermia (Odden *et al.*, 1999, Scholander *et al.*, 1942b), increased capillary density, increased mitochondrial density and decreased O_2 diffusion distances (Fuson *et al.*, 2003, Kerem and Elsner, 1973, Watson *et al.*, 2007), and changes in the concentration and function of the O_2 binding proteins cytoglobin and neuroglobin (Burmester and Hankeln, 2009, Burmester *et al.*, 2000, Williams *et al.*, 2008). In muscle, elevated myoglobin concentrations may also be protective in that myoglobin has peroxidase activity and can scavenge ROS as well as nitric oxide (Ordway and Garry, 2004).

Lastly, ischemic preconditioning may also occur in these animals. In this process, a tissue is conditioned by brief exposure to ischemia to better tolerate prolonged ischemia (Hausenloy and Yellon, 2011, Murry *et al.*, 1986). Ischemic pre-conditioning is associated with decreased rates of ATP depletion, glycogen breakdown, and lactate accumulation in the heart (Murry *et al.*, 1990). The intracellular mechanisms underlying such myocardial protection are complex and considered to include maintenance of calcium homeostasis through preservation of ryanodine receptors and Ca^{2+} pump ATPase in the sacroplasmic reticulum, activation of mitochondrial K^+ dependent ATPase, and reduced activation of the mitochondrial permeability transition pore (Chen *et al.*, 2006, Murphy and Steenbergen, 2008). Analogous processes of metabolic arrest and ion channel arrest also appear to occur in lower vertebrates (Bickler and Buck, 2007, Hochachka and Guppy, 1987). Other protective mechanisms may involve mitochondrial aldehyde dehydrogenase-2, the activation of which reduces infarct size in ischemic models (Chen *et al.*, 2008, 2010). If such protective mechanisms against ischemia are enhanced in any higher vertebrates, diving mammals and birds should be excellent candidates for investigative models.

13.2 Lung surfactant function

The routine collapse and re-expansion of marine mammal lungs during dives are considered the primary mechanism by which these animals avoid excess nitrogen absorption at depth. In elephant seals (*Mirounga angustirostris*), for example, compression-induced collapse of the lung alveoli is considered to occur by about 50 m depth (Kooyman *et al.*, 1973b). During the routine 400-m deep, 20-min long dives of these animals, the lungs remain collapsed. During the final 50 m of ascent of these dives, the alveoli must fully expand and regain function so that optimal gas exchange can occur during the two-min surface interval of this seal before it descends again. Such rapid re-expansion and resumption of alveolar function has led to the question as to whether there are adaptations in the pulmonary surfactant of diving mammals.

Pulmonary surfactant is a mixture of phospholipids, cholesterol, and proteins secreted by type II cells in the alveoli of mammalian lungs. The typical composition of mammalian surfactant is 90% lipid and 10% protein (Postle *et al.*, 2001). Cholesterol constitutes about 10% of surfactant by weight and is considered to contribute to the fluidity of surfactant by increasing the separation between phospholipid molecules (Daniels and Orgeig, 2003). Phospholipid, the primary component (80%) of surfactant, consists primarily of phosphatidyl choline (70–80% of phospholipid), phosphatidylglycerol (10–20%), and glycerolphosphatidylinositol (<5%) (Postle *et al.*, 2001). There are four surfactant proteins, A, B, C, and D. Surfactant protein A (SP-A) and SP-D are involved in immunological defense responses, while the hydrophobic SP-B and SP-C molecules interact with surfactant lipids, preventing surfactant absorption at the air–liquid interface. SP-B is essential for life; defects are also associated with fatal congenital alveolar proteinosis (Clark *et al.*, 1995, Lin *et al.*, 1998).

One of the primary functions of surfactant is to regulate the surface tension of the alveolar surface during inflation and deflation of the lung. As reviewed by Daniels and Orgeig (2003), the regulation of surface tension is based on compression of the surfactant layer during expiration, which "squeezes out" cholesterol and unsaturated phospholipids. This process concentrates disaturated phospholipids (i.e., phosphatidylcholine), tightly compressing their fully saturated fatty acid chains so that water is excluded and surface tension decreases. Surfactant has also been considered to have an "antiadhesive" function, especially in non-mammals. In this situation, as two surfaces, apposed against each other in a collapsed lung, begin to pull apart during lung expansion, surfactant lipids are considered to rise to the surface at the air–liquid interface, thereby lowering surface tension and decreasing the work required to expand the lung.

Surfactant abnormalities are associated with a variety of human pathologies. For example, absence of surfactant in premature infants results in neonatal respiratory distress syndrome. This severe lung disease requires ventilatory support and steroid treatments to increase surfactant production. Since the 1980s, outcomes have been improved with application of artificial surfactant (Jobe, 1993). In adults, acute respiratory distress syndrome (ARDS) can be secondary to pneumonia, chemical inhalation, drowning, burn injury, and multiple organ failure. Although there are significant deficiencies of SP-B and SP-A in ARDS, treatments with artificial surfactant have met with varied results (Hallman *et al.*, 1982, 2001, Spragg *et al.*, 2003).

Surfactant samples can be obtained by bronchoscopic collection of washings of the distal airways (bronchoalveolar lavage). Analysis of such fluid from young seals and sea lions revealed a greater concentration of phospholipid and more fluidic species of phosphatidylcholine than in terrestrial mammals (Spragg *et al.*, 2004). This was consistent with more rapid spreading during alveolar re-expansion. Notably, surfactant from elephant seals, the deepest divers, produced moderately elevated minimum surface tension measurements. Poor surface activity and minimal reduction of surface tension were also later observed in analyses of lavage fluid from excised lungs of other pinniped species (Miller *et al.*, 2006a). In addition, further analyses of surfactant composition of pinnipeds revealed a relative decrease in anionic phospholipids, an increase in short-chain phospholipids, and a decrease in surfactant protein B (Miller *et al.*, 2005, 2006b,). Based on all these findings, it has been suggested that the primary function of surfactant in deep-diving mammals may be an anti-adhesive function rather than only surface tension reduction (Foot *et al.*, 2006). This anti-adhesive function, as described above, has been considered to include the reduction of adhesive interactions between respiratory surfaces due to variation in surface tension.

Regulation of surfactant production has also been investigated in diving mammals. Phosphatidylcholine secretion of alveolar type II cells from California sea lions was found to be less sensitive to increased pressure than those from sheep (Miller *et al.*, 2004a). In addition, it has been demonstrated that, in contrast to adult terrestrial mammals, leptin is expressed in the lung tissue of adult gray seals (*Haliochoerus grypus*) and harbor seals (Hammond *et al.*, 2005). Because leptin plays a role in stretch-induced surfactant production by alveolar type II cells during fetal development (Torday and Rehan, 2002), it may be similarly involved in surfactant production in deep-diving mammals that undergo routine lung collapse and re-expansion. Thus, although changes in the composition of marine mammal surfactant have been relatively subtle, it may be that the production rate and quantity of surfactant available are also critical in marine mammals.

Surfactant composition and function in birds, which have tubular air capillaries in the lung, are different from those of surfactant in the mammalian alveoli (Bernhard *et al.*, 2001, 2004, Orgeig *et al.*, 2007). It has been suggested that its major function may be to facilitate airflow rather than to prevent collapse. Avian surfactant is more similar to the surfactant in the mammalian bronchial airway, another tubular structure. Low surface tensions cannot be achieved with dynamic compression studies with either mammalian airway or avian surfactant, whereas minimal tension is easily reached with alveolar surfactant. The composition of avian surfactant is also different; avian SP-B content is equal to or greater than that in mammals, while SP-A and SP-C are apparently absent. Furthermore, dipalmitoyl phosphatidylcholine is present in both alveolar and avian surfactant, but palmitoylmyristoyl phosphatidylcholine and palmitoylpalmitoleoyl phosphatidylcholine, both present in alveolar surfactant, are absent in avian surfactant. It is postulated that the latter two species of phosphatidylcholine are important in the dynamics of alveolar surfactant compression. Under dynamic cycling conditions, surface tension was not reduced significantly in either duck or chicken surfactant in comparison to that in the pig. These differences in function and composition between alveolar and avian surfactants have been attributed to the relatively rigid structure and lack of compressibility of the avian air capillary in contrast to the mammalian alveolus (Duncker, 1974, Watson *et al.*, 2008, West *et al.*, 2010). Investigation of surfactant from deep-diving birds such as emperor penguins might yield further insight into surfactant function of the penguin at depth.

13.3 Hypoxic pulmonary vasoconstriction

Hypoxic pulmonary vasoconstriction (HPV) is the process wherein a lowered O_2 concentration in the lung alveoli induces constriction of local pulmonary arteries (Moudgil *et al.*, 2005, Sommer *et al.*, 2008). HPV appears to occur in all mammals when alveolar P_{O2} is in the range of 20–60 mm Hg. In a healthy state, HPV represents a mechanism of matching ventilation and perfusion in the lung, minimizing intrapulmonary shunt, and optimizing gas exchange. Poorly ventilated alveoli with low O_2 content will receive less blood flow due to constriction of local branches of the pulmonary artery, while well-ventilated alveoli will receive more flow. However, global alveolar hypoxia, which can occur secondary to respiratory disease, sleep apnea, and high altitude, can result in widespread pulmonary arterial vasoconstriction and a significant increase in pulmonary vascular resistance (PVR). The elevated PVR increases the workload of the heart and can lead to right ventricular hypertrophy and eventually right heart failure.

The mechanisms underlying the HPV response are incompletely understood and considered to involve primarily the smooth muscle cells of the pulmonary artery (Aaronson *et al.*, 2006, Moudgil *et al.*, 2005, Sommer *et al.*, 2008). It has been hypothesized than an oxygen sensor (possibly mitochondria or nicotinamide adenine dinucleotide phosphate oxidases) produce either an increase or a decrease in reactive oxygen species (especially H_2O_2), which, in turn, inhibit voltage-gated K^+ channels, leading to membrane depolarization and calcium entry via voltage-dependent L-type channels. Increased cytoplasmic calcium secondary to calcium entry and/or calcium-induced release of stored calcium then leads to contraction and constriction of the pulmonary arterial wall. Many steps in this process are still undefined and debated.

In regard to diving mammals, the long breath holds and extremely low post-dive end tidal P_{O2}s of Weddell seals (Kooyman *et al.*, 1973a, Ponganis *et al.*, 1993a) raise the question of how the seal heart can adapt to an increase in pulmonary vascular resistance and right ventricular workload if HPV occurs during a dive. This would also be further complicated by elevated P_{CO2} and acidosis during a dive, which can also increase pulmonary vascular resistance (Barer and Shaw, 1971, Hyman and Kadowitz, 1975, Viles and Shepherd, 1968). Notably, pulmonary vascular resistance and right heart afterload would be maximal near the end of the dive, at a time when arterial hypoxemia is greatest and myocardial oxygen demand is additionally increased by the ascent tachycardia (increased work rate and decreased perfusion time as reviewed in Chapter 5).

The role of HPV in diving mammals is unclear. During sleep apnea of elephant seals, pulmonary artery pressures do not rise significantly despite the maintenance of cardiac output during these prolonged breath holds (Ponganis *et al.*, 2006a, 2006b). In addition, except for the hooded seal, the wall of the right ventricle in several seal species is not especially thick (not hypertrophied) (Drabek, 1975, 1977, Drabek and Burns, 2002). However, the right ventricular cavity has been reported to be large, and the ventricle has been hypothesized to dilate in response in an increase in pulmonary vascular resistance (Smodlaka *et al.*, 2008).

Recently, Olson and co-workers decided to extend their research on *in vitro* responses of pulmonary artery rings to California sea lions (*Zalophus caifornianus*) because of the debate over the mechanism of HPV and the question as to how significant increases in PVR might be avoided in a diving pinniped. Their results demonstrate novel responses in the sea lion and provide evidence for a new possible intracellular oxygen sensor (Olson *et al.*, 2010). On exposure to hypoxia, bovine pulmonary arteries demonstrated the expected constriction, whereas the pulmonary arteries of sea lions dilated. This hypoxic dilation had not been seen in the pulmonary arteries of any other mammal tested. Furthermore, exposure of the pulmonary artery rings to hydrogen sulfide (H_2S) resulted in the same responses – constriction in the cow and dilation in the sea lion. It was also found that H_2S was formed in the tissues of both animals in the absence of oxygen, but not in oxygen's presence. Thus, this study provides a mechanism by which pinnipeds may avoid the complications of HPV, and also, just as significantly, provides evidence for another intracellular oxygen sensor – O_2-dependent H_2S metabolism. More investigations are clearly needed, but this research again demonstrates how the physiological adaptations of a diving mammal are relevant to the understanding and potential treatment of human pathologies.

Another mechanism which could contribute to vascular dilation during hypoxia is an increase in the production of nitric oxide, which promotes vascular relaxation through its activation of guanyl cyclase and subsequent cGMP-mediated smooth muscle relaxation (Carvajal *et al.*, 2000, Ignarro *et al.*, 1987). Increased nitric oxide formation during hypoxemia has been postulated in diving mammals because of elevated blood nitrite levels found in porpoises (Soegaard *et al.*, 2012). Nitrite is a substrate for potential nitric oxide formation by deoxyhemoglobin and its nitrite reductase activity (Flögel *et al.*, 2010, Hill *et al.*, 2010, Lei *et al.*, 2013). The high concentrations of hemoglobin in elephant seals and the routine development of significant hemoglobin deoxygenation during their breath holds (Meir *et al.*, 2009, Stockard *et al.*, 2007) support this suggestion that elevated nitric oxide levels also contribute to the lack of pulmonary vasoconstriction in these animals. Elevated concentrations of nitric oxide in the exhalations of dolphins are also consistent with this hypothesis (Yeates *et al.*, 2014).

Carbon monoxide, which is elevated in some seals, is also a vasodilator. Prevention of pulmonary hypertension in llamas (*Lama glama*) exercising at high altitude has been attributed to production of carbon monoxide in their pulmonary vasculature (Herrera *et al.*, 2008). And although smoking is deleterious during pregnancy, the incidence of pre-eclampsia (hypertension of pregnancy) is lower in smokers than non-smokers, again suggesting that carbon monoxide can act as a vasodilator (Conde-Agudelo *et al.*, 1999). Thus, it is possible that carbon monoxide may also contribute to the lack of pulmonary hypertension during the long breath holds of these seals. In addition, in a pregnant mouse model, inhaled carbon monoxide resulted in increased uterine blood flow and increased size of uterine blood vessels (Venditti *et al.*, 2013). It is notable that uterine blood flow is maintained in pregnant Weddell seals even during the extreme bradycardia and peripheral vasoconstriction of forced submersion (Liggins *et al.*, 1980). In addition to differences in sympathetic nerve fiber distribution, the carbon monoxide in elephant seals as well as the possible elevations in nitric oxide may contribute to the maintenance of placental blood flow and oxygen delivery to the fetal seal during the long dives these seals routinely exhibit during their at-sea pregnancies.

13.4 Regulation of myoglobin production

The concentration of myoglobin in the best diving mammals and birds are 20–30 times the levels found in their non-diving counterparts (see Chapter 4). The molecular mechanisms responsible for the production of such extreme concentrations of myoglobin in divers are unknown. Apart from minor substitutions, the myoglobin genes in two marine mammals, the gray seal (*Halichoerus grypus*) and sperm whale (*Physeter macrocephalus*), are basically similar to those in other mammals (Blanchetot *et al.*, 1983, Garry *et al.*, 2003, Weller *et al.*, 1984). In mammals, the gene is composed of three exons and two entrons. In a terrestrial mammal such as the mouse, regulation of the production of myoglobin is considered to be dependent on the rate at which myoglobin messenger RNA (mRNA) is transcribed from the myoglobin gene (Kanatous *et al.*, 2009, Ordway and Garry, 2004, Underwood and Williams, 1987). In the developing mouse, it has been proposed that higher rates of myoglobin mRNA transcription result from increased neural activity (muscle maturation and innervation) and increased muscular activity (locomotion and muscle contraction). Both of these activities lead to elevated intracellular calcium concentrations, which in turn increase the production of the enzyme, calcineurin. Calcineurin then dephosphorylates the transcription factor, NFAT (nuclear factor of activated T cells), allowing it to migrate into the cell nucleus, where, in conjunction with other transcription factors, it regulates the expression of the myoglobin gene (Garry *et al.*, 2003, Chin *et al.*, 1998). This mechanism is considered to underlie the production of myoglobin during the early development of the neonatal mouse as its muscles mature and locomotory activity begins.

Hypoxia in combination with exercise has also been found to enhance intracellular calcium flux and increase myoglobin mRNA production via a similar pathway involving calcineurin (Kanatous *et al.*, 2009). It has been proposed (Kanatous *et al.*, 2009, Wittenberg, 2009) that this combination of hypoxia and exercise accounts for the increase in muscle myoglobin content in subjects undergoing hypoxic exercise training protocols (Terrados *et al.*, 1990, Vogt *et al.*, 2001). Such a mechanism involving both muscle activity and tissue hypoxia would also account for the fact that myoglobin content is elevated in cardiac muscle but not in skeletal muscle in the fetus, or in adult animals exposed only to hypoxia, but not exercise (Kanatous *et al.*, 2009, Longo *et al.*, 1973, Tipler *et al.*, 1978). This calcineurin-based mechanism would also be consistent with the gradual increase in myoglobin content in developing seal pups as they begin swimming and breath-holding during their initial diving activity (Thorson, 1993, Thorson and Le Boeuf, 1994).

Despite the remarkable advances in our understanding of the regulation of myoglobin production in terrestrial mammals, there has been little investigation of these mechanisms in diving animals. However, the expression of the myoglobin gene appears to be different in divers versus non-divers. In contrast to mice, which have little skeletal muscle myoglobin at birth, muscle myoglobin content in marine mammal neonates is already elevated at levels four to five times that in adult terrestrial mammals (Burns and Castellini, 1996, Burns *et al.*, 2005, 2007, Noren *et al.*, 2001, 2005, Petrov, 1985, Thorson, 1993, Thorson and Le Boeuf, 1994). In marine mammals, the relative tissue hypoxia *in utero* during dives of pregnant seals may provide the stimulus for myoglobin mRNA transcription during fetal development. However, the molecular mechanism underlying that production of myoglobin in the fetal pinniped is probably different from that described in mice by Kanatous and co-workers (2009) in that they found that hypoxia alone did not increase myoglobin production in mouse skeletal muscle. Muscle exercise was required, and the addition of hypoxia accentuated myoglobin production.

The Kanatous research team has further explored the early development of myoglobin production in cultured Weddell seal muscle cells, and found that an initial increase in myoglobin concentration was achieved with hypoxic exposure and lipid supplementation (De Miranda *et al.*, 2012). Exercise was not required. It was concluded that early regulation of myoglobin production in Weddell seal muscle was different than that in the mouse, but that another stimulus (perhaps exercise) was still needed to elevate myoglobin concentrations to adult levels.

One other potential factor in regulation of the fetal myoglobin production in seals is the degree of hypoxic exposure in the pinniped fetus. Oxygen levels in the fetal seal may be much lower than those in hypoxic terrestrial models. Although placental blood flow is maintained during dives, and the greater O_2 affinity of hemoglobin in the fetal seal optimizes O_2 transport, adult arterial P_{O2} and hemoglobin saturation during dives reach exceptionally low levels (Elsner, 1965, Elsner *et al.*, 1970a, Liggins *et al.*, 1980, Meir *et al.*, 2009, Qvist *et al.*, 1981, 1986). The effects of such low P_{O2}s on molecular regulatory mechanisms during fetal development remain to be investigated.

In contrast to pinnipeds and dolphins, significant fetal skeletal muscle myoglobin production does not occur in another excellent diver, the emperor penguin. Notably, and also similar to a terrestrial mammal, the developing penguin inside its egg does not experience the hypoxia of diving. In penguin chick carcasses with body masses near birth weight, the pectoral muscle is as white as that of chicken pectoral muscle (Ponganis, personal observation). By three months of age, chicks have myoglobin levels equivalent to that in non-diving birds, but only about one-quarter that found in the neonatal seal (Ponganis *et al.*, 1999b). These mechanisms for increased Mb production may also apply to other birds. Thus, the elevated myoglobin levels in fetal and newborn pinnipeds contrast with the extremely low levels found in both newborn terrestrial mammals and newly hatched penguin chicks.

In light of the calcineurin-based regulation of myoglobin production discovered in mice, it is remarkable that pectoral muscle myoglobin content of the emperor penguin chick continues to increase during the first six months of life without undergoing the exercise and hypoxia of diving. Emperor penguins do not enter the water until the time of fledging, at about six months of age, at which time the pectoral myoglobin concentration is about one-third the adult level (Ponganis *et al.*, 1999b). This is equivalent to that in newborn pinnipeds, and is four to five times that in non-diving birds and mammals. Certainly, the innervation and differentiation/maturation of muscle fibers in early development may contribute to a calcineurin-based increase in myoglobin. But the only pectoral muscle activity of chicks prior to fledging is occasional wing flapping.

Non-shivering thermogenesis has been proposed as a mechanism for increased contractile activity and eventual myoglobin production in penguin chicks (Noren *et al.*, 2001). Non-shivering thermogenesis occurs in experimentally cold-adapted 6–9-month-old king penguin (*Aptenodytes patagonicus*) chicks and is associated with increased ryanodine receptors and Ca^{2+} ATPase activity in the sacroplasmic reticulum (Bicudo *et al.*, 2001, Duchamp *et al.*, 1989, 1991, 1993, Dumonteil *et al.*, 1993, Meis *et al.*, 2005, Mozo *et al.*, 2005). In addition, once the chick leaves the relative warmth of the parent's brood patch, thermoregulation is initially achieved through shivering thermogenesis and increased insulation (Duchamp *et al.*, 2002). Thus, shivering, non-shivering thermogenesis, and wing flapping during different phases of development may provide at least some of the necessary contractile activity to elevate calcineurin levels and increase myoglobin production in penguin chicks. Overall, however, production of such high concentrations of myoglobin in chicks of emperor and other penguin species (Noren *et al.*, 2001, Ponganis *et al.*, 1999b, Weber *et al.*, 1974) suggests additional regulatory mechanisms. Similar regulatory processes may also occur in another marine bird, the pigeon guillemot (*Cepphus columbia*). Although adult birds have a pectoral Mb concentration of 2.2 g 100 g^{-1} (Haggblom *et al.*, 1988), Mb content is non-detectable in the chick, but increases to 0.5 g 100 g^{-1} in the fledgling (prior to any flight activity).

Little is known about the life history, let alone muscle myoglobin development, of juvenile emperor penguins (*A. forsteri*) once they go to sea (Kooyman and Ponganis, 2008, Kooyman *et al.*, 1996, Ponganis *et al.*, 1999b). The young penguins do not return to the colonies until they are young adults, about five years of age. During this period, however, myoglobin increases three-fold to adult levels (Ponganis *et al.*, 1997a). Biopsy of adult emperor penguins that had been in captivity since the pre-fledging period and of a three-year-old juvenile born in captivity revealed that pectoral myoglobin concentrations were only 60% of that of adult birds in the wild (Ponganis *et al.*, 2010b). It thus appears that exceptional levels of myoglobin, higher than that in many other divers, can be achieved in emperor penguins without exposure to the exercise, hypoxia, and the increased ambient pressure associated with diving in the wild. However, the final 40% increase to adult levels may well be dependent on dive activity and the exercise-hypoxia calcineurin-based mechanisms proposed by Kanatous *et al.* (2009).

In emperor penguins, myoglobin production is at least partially regulated at the level of transcription. Examination of myoglobin and myoglobin mRNA concentrations in developing emperor penguin chicks has revealed that both increase throughout development into adulthood, but not in a 1:1 fashion (Ponganis *et al.*, 2010b). For example, a 15-fold difference in myoglobin concentration between a 3.5-month-old chick and an adult is only associated with about a three-fold difference in myoglobin mRNA concentrations. This lack of a 1:1 relationship has also been seen in mice and in notothenoid fish (Moylan and Sidell, 2000, Weller *et al.*, 1986). This contrasts with findings in humans and gray seals, where differences in myoglobin concentrations appeared proportional to the size of the myoglobin mRNA pool (Weller *et al.*, 1986). It has been suggested that nutritional limitations in chicks, more efficient translation, and differences in myoglobin degradation rate (half-life) may account for the lack of proportionality between myoglobin and its mRNA pool in emperor penguins (Ponganis *et al.*, 2010b).

In summary, the regulation of the production of the exceptional myoglobin levels in diving birds and mammals remains to be resolved at the molecular level. Although myoglobin-less mice can be produced (Garry *et al.*, 1998, Godecke *et al.*, 1999), the opposite extreme (a diving mouse with high myoglobin concentrations) has not been achieved. What exactly controls the level of myoglobin in a given marine mammal or penguin species is not known. Furthermore, marine mammals contrast with terrestrial mammals in that they are born with already significantly elevated myoglobin levels in skeletal muscle. In addition, penguins, prior to fledging, develop significant myoglobin concentrations without any exposure to hypoxia or obvious exercise. These findings are inconsistent with available regulatory models of myoglobin production; additional regulatory mechanisms may be involved in diving animals.

13.5 Myoglobin "biophysics"

The high concentrations of myoglobin (Mb) in marine mammals and penguins have long been considered to serve as an O_2 store during diving (Scholander, 1940). Even the low Mb concentrations in terrestrial mammals serve as O_2 stores during muscle contractions (Millikan, 1939, Wittenberg, 1970). In addition, as will be reviewed below, Mb is considered to facilitate the intracellular diffusion of O_2 to the mitochondrion, the site of aerobic energy production in the cell (Wittenberg, 1970). Facilitated diffusion has been considered especially important in the maintenance of O_2 flux in exercising muscle (Wagner, 1996, Wittenberg and Wittenberg, 1989).

In terrestrial mammals exercising at maximal O_2 consumption, the P_{O2} of effluent venous blood from muscle is usually never below 15–20 mm Hg (1.99–2.66 kPa)

(Roca *et al.*, 1992, Wagner, 1996). More O_2 cannot be extracted. This has been considered to be secondary to an O_2 diffusion limitation at the capillary–muscle membrane interface. In contrast, venous P_{O2} values are frequently less than 10 mm Hg and even approach zero during dives of elephant seals, sea lions, and emperor penguins (McDonald and Ponganis, 2013, Meir and Ponganis, 2009, Meir *et al.*, 2009, Ponganis *et al.*, 2007, 2009). Given the potential for maintenance of muscle blood flow during sleep apnea and some dives (Guyton *et al.*, 1995, Ponganis *et al.*, 2006b, 2008, Williams *et al.*, 2011a), the elevated Mb concentrations of diving animals may also play an important role in blood-to-muscle O_2 transfer during the hypoxemia of diving. Such high Mb concentrations may contribute to greater blood O_2 extraction and more complete utilization of the blood O_2 store during the breath holds of these animals. This concept is supported by the maintenance of some muscle blood flow, slower rates of muscle oxygen desaturation, and lower venous P_{O2}s during trained versus naïve submersions of harbor seals (*Phoca vitulina*) (Jobsis *et al.*, 2001). It is for these reasons that it is important to understand the "biophysics" of the Mb molecule, especially at the high concentrations found in the muscles of marine mammals and penguins.

The movement or flux of oxygen (O_2) from the surface membrane of a muscle cell to the mitochondria may occur by diffusion of the O_2 molecule through the cytoplasm or by facilitated diffusion via the binding and transport of O_2 by the myoglobin (Mb) molecule (Wittenberg, 1970). Given the low water solubility of O_2 (Sendroy *et al.*, 1934), and the greater transport capacity of Mb (1.34 ml O_2 g^{-1} Mb), one might consider Mb to be the major pathway. Indeed, the first evidence for facilitated diffusion of O_2 through membranes consisted of the comparison of O_2 transport rates in the absence or presence of either hemoglobin (Hb) or Mb (Hemmingsen, 1963, Hemmingsen and Scholander, 1960, Scholander, 1960, Wittenberg, 1959).

However, the magnitude of the Mb-facilitated flux *in vivo* is considered to be dependent on several factors including the concentration of Mb, the translational diffusion of the Mb molecule through the cell cytoplasm, and the difference in bound oxygen concentrations at the cell surface and the mitochondrion (Wittenberg, 1970). Because the Mb-bound O_2 concentrations at the cell surface and mitochondrion are a function of the O_2–Mb dissociation curve and the local intracellular P_{O2}, the ratio of the Mb-based flux and total O_2 flux can be expressed with the following modification of the classic Fick equation (Lin *et al.*, 2007):

$$F_{O2Mb}/F_{O2} = (D_{Mb} \times C_{Mb})/K_0(P_{O2} \times P_{50})$$

where F_{O2Mb} is the O_2 flux due to Mb, F_{O2} is total O_2 flux, D_{Mb} is the Mb diffusion coefficient, C_{Mb} is the concentration of Mb, K_0 is Krogh's diffusion constant for free O_2, P_{O2} is the O_2 partial pressure at the cell surface, and P_{50} is the P_{O2} at which Mb is 50% saturated. The P_{O2} at the mitochondrion is sufficiently low that it is assumed to be zero; therefore, it is not included in this simplified equation.

D_{Mb} has been one of the most difficult to determine variables in the O_2 flux equation due to the potential effect of viscosity on the diffusion of the Mb molecule within the cell. The retardant effect of increased viscosity on Mb-facilitated diffusion had been known since Scholander's original experiment (Scholander, 1960). Addition of 10%

gelatin to the Hb solution halved the O_2 transport rate, but had no effect on N_2 transfer. Hence, the movement of Mb was considered to be restricted in the viscous gelatin.

D_{Mb} values, based on Hb and Mb solution studies, had ranged from 5×10^{-7} to 23×10^{-7} cm^2 s^{-1} (Jurgens *et al.*, 2000, Lin *et al.*, 2007). In contrast, *in vivo* studies have yielded much lower values near 1×10^{-7} to 2×10^{-7} cm^2 s^{-1} (Jurgens *et al.*, 1994, 2000, Papadopoulos *et al.*, 1995). With the low D_{Mb} values measured *in vivo*, the flux equation would predict that only 4% of total O_2 flux is secondary to facilitated diffusion in the exercising muscle of a terrestrial mammal (Jurgens *et al.*, 2000). These predictions contrasted with previous spectrophotometric investigations of Mb saturation that supported a significant role for Mb-facilitated diffusion in exercising muscle (Gayeski and Honig, 1986). On the other hand, questions have been raised regarding the *in vivo* studies and the potential effects of tissue manipulation on the results.

Determination of both the value of D_{Mb} and the role of facilitated diffusion in intracellular O_2 flux have been further refined through the application of pulsed field gradient NMR (nuclear magnetic resonance) methods to muscle tissue (Lin *et al.*, 2007). In intact heart muscle, D_{Mb} was 4.2×10^{-7} cm^2 s^{-1}; the surface P_{O2} at which the contribution of free diffusion and Mb-facilitated diffusion were equal (the equipoise P_{O2}) was calculated to be 1.77 mm Hg (0.24 kPa). Although the flux equation indicated that facilitated diffusion occurred in rat myocardium, it was concluded that Mb did not dominate O_2 flux unless P_{O2} was extremely low.

In marine mammals and penguins, the magnitude of D_{Mb} and the role of the exceptionally elevated Mb concentration in O_2 flux from the muscle surface to the interior are further complicated by the potential effect of high Mb concentrations (C_{Mb}) on intracellular viscosity. Although the increased C_{Mb} in the numerator of the flux equation should increase F_{O2Mb}, the potential viscosity effect on D_{Mb} would do the opposite. Application of the ^{1}H NMR techniques developed by Jue and colleagues to seal muscle revealed a D_{Mb} of 4.5×10^{-7} cm^2 s^{-1}, similar to the value determined with the same technique in rat myocardium (Ponganis *et al.*, 2008). Surprisingly, D_{Mb} was not affected by the elevated Mb concentration within the cell. In addition, because of that high concentration of Mb in the seal, the calculated equipoise P_{O2} was 67 mm Hg (8.93 kPa). Based on blood P_{O2} determinations during sleep apnea and during dives (Meir *et al.*, 2009, Stockard *et al.*, 2007), Mb-facilitated diffusion should dominate O_2 flux within seal muscle under most physiological conditions during both eupnea and apnea.

In regard to the flux of O_2 from the capillary to the muscle mitochondrion, ^{1}H NMR spectroscopy has revealed a predominant role for facilitated diffusion of O_2 in seal muscle in contrast to its minimal role in the rat myocardium. It should be noted that due to potential diffusion limitation between the capillary and the muscle surface, the value of P_{O2} at the surface of the muscle is difficult to determine from the capillary or effluent venous P_{O2}. Consequently, especially during exercise of terrestrial mammals with lower Mb contents, it is still difficult to define the relative contribution of facilitated diffusion to total O_2 flux at different muscle workloads. Nonetheless, the difference in the equipoise P_{O2} (facilitated diffusion = free diffusion) between terrestrial mammals and the seal is primarily due to the high concentration of Mb in seal muscle. Translational diffusion, as reflected by the D_{Mb}, is not affected by the 10–20-fold higher concentration of Mb in seal muscle. This lack of effect of Mb concentration on translational diffusion may be secondary to a higher net surface charge of Mb molecules in the seal than in terrestrial mammals (Mirceta *et al.*, 2013). It has been postulated that higher net surface charges of Mbs in diving mammals promote electrostatic repulsion, minimizing protein self-association which might limit diffusion (Mirceta *et al.*, 2013).

The similarity of D_{Mb} values in both rat myocardium and seal muscle do not support the low D_{Mb} values determined from labeled microinjection techniques (Jurgens *et al.*, 1994, Papadopoulos *et al.*, 1995). As emphasized by Linn *et al.* (2007), the NMR-determined D_{Mb} values support the D_{Mb} of $5–7 \times 10^{-7}$ cm^2 s^{-1} obtained from studies of Mb solutions and used in many modeling studies (Riveros-Moreno and Wittenberg, 1972). The relative contribution of facilitated diffusion appears dependent on the muscle membrane P_{O2} and the Mb concentration. In terrestrial mammals, facilitated diffusion does not dominate O_2 flux until the muscle membrane surface P_{O2} is extremely low. In contrast, the higher Mb content of divers allows for greater facilitated diffusion at any given surface P_{O2}, including the higher P_{O2}s which occur during eupnea and the early portions of dives.

References

Aakhus, T. & Johansen, K. 1964. Angiocardiography of the duck during submersion asphyxia. *Acta Physiologica Scandinavica*, **62**, 10–17.

Aaronson, P. I., Robertson, T. P., Knock, G. A., *et al.* 2006. Hypoxic pulmonary vasoconstriction: mechanisms and controversies. *Journal of Physiology*, **570**, 53–58.

Adachi, H., Strauss, W., Ochi, H., & Wagner, H. N. 1976. The effect of hypoxia on the regional distribution of cardiac output in the dog. *Circulation Research*, **39**, 314–319.

Adachi, T., Maresh, J. L., Robinson, P. W., *et al.* 2014. The foraging benefits of being fat in a highly migratory marine mammal. *Proceedings Royal Society B*, **281**, DOI: 10.1098/rspb.2014.2120.

Aguilar De Soto, N., Johnson, M. P., Madsen, P. T., *et al.* 2008. Cheetahs of the deep sea: deep foraging sprints in short-finned pilot whales off Tenerife (Canary Islands). *Journal of Animal Ecology*, **77**, 936–947.

Aguilar, J. S., Benvenuti, S., Dall'antonia, L., McMinn-Grive, M., & Mayol-Serra, J. 2003. Preliminary results on the foraging ecology of Balearic shearwaters (*Puffinus mauretanicus*) from bird-borne data loggers. *Scientia Marina*, **67**, 129–134.

Ainley, D. G., Strong, C., Penniman, T., & Boekelheide, T. 1990. The feeding ecology of Farallon seabirds. In: Ainley, D. G. & Boekelheide, R. J. (eds.) *Seabirds of the Farallon Islands: Ecology, Dynamics, and Structure of an Upwelling-System Community*. Stanford, CA: Stanford University Press.

Akamatsu, T., Wang, D., Wang, K., Li, S., & Dong, S. 2010. Scanning sonar of rolling porpoises during prey capture dives. *Journal of Experimental Biology*, **213**, 146–152.

Alexander, R. S. 1963. The peripheral venous system. In: Hamilton, W. F. & Dow, P. (eds.) *Handbook of Physiology: Circulation*. Washington, DC: American Physiological Society.

Alves, F., Dinis, A., Cascão, I., & Freitas, L. 2010. Bryde's whale (*Balaenoptera brydei*) stable associations and dive profiles: New insights into foraging behavior. *Marine Mammal Science*, **26**, 202–212.

Ancel, A., Starke, L. N., Ponganis, P. J., Van Dam, R., & Kooyman, G. L. 2000. Energetics of surface swimming in Brandt's cormorants (*Phalacrorax penicillatus Brandt*). *Journal of Experimental Physiology*, **203**, 3727–3731.

Andersen, H. T. 1959. A note on the composition of alveolar air in the diving duck. *Acta Physiologica Scandinavica*, **46**, 240–243.

Andersen, H. T. 1963. The reflex nature of the physiological adjustments and their afferent pathway. *Acta Physiologica Scandanavica*, **58**, 263–273.

Andersen, H. T. 1966. Physiological adaptations in diving vertebrates. *Physiological Reviews*, **40**, 212–243.

Andersson, J. P. A., Biasoletto-Tjellstr, O. M., & Schagatay, E. K. A. 2008. Pulmonary gas exchange is reduced by the cardiovascular diving response in resting humans. *Respiratory Physiology and Neurobiology*, **160**, 320–324.

Andersson, J. P. A. & Evaggelidis, L. 2009. Arterial oxygen saturation and diving response during dynamic apneas in breath-hold divers. *Scandinavian Journal of Medicine & Science in Sports*, **19**, 87–91.

Andersson, J. P. A., Liner, M. H., & Jonsson, H. 2009a. Asystole and increased serum myoglobin levels associated with "packing blackout" in a competitive breath-hold diver. *Clinical Physiology and Functional Imaging*, **29**, 458–461.

Andersson, J. P. A., Liner, M. H., & Jonsson, H. 2009b. Increased serum levels of the brain damage marker S100B after apnea in trained breath-hold divers: a study including respiratory and cardiovascular observations. *Journal of Applied Physiology*, **107**, 809–815.

Andersson, J. P. A., Liner, M. H., Runow, E., & Schagatay, E. K. A. 2002. Diving responses and arterial oxygen saturation during apnea and exercise in breathhold divers. *Journal of Applied Physiology*, **93**, 882–886.

Andrews, R. D., Costa, D. P., Le Boeuf, B. J., & Jones, D. R. 2000. Breathing frequencies of northern elephant seals at sea and on land revealed by heart rate spectral analysis. *Respiration Physiology*, **123**, 71–85.

Andrews, R. D., Jones, D. R., Williams, J. D., *et al.* 1997. Heart rates of northern elephant seals diving at sea and resting on the beach. *Journal of Experimental Biology*, **200**, 2083–2095.

Andrews, R. D., Pitman, R. L., & Ballance, L. T. 2008. Satellite tracking reveals distinct movement patterns for Type B killer whales in the southern Ross Sea, Antarctica. *Polar Biology*, **31**, 1461–1468.

Angell-James, J. E., De Burgh Daly, M., & Elsner, R. 1978. Arterial baroreceptor reflexes in the seal and their modification during experimental dives. *American Journal of Physiology. Heart and Circulatory Physiology*, **234**, H730–H739.

Angell-James, J. E., Elsner, R., & De Burgh Daly, M. 1981. Lung inflation: effects on heart rate, respiration, and vagal afferent activity in seals. *American Journal of Physiology: Heart and Circulatory Physiology*, **240**, H190–H198.

Antonini, E. 1965. Interrelationship between structure and function in hemoglobin and myoglobin. *Physiological Reviews*, **45**, 123–170.

Arad, Z., Midtgard, U., & Bernstein, M. H. 1989. Thermoregulation in turkey vultures: vascular anatomy, arteriovenous heat exchange, and behavior. *Condor*, **91**, 505–514.

Arai, A. E., Grauer, S. E., Anselone, C. G., Pantely, G. A., & Bristow, J. D. 1995. Metabolic adaptation to a gradual reduction in myocardial blood flow. *Circulation*, **92**, 244–252.

Arai, A. E., Pantely, G. A., Anselone, C. G., Bristow, J., & Bristow, J. D. 1991. Active downregulation of myocardial energy requirements during prolonged moderate ischemia in swine. *Circulation Research*, **69**, 1458–1469.

Armstrong, R. B., Delp, M. D., Goljan, G. F., & Laughlin, M. H. 1987. Distribution of blood flow in muscles of miniature swine during exercise. *Journal of Applied Physiology*, **62**, 1285–1298.

Armstrong, R. B., Essen-Gustavsson, B., Hoppeler, H., *et al.* 1992. O_2 delivery at V_{O2max} and oxidative capacity in muscles of Standardbred horses. *Journal of Applied Physiology*, **73**, 2274–2282.

Armstrong, R. B. & Laughlin, M. H. 1984. Exercise blood flow patterns within and among rat muscles after training. *American Journal of Physiology: Heart and Circulatory Physiology*, **246**, H59–H68.

Arnould, J. P. & Hindell, M. 2001. Dive behaviour, foraging locations, and maternal-attendance patterns of Australian fur seals (*Arctocephalus pusillus doriferus*). *Canadian Journal of Zoology*, **79**, 35–48.

Arsham, A. M., Howell, J. J., & Simon, M. C. 2003. A novel hypoxia-inducible factor-independent hypoxic response regulating mammalian target of rapamycin and its targets. *Journal of Biological Chemistry*, **278**, 29655–29660.

Aschoff, S. & Pohl, H. 1970. Rhythmic variation in energy metabolism. *Federation Proceedings, Federation of American Societies for Experimental Biology*, **29**, 1541–1552.

Austin, D., Bowen, W. D., McMillan, J. I., & Boness, D. J. 2006. Stomach temperature telemetry reveals temporal patterns of foraging success in a free-ranging marine mammal. *Journal of Animal Ecology*, **75**, 408–420.

Bagge, L. E., Koopman, H. N., Rommel, S. A., Mclellan, W. A., & Pabst, D. A. 2012. Lipid class and depth-specific thermal properties in the blubber of the short-finned pilot whale and the pygmy sperm whale. *Journal of Experimental Biology*, **215**, 4330–4339.

Bagnoli, P., Cozzi, B., Zaffora, A., *et al.* 2011. Experimental and computational biomechanical characterisation of the tracheo-bronchial tree of the bottlenose dolphin (*Tursiops truncatus*) during diving. *Journal of Biomechanics*, **44**, 1040–1045.

Baird, R. W. 1998. Studying diving behavior of whales and dolphins uising suction-cup attached tags. *Whale Watcher*, **Spring/Summer**, 3–6.

Baird, R. W., Borsani, J. F., Hanson, M. B., & Tyack, P. L. 2002. Diving and night-time behavior of long-finned pilot whales in the Ligurian Sea. *Marine Ecology Progress Series*, **237**, 301–305.

Baird, R. W., Hanson, M. B., & Dill, L. M. 2006a. Factors influencing the diving behaviour of fish-eating killer whales: sex differences and diel and interannual variation in diving rates. *Canadian Journal of Zoology*, **83** 257–267.

Baird, R. W., Ligon, A. D., Hooker, S. K., & Gorgone, A. M. 2001. Subsurface and nighttime behaviour of pantropical spotted dolphins in Hawaii. *Canadian Journal of Zoology*, **79**, 988–996.

Baird, R. W., Webster, D. L., Mcsweeney, D. J., *et al.* 2006b. Diving behavior of Cuvier's (*Ziphius cavirostris*) and Blaineville's (*Mesoplodon densirostris*) beaked whales in Hawaii. *Canadian Journal of Zoology*, **84**, 1120–1128.

Baird, R. W., Webster, D. L., Schorr, G. S., Mcsweeney, D. J., & Barlow, J. 2008. Diel variation in beaked whale diving behavior. *Marine Mammal Science*, **24**, 630–642.

Baker, D. A. & Gough, D. A. 1995. A continuous, implantable lactate sensor. *Analytical Chemistry*, **67**, 1536–1540.

Baker, J., Yost, C. S., & Niemann, C. U. 2005. Organ transplantation. In: Miller, R. D. (ed.) *Miller's Anesthesia*. 6th ed. Philadelphia, PA: Elsevier.

Baldwin, J. 1988. Prediciting the swimming and diving behavior of penguins from muscle biochemistry. *Hydrobiologia*, **165**, 255–261.

Baldwin, J., Jardel, J.-P., Montague, T., & Tomkin, R. 1984. Energy metabolism in penguin swimming muscles. *Molecular Physiology*, **6**, 33–42.

Balmer, B. C., Wells, R. S., Howle, L. E., *et al.* 2014. Advances in cetacean telemetry: a review of single-pin transmitter attachment techniques on small cetaceans and development of a new satellite-linked transmitter design. *Marine Mammal Science*, **30**, 656–673.

Barer, G. R. & Shaw, J. W. 1971. Pulmonary vasodilatory and vasoconstrictor actions of carbon dioxide. *Journal of Physiology*, **213**, 633–645.

Barnas, G. M., Mather, F. B., & Fedde, M. R. 1978. Response of avian intrapulmonary smooth muscle to changes in carbon dioxide concentration. *Poultry Science*, **57**, 1400–1407.

Barre, H. & Roussel, B. 1986. Thermal and metabolic adaptation to first cold-water immersion in juvenile penguins. *American Journal of Physiology*, **251**, R456–R462.

Bartholomew, G. A. 1954. Body temperature and respiratory and heart rates in the northern elephant seal. *Journal of Mammalogy*, **35**, 211–218.

Bashir, T., Khan, A., Behera, S. K., & Gautam, P. 2013. Time dependent activity pattern of Ganges River dolphin *Platanista gangetica gangetica* and its response to human presence in Upper Ganges River, India. *Mammal Study*, **38**, 9–17.

Batulis, J. C. & Bongiorno, S. F. 1972. Effect of water depth on diving times in the American coot (*Fulica americana*). *The Auk*, **89**, 665–667.

Baudinette, R., Loveridge, J., Wilson, K., Mills, C., & Schmidt-Nielsen, K. 1976. Heat loss from feet of herring gulls at rest and during flight. *American Journal of Physiology: Legacy Content*, **230**, 920–924.

Baudinette, R. V. & Gill, P. 1985. The energetics of "flying" and "paddling" in water: locomotion in penguins and ducks. *Journal of Comparative Physiology B*, **155**, 373–380.

Baumgartner, M. F. & Mate, B. R. 2003. Summertime foraging ecology of North Atlantic right whales. *Marine Ecology Progress Series*, **264**, 123–135.

Bayly, W. M., Hodgson, D. R., Schulz, D. A., Dempsey, J. A., & Gollnick, P. D. 1989. Exercise-induced hypercapnia in the horse. *Journal of Applied Physiology*, **67**, 1958–1966.

Bech, C. & Praesteng, K. E. 2004. Thermoregulatory use of heat increment of feeding in the tawny owl (*Strix aluco*). *Journal of Thermal Biology*, **29**, 649–654.

Beck, C. A., Bowen, W. D., McMillan, J. I., & Iverson, S. J. 2003. Sex differences in the diving behaviour of a size-dimorphic capital breeder: the grey seal. *Animal Behaviour*, **66**, 777–789.

Behrman, R. E. & Lees, M. H. 1971. Organ blood flows of the fetal, newborn and adult rhesus monkey. *Neonatology*, **18**, 330–340.

Belanger, L. F. 1940. A study of the histological structure of the respiratory portion of the lungs of aquatic mammals. *American Journal of Anatomy*, **67**, 437–461.

Bello, M. A., Roy, R. R., Martin, T. P., Goforth, H. W., & Edgerton, V. R. 1985. Axial musculature in the dolphin (*Tursiops truncatus*): some architectural and histochemical characteristics. *Marine Mammal Science*, **1**, 324–336.

Bengtson, J. L., Croll, D. A., & Goebel, M. E. 1993. Diving behavior of chinstrap penguins at Seal Island. *Antarctic Science*, **5**, 9–15.

Bennett, A. F. 1984. Thermal dependence of muscle function. *American Journal of Physiology: Regulatory, Integrative and Comparative Physiology*, **247**, R217–R229.

Bennett, J. B. & Rostain, J. C. 2003. High pressure nervous syndrome. In: Brubakk, A. O. & Neuman, T. S. (eds.) *Bennett and Elliott's Physiology and Medicine of Diving*. Cornwall: Saunders.

Bennett, K., Mcconnell, B., & Fedak, M. A. 2001. Diurnal and seasonal variations in the duration and depth of the longest dives in Southern Elephant seals (*Mirounga leonina*): possible physiological and behavioural constraints. *Journal of Experimental Biology*, **204**, 649–662.

Benoit-Bird, K. J., Würsig, B., & Mfadden, C. J. 2004. Dusky dolphin (*Lagenorhynchus obscurus*) foraging in two different habitats: active acoustic detection of dolphins and their prey. *Marine Mammal Science*, **20**, 215–231.

Benvenuti, S., Dall'antonia, L., & Lyngs, P. 2001. Foraging behaviour and time allocation of chick-rearing Razorbills *Alca torda* at Graesholmen, central Baltic Sea. *Ibis*, **143**, 402–412.

Bernaldo De Quirós, Y., González-Díaz, Ó., Arbelo, M., *et al.* 2012. Decompression vs. decomposition: distribution, amount and gas composition of bubbles in stranded marine mammals. *Frontiers in Physiology*. DOI: 10.3389/fphys.2012.00177.

Bernaldo De Quirós, Y., González-Díaz, Ó., Saavedra, P., *et al.* 2011. Methodology for in situ gas sampling, transport and laboratory analysis of gases from stranded cetaceans. *Scientific Reports*. DOI: 10.1038/srep00193.

Bernard, S. A., Gray, T. W., Buist, M. D., *et al.* 2002. Treatment of comatose survivors of out-of-hospital cardiac arrest with induced hypothermia. *New England Journal of Medicine*, **346**, 557–563.

Berne, R. M. & Levy, M. N. 1998. *Physiology*, St. Louis, MO: Mosby.

Bernhard, W., Gebert, A., Vieten, G., *et al.* 2001. Pulmonary surfactant in birds: coping with surface tension in a tubular lung. *American Journal of Physiology: Regulatory, Integrative and Comparative Physiology*, **281**, R327–R337.

Bernhard, W., Haslam, P. L., & Floros, J. 2004. From birds to humans: new concepts on airways relative to alveolar surfactant. *American Journal of Respiratory Cell and Molecular Biology*, **30**, 6–11.

Bernstein, M. H., Curtis, M. B., & Hudson, D. M. 1979. Independence of brain and body temperatures in flying American kestrels, *Falco sparverius*. *American Journal of Physiology: Regulatory, Integrative and Comparative Physiology*, **237**, R58–R62.

Bernstein, M. H., Duran, H. L., & Pinshow, B. 1984. Extrapulmonary gas exchange enhances brain oxygen in pigeons. *Science*, **226**, 564–566.

Bert, P. 1870. *Lecons Sur la Physiologie Comparee de la Respiration*. Paris: Bailliere.

Berta, A., Sumich, J. L., & Kovacs, K. M. 2006. *Marine Mammals Evolutionary Biology*. San Diego, CA: Academic Press.

Bethge, P., Munks, S., Otley, H., & Nicol, S. 2003. Diving behaviour, dive cycles and aerobic dive limit in the platypus *Ornithorhynchus anatinus*. *Comparative Biochemistry and Physiology Part A* **136**, 799–809.

Bethge, P., Nicol, S., Culik, B. M., & Wilson, R. P. 1997. Diving behaviour and energetics in breeding little penguins (*Eudyptula minor*). *Journal of Zoology, London*, **242**, 483–502.

Bethke, R. W. & Thomas, V. G. 1988. Differences in flight and heart muscle mass among geese, dabbling ducks, and diving ducks relative to habitat use. *Canadian Journal of Zoology*, **66**, 2024–2028.

Bevan, R., Speakman, J. R., & Butler, P. 1995a. Daily energy expenditure of tufted ducks: a comparison between indirect calorimetry, doubly labelled water and heart rate. *Functional Ecology*, **9**, 40–47.

Bevan, R. M., Boyd, I. L., Butler, P. J., *et al.* 1997. Heart rates and abdominal temperatures of free-ranging South Georgian shags, *Phalacrocorax georgianus*. *Journal of Experimental Biology*, **200**, 661–675.

Bevan, R. M. & Butler, P. J. 1992a. Cardiac output and blood flow distribution during swimming and voluntary diving of the tufted duck (*Aythya fuligula*). *Journal of Experimental Biology*, **1668**, 199–217.

Bevan, R. M. & Butler, P. J. 1992b. The effects of temperature on the oxygen consumption, heart rate and deep body temperature during diving in the tufted duck *Aythya fuligula*. *Journal of Experimental Biology*, **163**, 139–151.

Bevan, R. M., Butler, P. J., Woakes, A. J., & Boyd, I. L. 2002. The energetics of gentoo penguins, *Pygoscelis papua*, during the breeding season. *Functional Ecology*, **16**, 175–190.

Bevan, R. M., Keijer, E., & Butler, P. J. 1992. A method for controlling the feeding behaviour of aquatic birds: heart rate and oxgyen consumption during dives of different duration. *Journal of Experimental Biology*, **162**, 91–106.

Bevan, R. M., Woakes, A. J., Butler, P. J., & Croxall, J. P. 1995b. Heart rate and oxygen consumption of exercising gentoo penguins. *Physiological Zoology*, **68**, 855–877.

Bickler, P. E. & Buck, L. T. 2007. Hypoxia tolerance in reptiles, amphibians, and fishes: life with variable oxygen availability. *Annual Review of Physiology*, **69**, 145–170.

Bicudo, J. E., Vianna, C. R., & Chaui-Berlinck, J. 2001. Thermogenesis in birds. *Bioscience Reports*, **21**, 181–188.

Bisaillon, A., Martineau, D., & St-Pierre, M. A. 1987. Anatomy of the heart of the beluga whale (*Delphinapterus leucas*). *Journal of Morphology*, **191**, 89–100.

Biuw, M., Mcconnell, B., Bradshaw, C. J. A., Burton, H., & Fedak, M. 2003. Blubber and buoyancy: monitoring the body condition of free-ranging seals using simple dive characteristics. *Journal of Experimental Biology*, **206**, 3405–3423.

Black, C. P. & Tenney, S. M. 1980. Oxygen transport during progressive hypoxia in high-altitude and sea-level waterfowl. *Respiration Physiology*, **39**, 217–239.

Blackstone, E., Morrison, M., & Roth, M. B. 2005. H_2S induces a suspended animation-like state in mice. *Science*, **308**, 518.

Blackwell, S. B. & Le Boeuf, B. J. 1993. Developmental aspects of sleep apnea in northern elephant seals, *Mirounga angustirostris*. *Journal of Zoology, London*, **231**, 437–447.

Blanchetot, A., Wilson, V., Wood, D., & Jeffreys, A. J. 1983. The seal myoglobin gene: an unusually long globin gene. *Nature*, **301**, 732–734.

Blei, M. L., Conley, K. E., & Kushmerick, M. J. 1993. Separate measures of ATP utilization and recovery in human skeletal muscle. *Journal of Physiology (Cambridge)*, **465**, 203–222.

Blessing, M. H. 1972a. Myoglobin concentration in *Platanista indi*. *Investigations of Cetacea*, **4**, 91–92.

Blessing, M. H. 1972b. Studies on the concentration of myoglobin in the sea-cow and porpoise. *Comparative Physiology and Biochemistry*, **41A**, 475–480.

Blix, A. S. 1971. Creatine in diving animals: a comparative study. *Comparative Biochemistry and Physiology A*, **40**, 805–807.

Blix, A. S. 2011. The venous system of seals, with new ideas on the significance of the extradural intravertebral vein. *Journal of Experimental Biology*, **214**, 3507–3510.

Blix, A. S., Berg, T., & Fyhn, H. J. 1970. Lactate dehydrogenase in a diving mammal, the common seal (*Phoca vitulina vitulina l.*). *International Journal of Biochemistry*, **1**, 292–294.

Blix, A. S., Elsner, R. W., & Kjekhus, J. K. 1983. Cardiac output and its distribution through capillaries and A–V shunts in diving seals. *Acta Physiologica Scandanavica*, **118**, 109–116.

Blix, A. S. & Folkow, B. 1983. Cardiovascular adjustments to diving in mammals and birds. In: *Handbook of Physiology: The Cardiovascular System. Peripheral Circulation and Organ Blood Flow*. Bethesda, MD: American Physiology Society.

Blix, A. S. & Folkow, L. P. 1995. Daily energy expenditure in free living minke whales. *Acta Physiologica Scandinavica*, **153**, 61–66.

Blix, A. S. & From, S. H. J. 1971. Lactate dehydrogenase in diving animals: a comparative study with special reference to the eider (*Somateria mollissima*). *Comparative Biochemistry and Physiology Part B: Comparative Biochemistry*, **40**, 579–584.

Blix, A. S., Gautvik, E. L., & Refsum, H. 1974. Aspects of the relative roles of peripheral vasoconstriction and vagal bradycardia in the establishment of the "diving reflex" in ducks. *Acta Physiologica Scandinavica*, **90**, 289–296.

Blix, A. S., Grav, H. J., & Ronald, K. 1975. Brown adipose tissue and the significance of the venous plexuses in pinnipeds. *Acta Physiologica Scandinavica*, **94**, 133–135.

Blix, A. S. & Hol, R. 1973. Ventricular dilatation in the diving seal. *Acta Physiologica Scandinavica*, **87**, 431–432.

Blix, A. S. & Nordoy, E. S. 2007. Ross seal (*Ommatophoca rossii*) annual distribution, diving behaviour, breeding and moulting, off Queen Maud Land, Antarctica. *Polar Biology*, **30**, 1449–1458.

Blix, A. S., Walloe, L., & Messelt, E. 2013. On how whales avoid decompression sickness and why they sometimes strand. *Journal of Experimental Biology*, **216**, 3385–3387.

Blix, A. S., Walloe, L., Messelt, E. B., & Folkow, L. P. 2010. Selective brain cooling and its vascular basis in diving seals. *Journal of Experimental Biology*, **213**, 2610–2616.

Bocher, P., Cherel, Y., & Hobson, K. A. 2000a. Complete trophic segregation between South Georgian and common diving petrels during breeding at Iles Kerguelen. *Marine Ecology Progress Series*, **208**, 249–264.

Bocher, P., Labidoire, B., & Cherel, Y. 2000b. Maximum dive depths of common diving petrels (*Pelecanoides urinatrix*) during the annual cycle at Mayes Island, Kerguelen. *Journal of Zoology*, **251**, 517–524.

Boersma, P. D. 1976. An ecological and behavioral study of the Galapagos penguin. *Living Bird*, **15**, 43–93.

Boggs, D. F. 1997. Coordinated control of respiratory pattern during locomotion in birds. *American Zoologist*, **37**, 41–53.

Boggs, D. F., Baudinette, R. V., Frappell, P. B., & Butler, P. 2001. The influence of locomotion on air-sac pressures in little penguins. *Journal of Experimental Biology*, **204**, 3581–3586.

Bond, C. F. & Gilbert, P. W. 1958. Comparative study of blood volume in representative aquatic and nonaquatic birds. *American Journal of Physiology: Legacy Content*, **194**, 519–521.

Born, E. W. & Knutsen, L. O. 1997. Haul-out and diving activity of male Atlantic walruses (*Odobenus rosmarus*) in NE Greenland. *Journal of Zoology London*, **243**, 381–396.

Born, E. W., Teilmann, J., Acquarone, M., & Riget, F. F. 2004. Habitat use of ringed seals (*Phoca hispida*) in the North Water Area (North Baffin Bay). *Arctic*, **7**, 147–152.

Bostrom, B. L., Falhlman, A., & Jones, D. R. 2008. Tracheal compression delays alveolar collapse during deep diving in marine mammals. *Respiration Physiology and Neurobiology*, **161**, 298–305.

Bouverot, P. 1978. Control of breathing in birds compared with mammals. *Physiological Reviews*, **58**, 604–655.

Bowen, W. D., Boness, D. J., & Iverson, S. J. 1999. Diving behaviour of lactating harbour seals and their pups during maternal foraging trips. *Canadian Journal of Zoology*, **77**, 978–988.

Boyd, I. L. 2000. Skin temperatures during free-ranging swimming and diving in Antarctic fur seals. *Journal of Experimental Biology*, **203**, 1907–1914.

Boyd, I. L., Bevan, R. M., Woakes, A. J., & Butler, P. J. 1999. Heart rate and behavior of fur seals: implications for measurement of field energetics. *American Journal of Physiology*, **276**, H844–H857.

Boyd, I. L. & Croxall, J. P. 1992. Diving behaviour of lactating Antarctic fur seals. *Canadian Journal of Zoology*, **70**, 919–928.

Boyd, I. L. & Croxall, J. P. 1996. Dive durations in pinnipeds and seabirds. *Canadian Journal of Zoology*, **74**, 1696–1705.

Boyd, I. L., Woakes, A. J., Butler, P. J., Davis, R. W., & Williams, T. M. 1995. Validation of heart rate and doubly labelled water as measures of metabolic rate during swimming in California sea lions. *Functional Ecology*, **9**, 151–160.

Braunwald, E. 1971. Control of myocardial oxygen consumption: physiological and clinical considerations. *American Journal of Cardiology*, **27**, 416–432.

Bried, J. 2005. Diving ability of the Madeiran storm petrel. *Waterbirds*, **28**, 162–166.

Brierley, A. S. & Fernandes, P. G. 2001. Diving depths of Northern gannets: acoustic observations of *Sula bassana* from an autonomous underwater vehicle. *The Auk*, **118**, 529–534.

Bron, K. M., Murdaugh, J. H. V., Millen, J. E., *et al.* 1966. Arterial constrictor response in a diving mammal. *Science*, **152**, 540–543.

Brown, R. G. B., Bourne, W. R. P., & Wahl, T. R. 1978. Diving by shearwaters. *Condor*, **80**, 123–125.

Brown, Z. W., Welcker, J., Harding, A. M. A., Walkusz, W., & Karnovsky, N. J. 2012. Divergent diving behavior during short and long trips of a bimodal forager, the little auk *Alle alle*. *Journal of Avian Biology*, **43**, 215–226.

Brubakk, A. O. & Neuman, T. S. 2003. *Bennett and Elliott's Physiology and Medicine of Diving*. Edinburgh: Saunders.

Bryden, M. M. 1972. Body size and composition of elephant seals (*Mirounga leonina*): absolute measurements and estimates from bone dimensions. *Journal of Zoology (London)*, **167**, 265–276.

Bryden, M. M. & Erickson, A. W. 1976. Body size and composition of crabeater seals (*Lobodon carcinophagus*), with observations on tissue and organ size in Ross seals (*Ommatophoca rossi*). *Journal of Zoology (London)*, **179**, 235–247.

Bryden, M. M. & Felts, J. L. 1974. Quantitative anatomical observations on the skeletal and muscular systems of four species of Antarctic seals. *Journal of Anatomy*, **118**, 589–600.

Bryden, M. M. & Lim, G. H. K. 1969. Blood parameters of the southern elephant seal (*Mirounga leonina*) in relation to diving. *Comparative Biochemistry and Physiology*, **28**, 139–148.

Bryden, M. M. & Molyneux, G. S. 1978. Arteriovenous anastomoses in the skin of seals II. The California sea lion (*Zalopohus californianus*) and the northern fur seal (*Callorhinus ursinus*) (Pinnipedia: Otariidae). *Anatomical Record*, **191**, 253–260.

Burger, A. E. 2001. Diving depths of shearwaters. *The Auk*, **118**, 755–759.

Burger, A. E. & Powell, D. W. 1990. Diving depths and diet of Cassin's Auklet at Reef Island, British Columbia. *Canadian Journal of Zoology*, **68**, 1572–1577.

Burger, A. E. & Simpson, M. 1986. Diving depths of Atlantic puffins and common murres. *The Auk*, **103**, 828–830.

Burger, A. E., Wilson, R. P., Garnier, D., & Wilson, M.-P. T. 1993. Diving depths, diet, and underwater foraging of Rhinoceros Auklets in British Columbia. *Canadian Journal of Zoology*, **71**, 2528–2540.

Burmester, T. & Hankeln, T. 2009. What is the function of neuroglobin? *Journal of Experimental Biology*, **212**, 1423–1428.

Burmester, T., Weich, B., Reinhardt, S., & Hankein, T. 2000. A vertebrate globin expressed in the brain. *Nature*, **407**, 520–523.

Burns, J. M. 1999. The development of diving behavior in juvenile Weddell seals: pushing physiological limits in order to survive. *Canadian Journal of Zoology*, **77**, 737–747.

Burns, J. M. & Castellini, M. A. 1996. Physiological and behavioral determinants of the aerobic dive limit in Weddell seal (*Letonychotes weddellii*) pups. *Journal of Comparative Physiology B*, **166**, 473–483.

Burns, J. M., Costa, D. P., Fedak, M. A., *et al.* 2004. Winter habitat use and foraging behavior of crabeater seals along the Western Antarctic Peninsula. *Deep Sea Research Part II: Topical Studies in Oceanography*, **51**, 2279–2303.

Burns, J. M., Costa, D. P., Frost, K., & Harvey, J. T. 2005. Physiological development in juvenile harbor seals. *Physiological and Biochemical Zoology*, **78**, 1057–1068.

Burns, J. M., Lestyk, K. C., Hammill, M. O., Folkow, L. P., & Blix, A. S. 2007. Size and distribution of oxygen stores in harp and hooded seals from birth to maturity. *Journal of Comparative Physiology B*, **177**, 687–700.

Burns, J. M., Skomp, N., Bishop, N., Lestyk, K., & Hammill, M. 2010. Development of aerobic and anaerobic metabolism in cardiac and skeletal muscles from harp and hooded seals. *Journal of Experimental Biology*, **213**, 740–748.

Butler, P. J. 1982. Respiratory and cardiovascular control during diving in birds and mammals. *Journal of Experimental Biology*, **100**, 195–221.

Butler, P. J. 1991. Exercise in birds. *Journal of Experimental Biology*, **160**, 233–262.

Butler, P. J. 1994. To what extent can heart rate be used as an indicator of metabolic rate in free-living marine mammals. In: Boyd, I. L. (ed.) *Marine Mammals: Advances in Behavioural and Population Biology*. London: Zoological Society of London.

Butler, P. J. 2000. Energetic costs of surface swimming and diving of birds. *Physiolgical and Biochemical Zoology*, **73**, 699–705.

Butler, P. J. 2004. Metabolic regulation in diving birds and mammals. *Respiratory Physiology and Neurobiology*, **141**, 297–315.

Butler, P. J. 2006. Aerobic dive limit. What is it and is it always used properly? *Comparative Biochemistry and Physiology A*, **145**, 1–6.

Butler, P. J. & Bishop, C. M. 2000. Flight. In: Whittow, G. C. (ed.) *Sturkie's Avian Physiology*. San Diego, CA: Academic Press.

Butler, P. J., Green, J. A., Boyd, I. L., & Speakman, J. R. 2004. Measuring metabolic rate in the field: the pros and cons of the doubly labelled water and heart rate methods. *Functional Ecology*, **18**, 168–183.

Butler, P. J. & Jones, D. R. 1968. Onset of and recovery from diving bradycardia in ducks. *Journal of Physiology (London)*, **196**, 255–272.

Butler, P. J. & Jones, D. R. 1997. The physiology of diving of birds and mammals. *Physiological Reviews*, **77**, 837–899.

Butler, P. J. & Taylor, E. W. 1973. The effect of hypercapnic hypoxia, accompanied by different levels of lung ventilation, on heart rate in the duck. *Respiration Physiology*, **19**, 176–187.

Butler, P. J. & Taylor, E. W. 1983. Factors affecting the respiratory and cardiovascular responses to hypercapnic hypoxia in mallard ducks. *Respiration Physiology*, **53**, 109–127.

Butler, P. J., Turner, D. L., Al-Wassiaf, A., & Bevan, R. M. 1988. Regional distribution of blood flow during swimming in the tufted duck (*Aythya fuligula*). *Journal of Experimental Biology*, **135**, 461–472.

Butler, P. J. & Woakes, A. J. 1979. Changes in heart rate and respiratory frequency during natural behaviour of ducks, with particular reference to diving. *Journal of Experimental Biology*, **79**, 283–300.

Butler, P. J. & Woakes, A. J. 1982. Control of heart rate by carotid body chemoreceptors during diving in tufted ducks. *Journal of Applied Physiology*, **53**, 1405–1410.

Butler, P. J. & Woakes, A. J. 1984. Heart rate and aerobic metabolism in Humboldt penguins (*Spheniscus humboldti*) during voluntary dives. *Journal of Experimental Biology*, **108**, 419–428.

Cabanac, A., Folkow, L. P., & Blix, A. S. 1997. Volume capacity and contraction control of the seal spleen. *Journal of Applied Physiology*, **82**, 1989–1994.

Cabanac, A. J. 2002. Contracted spleen in seals, estimates of dilated organs, and dive capacity. *Polar Biology*, **25**, 1–4.

Cabanac, A. J., Messelt, E. B., Folkow, L. P., & Blix, A. S. 1999. The structure and blood-storing function of the spleen of the hooded seal (*Cystophora cristata*). *Journal of Zoology, London*, **248**, 75–81.

Cabanac, A., Folkow, L. P., & Blix, A. S. 1998. Effects of adrenergic and cholinergic drugs on splenic arteries and veins from hooded seals. *Comparative Biochemistry and Physiology A*, **120**, 277–281.

Camporesi, E. M. & Bosco, G. 2003. Ventilation, gas exchange and exercise under pressure. In: Brubakk, A. O. & Neuman, T. S. (eds.) *Bennett and Elliott's Physiology and Medicine of Diving*. Edinburgh: Saunders.

Caruso, J. L. 2003. Pathology of diving accidents. In: Brubakk, A. O. & Neuman, T. S. (eds.) *Bennett and Elliott's Physiology and Medicine of Diving*. Edinburgh: Saunders.

Carvajal, J. A., Germain, A. M., Huidobro-Toro, J. P., & Weiner, C. P. 2000. Molecular mechanism of cGMP-mediated smooth muscle relaxation. *Journal of Cellular Physiology*, **184**, 409–420.

Casadei, B. 2001. Vagal control of myocardial contractility in man. *Experimental Physiology*, **86**.6, 817–823.

Casaux, R. 2004. Diving patterns in the Antarctic shag. *Waterbirds*, **27**, 382–387.

Casaux, R. & Barrera-Oro, E. 2006. Shags in Antarctica: their feeding behaviour and ecological role in the marine food web. *Antarctic Science*, **18**, 3–14.

Casaux, R., Favero, M., Silva, P., & Baroni, A. 2001. Sex differences in diving depths and diet of Antarctic shags at the South Shetland Islands. *Journal of Field Ornithology*, **72**, 22–29.

Casey, D. P. & Joyner, M. J. 2011. Local control of skeletal muscle blood flow during exercise: influence of available oxygen. *Journal of Applied Physiology*, **111**, 1527–1538.

Casey, D. P., Madery, B. D., Curry, T. B., *et al.* 2010. Nitric oxide contributes to the augmented vasodilatation during hypoxic exercise. *Journal of Physiology*, **588**.2, 373–385.

Casler, C. L. 1973. The air-sac systems and buoyancy of the anhinga and double-crested cormorant. *The Auk*, **90**, 324–340.

Castellini, J. M. & Castellini, M. A. 1993. Estimation of splenic volume and its relationship to long-duration apnea in seals. *Physiological Zoology*, **66**, 619–627.

Castellini, M. A. 1994. Apnea tolerance in the elephant seal during sleeping and diving: physiological mechanisms and correlations. In: Le Boeuf, B. J. & Laws, R. M. (eds.) *Elephant Seals: Population Ecology, Behavior, and Physiology*. Berkeley, CA: University of California Press.

Castellini, M. A., Baskurt, O. K., Castellini, J. M., & Meiselman, H. J. 2010. Blood rheology in marine mammals. *Frontiers in Physiology* [Online], **1**:146.

Castellini, M. A., Castellini, J. M., & Rivera, P. M. 2001. Adaptations to pressure in the RBC metabolism of diving mammals. *Comparative Biochemistry and Physiology Part A: Molecular & Integrative Physiology*, **129**, 751–757.

Castellini, M. A., Costa, D. P., & Huntley, A. 1986. Hematocrit variation during sleep apnea in elephant seal pups. *American Journal of Physiology*, **251**, R429–R431.

Castellini, M. A., Costa, D. P., & Huntley, A. 1987. Fatty acid metabolism in fasting elephant seal pups. *Journal of Comparative Physiology*, **157B**, 445–448.

Castellini, M. A., Davis, R. W., & Kooyman, G. L. 1988. Blood chemistry regulation during repetitive diving in Weddell seals. *Physiological Zoology*, **61**, 379–386.

Castellini, M. A., Davis, R. W., & Kooyman, G. L. 1992a. *Diving Behavior of the Weddell Seal: Annual Cycles*, Berkeley, CA: University of California Press.

Castellini, M. A., Elsner, R., Baskurt, O. K., Wenby, R. A., & Meiselman, H. J. 2006. Blood rheology of Weddell seals and bowhead whales. *Biorheology*, **43**, 57–69.

Castellini, M. A., Kooyman, G. L., & Ponganis, P. J. 1992b. Metabolic rates of freely diving Weddell seals: correlations with oxygen stores, swim velocity, and diving duration. *Journal of Experimental Biology*, **165**, 181–194.

Castellini, M. A. & Mellish, J. E. (eds.) 2015. *Marine Mammal Physiology: Requisites for Ocean Living*. Boca Raton, FL: CRC Press.

Castellini, M. A., Milsom, W. K., Berger, R. J., *et al.* 1994a. Patterns of respiration and heart rate during wakefulness and sleep in elephant seal pups. *American Journal of Physiology*, **266**, R863–R869.

Castellini, M. A., Murphy, B. J., Fedak, M. A., *et al.* 1985. Potentially conflicting demands of diving and exercise in seals. *Journal of Applied Physiology*, **58**, 392–399.

Castellini, M. A., Rea, L. D., Sanders, J. L., Castellini, J. M., & Zenteno-Savin, T. 1994b. Developmental changes in cardiorespiratory patterns of sleep-associated apnea in northern elephant seals. *American Journal of Physiology*, **267**, R1294–R1301.

Castellini, M. A., Rivera, P. M., & Castellini, J. M. 2002. Biochemical aspects of pressure tolerance in marine mammals. *Comparative Biochemistry and Physiology Part A: Molecular & Integrative Physiology*, **133**, 893–899.

Castellini, M. A. & Somero, G. N. 1981. Buffering capacity of vertebrate muscle: correlations with potentials for anaerobic function. *Journal of Comparative Physiology B*, **143**, 191–198.

Castellini, M. A., Somero, G. N., & Kooyman, G. L. 1981. Glycolytic enzyme activities in tissues of marine and terrestrial mammals. *Physiological Zoology*, **54**, 242–252.

Chandler, A. C. 1916. A study of the structure of feathers with reference to their taxonomic significance. *University of California Publications in Zoology*, **13**, 243–446.

Chapla, M. E., Nowacek, D. P., Rommel, S. A., & Sadler, V. M. 2007. CT scans and 3D reconstructions of Florida manatee (*Trichechus manatus latirostris*) heads and ear bones. *Hearing Research*, **228**, 123–135.

Chappell, M. A., Shoemaker, V. A., Jones, D. N., & Maloney, S. K. 1993. Diving behavior during foraging in breeding Adelie penguins. *Ecology*, **74**, 1204–1215.

Charrassin, J.-B., Kato, A., Handrich, Y., *et al.* 2001. Feeding behaviour of free-ranging penguins determined by oesophageal temperature. *Proceedings of the Royal Society London B*, **268**, 151–157.

Charrassin, J. B. & Bost, C.-A. 2001. Utilisation of the oceanic habitat by king penguins over the annual cycle. *Marine Ecology Progress Series*, **221**, 285–297.

Chastel, O. 1994. Maximum diving depths of common diving petrels *Pelecanoides urinatrix* at Kerguelen Islands. *Polar Biology*, **14**, 211–213.

Chastel, O. & Bried, J. 1996. Diving ability of blue petrels and thin-billed prions. *The Condor*, **98**, 621–629.

Chen, C.-H., Sun, L., & Mochly-Rosen, D. 2010. Mitochondrial aldehyde dehydrogenase and cardiac diseases. *Cardiovascular Research*, **88**, 51–57.

Chen, C., Budas, G. R., Churchill, C. N., *et al.* 2008. Activation of aldehyde dehydrogenase-2 reduced ischemic damage to the heart. *Science*, **321**, 1493–1495.

Chen, L., Lu, X.-Y., Li, J., *et al.* 2006. Intermittent hypoxia protects cardiomyocytes against ischemia-reperfusion injury-induced alterations in Ca^{2+} homeostasis and contraction via the sarcoplasmic reticulum and Na^+/Ca^{2+} exchange mechanisms. *American Journal of Physiology Cell Physiology*, **290**, C1221–C1229.

Cherel, Y., Bocher, P., Broyer, C. D., & Hobson, K. A. 2002. Food and feeding ecology of the sympatric thinbilled *Pachyptila belcheri* and Antarctic *P. desolata* prions at Iles Kerguelen, Southern Indian Ocean. *Marine Ecology Progress Series*, **228**, 263–281.

Cherel, Y., Charassin, J.-B., & Handrich, Y. 1993. Comparison of body reserve buildup in prefasting chicks and adults of king penguins (*Aptenodytes patagonicus*). *Physiological Zoology*, **66**, 750–770.

Cherel, Y., Tremblay, Y., Guinard, E., & Georges, J.-Y. 1999. Diving behaviour of female northern rockhopper penguins, *Eudyptes chrysocome moseleyi*, during the brooding period at Amsterdam Island (Southern Indian Ocean). *Marine Biology*, **134**, 375–385.

Cherepanova, V., Neshumova, T. V., & Elsner, R. 1993. Muscle blood flow in diving mammals. *Comparative Biochemistry and Physiology A*, **106**, 1–6.

Chicco, A. J., Le, C. H., Schlater, A., *et al.* 2014. High fatty acid oxidation capacity and phosphorylation control despite elevated leak and reduced respiratory capacity in northern elephant seal muscle mitochondria. *Journal of Experimental Biology*, **217**, 2947–2955.

Chilvers, B. L. 2008. New Zealand sea lions *Phocarctos hookeri* and squid trawl fisheries: bycatch problems and management options. *Endangered Species Research*, **5**, 193–204.

Chilvers, B. L., Delean, S., Gales, N. J., *et al.* 2004. Diving behavior of dugongs, *Dugong dugon*. *Journal of Experimental Marine Biology and Ecology*, **304**, 203–224.

Chilvers, B. L., Wilkinson, I. S., Duignan, P. J., & Gemmell, N. J. 2006. Diving to extremes: are New Zealand sea lions (*Phocarctos hookeri*) pushing their limits in a marginal habitat? *Journal of Zoology*, **269**, 233–240.

Chin, E. R., Olson, E. N., Richardson, J. A., et al. 1998. A calcineurin dependent transcriptional pathway controls skeletal muscle fiber type. *Genes & Development*, **12**, 2499–2509.

Clark, B. D. & Bemis, W. 1979. Kinematics of swimming of penguins at the Detroit Zoo. *Journal of Zoology*, **188**, 411–428.

Clark, J. C., Wert, S. E., Bachurski, C. J., *et al.* 1995. Targeted disruption of the surfactant protein B disrupts surfactant homeostasis, causing respiratory failure in newborn mice. *Proceedings of the National Academy of Sciences*, **92**, 7794–7798.

Clarke, J. & Nicol, S. 1993. Blood viscosity of the little penguin, *Eudyptula minor*, and the Adélie penguin, *Pygoscelis adeliae*: effects of temperature and shear rate. *Physiological Zoology*, **66**, 720–731.

Clowater, J. S. & Burger, A. E. 1994. The diving behaviour of pigeon guillemots (*Cepphus columba*) off southern Vancouver Island. *Canadian Journal of Zoology*, **72**, 863–872.

Colin, P., Ghaleh, B., Monnet, X., *et al.* 2003. Contributions of heart rate and contractility to myocardial oxygen balance during exercise. *American Journal of Physiology: Heart and Circulatory Physiology*, **284**, H676–H682.

Comellas, A. P., Dada, L. A., Lecuona, E., *et al.* 2006. Hypoxia-mediated degradation of Na,K-ATPase via mitochondrial reactive oxygen species and the ubiquitin-conjugating system. *Circulation Research*, **98**, 1314–1322.

Conde-Agudelo, A., Althabe, F., Belizán, J. M., & Kafury-Goeta, A. C. 1999. Cigarette smoking during pregnancy and risk of preeclampsia: a systematic review. *American Journal of Obstetrics and Gynecology*, **181**, 1026–1035.

Conroy, J. W. & Jenkins, D. 1986. Ecology of otters in northern Scotland: VI. Diving times and hunting success of otters (*Lutra lutra*) at Dinnet Lochs, Aberdeenshire and in Yell Sound, Shetland. *Journal of Zoology London*, **209**, 341–346.

Cook, T. R., Kato, A., Tanaka, H., Ropert-Coudert, Y., & Bost, C.-A. 2010. Buoyancy under control: underwater locomotor performance in a deep diving seabird suggests respiratory strategies for reducing foraging effort. *PloS ONE*, **5**, e9839.

Cook, T. R., Lescroel, A., Tremblay, Y., & Bost, C.-A. 2008. To breathe or not to breathe? Optimal breathing, aerobic dive limit, and oxygen stores in deep-diving blue-eyed shags. *Animal Behaviour*, **76**, 565–576.

Corsolini, M., Nigro, M., Olmastroni, S., Focardi, S., & Regoli, F. 2001. Susceptibility to oxidative stress in Adelie and emperor penguin. *Polar Biology*, **24**, 365–368.

Costa, D. P., Gales, N. J., & Crocker, D. E. 1998. Blood volume and diving ability of the New Zealand sea lion. *Physiological Zoology*, **71**, 208–213.

Costa, D. P., Gales, N. J., & Goebel, M. E. 2001. Aerobic dive limit: how often does it occur in nature? *Comparative Biochemistry and Physiology A*, **129**, 771–783.

Costa, D. P., Huckstadt, L. A., Crocker, D. E., *et al.* 2010. Approaches to studying climatic change and its role on the habitat selection of Antarctic pinnipeds. *Integrative and Comparative Biology*, **50**, 1018–1030.

Costa, D. P. & Kooyman, G. L. 1984. Contribution of specific dynamic action to heat balance and thermoregulation in the sea otter *Enhydra lutris*. *Physiological Zoology*, **57**, 199–203.

Costa, D. P., Kuhn, C. E., Weise, M. J., Shaffer, S. A., & Arnould, J. P. Y. 2004. When does physiology limit the foraging behaviour of freely diving mammals? *International Congress Series*, **1275**, 359–366.

Costidis, A. & Rommel, S. A. 2012. Vascularization of air sinuses and fat bodies in the head of the bottlenose dolphin (*Tursiops truncatus*): morphological implications on physiology. *Frontiers in Physiology*. DOI: 10.3389/fphys.2012.00243.

Coulombe, H. N., Ridgway, S. H., & Evans, W. E. 1965. Respiratory water exchange in two species of porpoise. *Science*, **149**, 86–88.

Cowan, D. F. & Curry, B. E. 2008. Histopathology of the alarm reaction in small odontocetes. *Journal of Comparative Pathology*, **139**, 24–33.

Cox, T. M., Ragen, T. J., Read, A. J., *et al.* 2006. Understanding the impacts of anthropogenic sound on beaked whales. *Journal of Cetacean Research Management*, **7**, 177–187.

Cozzi, B., Bagnoli, P., Acocella, F., & Costantino, M. L. 2005. Structure and biomechanical properties of the trachea of the striped dolphin *Stenella coeruleoalba*: evidence for evolutionary adaptations to diving. *Anatomical Record*, **284**, 500–510.

Craig, Jr, A. B. 1968. Depth limits of breath hold diving (an example of fennology). *Respiration Physiology*, **5**, 14–22.

Craig, Jr, A. B. & Påsche, A. 1980. Respiratory physiology of freely diving harbor seals (*Phoca vitulina*). *Physiological Zoology*, **53**, 419–432.

Cranford, T. W., Amundin, M., & Norris, K. S. 1996. Functional morphology and homology in the odontocete nasal complex: implications for sound generation. *Journal of Morphology*, **228**, 223–285.

Cranford, T. W. & Krysl, P. 2015. Fin whale sound reception mechanisms: skull vibration enables low-frequency hearing. *PloS ONE*, **10**, e0116222.

Cranford, T. W., Krysl, P., & Amundin, M. 2010. A new acoustic portal into the odontocete ear and vibrational analysis of the tympanoperiotic complex. *PloS ONE*, **5**.8, e11927. DOI:10.371/journal.pone.0011927.

Cranford, T. W., Mckenna, M. F., Soldevilla, M. S., *et al.* 2008. Anatomic geometry of sound transmission and reception in Cuvier's beaked whale (*Ziphius cavirostris*). *Anatomical Record*, **29**, 353–378.

Cranford, T. W., Van Bonn, W. G., Chaplin, M. S., *et al.* 1997. Visualizing dolphin sonar signal generation using high-speed video endoscopy. *Journal of the Acoustical Society of America*, **102**, 3123.

Creed, J. C. 2004. Capybara (*Hydrochaeris hydrochaeris* Rodentia: Hydrochaeridae): a mammalian seagrass herbivore. *Estuaries*, **27**, 197–200.

Crespo, F. A. & Lauria De Cidre, L. 2005. Functional significance of bronchial sphincters in two Southwestern Atlantic dolphins: *Pontoporia blainvillei* and *Lagenorhynchus obscurus* – a comparative approach. *Mammalia*, **69**, 233–238.

Crocker, D. E., Le Boeuf, B. J., & Costa, D. P. 1997. Drift diving in female northern elephant seals. *Canadian Journal of Zoology*, **75**, 27–39.

Croll, D. A., Acevedo-Gutiérrez, A., Tershy, B. R., & Urbán-Ramírez, J. 2001. The diving behavior of blue and fin whales: is dive duration shorter than expected based on oxygen stores? *Comparative Biochemistry and Physiology Part A: Molecular & Integrative Physiology*, **129**, 797–809.

Croll, D. A., Gaston, A. J., Burger, A. E., & Konnoff, D. 1992a. Foraging behavior and physiological adaptation for diving in thick-billed murres. *Ecology*, **73**, 344–356.

Croll, D. A. & McLaren, E. 1993. Diving metabolism and thermoregulation in common and thick-billed murres. *Journal of Comparative Physiology B: Biochemical, Systemic, and Environmental Physiology*, **163**, 160–166.

Croll, D. A., Nishiguchi, M. K., & Kaupp, S. 1992b. Pressure and lactate dehydrogenase function in diving mammals and birds. *Physiological Zoology*, **65**, 1022–1027.

Croll, D. A., Tershy, B. R., Hewitt, R. P., *et al.* 1998. An integrated approach to the foraging ecology of marine birds and mammals. *Deep-sea Research II*, **45**, 1353–1371.

Cross, E. R. 1965. Taravana: diving syndrome in the Tuamotu diver. In: Rahn, H. & Yokoyama, T. (eds.) *Physiology of breath-hold diving and the Ama of Japan.* Washington, DC: National Academy of Sciences, National Research Council.

Cross, J. P., Mackintosh, C. G., & Griffin, J. F. T. 1988. Effect of physical restraint and xylazine sedation of haemotological values in red deer (*Cervus elaphus*). *Research in Veterinary Science*, **45**, 281–286.

Croxall, J. P. (ed.) 1987. *Seabirds: Feeding Ecology and Role in Marine Ecosystems*, Cambridge: Cambridge University Press.

Croxall, J. P., Briggs, D. R., Kato, A., *et al.* 1993. Diving pattern and performance in the macaroni penguin *Eudyptes chrysolophus*. *Journal of Zoology, London*, **230**, 31–47.

Croxall, J. P., Naito, Y., Kato, A., Rothery, P., & Briggs, D. 1991. Diving patterns and performance in the Antarctic blue-eyed shag *Phalacrocorax atriceps*. *Journal of Zoology London*, **225**, 177–199.

Crum, L. A., Bailey, M. R., Guan, J., *et al.* 2005. Monitoring bubble growth in supersaturated blood and tissue ex vivo and the relevance to marine mammal bioeffects. *Acoustics Research Letters Online*, **6**, 214–220.

Crum, L. A. & Mao, Y. 1996. Acoustically enhanced bubble growth at low frequencies and its implications for human diver and marine mammal safety. *Journal of the Acoustic Society of America*, **99**, 2898–2907.

Crush, K. G. 1970. Carnosine and related substances in animal tissues. *Comparative Biochemistry and Physiology*, **34**, 3–30.

Culik, B. 1992. Diving heart rates in Adelie penguins (*Pygoscelis adeliae*). *Comparative Biochemistry and Physiology*, **102A**, 487–490.

Culik, B. 2001. Finding food in the open ocean: foraging strategies in Humboldt penguins. *Zoology*, **104**, 327–338.

Culik, B. M., Hennicke, J., & Martin, T. 2000. Humboldt penguins outmanoeuvring El Nino. *Journal of Experimental Biology*, **203**, 2311–2322.

Culik, B. M., Putz, K., Wilson, R. P., *et al.* 1996a. Diving energetics in king penguins (*Aptenodytes patagonicus*). *Journal of Experimental Biology*, **199**, 973–983.

Culik, B. M., Putz, K., Wilson, R. P., *et al.* 1996b. Core temperature variability in diving king penguins (*Aptenodytes patagonicus*): a preliminary analysis. *Polar Biology*, **16**, 371–378.

Culik, B. M., Wilson, R. P., & Bannasch, R. 1994. Under-water swimming at low energetic cost by Pygoscelid penguins. *Journal of Experimental Biology*, **197**, 65–78.

Culik, B. M., Wilson, R. P., Dannfeld, R., *et al.* 1991. Pygoscelid penguins in a swim canal. *Polar Biology*, **11**, 277–282.

Cuyler, L. C., Wiulsrød, R., & Øritsland, N. A. 1992. Thermal infrared radiation from free living whales. *Marine Mammal Science*, **8**, 120–134.

Damant, G. C. C. 1934. Physiology of deep diving in the whale. *Nature*, **133**, 874.

Daniels, C. B. & Orgeig, S. 2003. Pulmonary surfactant: the key to the evolution of air breathing. *News Physiological Sciences*, **18**, 151–157.

Daniels, D. & Grossman, Y. 2003. Biological effects of pressure. In: Brubakk, A. O. & Neuman, T. S. (eds.) *Bennett and Elliott's Physiology and Medicine of Diving*. Edinburgh: Saunders.

Davenport, J., Cotter, L., Rogan, E., Kelliher, D., & Murphy, C. 2013. Structure, material characteristics and function of the upper respiratory tract of the pygmy sperm whale. *The Journal of Experimental Biology*, **216**, 4639–4646.

Davenport, J., Hughes, R. N., Shorten, M., & Larsen, P. S. 2011. Drag reduction in emperor penguins: a novel hypothesis. *Marine Ecology Progress Series*, **430**, 171–182.

Davis, L. S. & Darby, J. T. (eds.) 1990. *Penguin Biology*. London: Academic Press.

Davis, M. B. & Guderley, H. 1987. Energy metabolism in the locomotor muscles of the common nurre (*Uria aalge*) and the atlantic puffin (*Fratercula arctica*). *The Auk*, **104**, 733–739.

Davis, M. B. & Guderley, H. 1990. Biochemical adpatations to diving in the common murre (*Uria aalge*) and the atlantic puffin (*Fratercula arctica*). *Journal of Experimental Zoology*, **253**, 235–244.

Davis, R. W. 2014. A review of the multi-level adaptations for maximizing aerobic dive duration in marine mammals: from biochemistry to behavior. *Journal of Comparative Physiology B*, **184**, 23–53.

Davis, R. W., Castellini, M. A., Kooyman, G. L., & Maue, R. 1983. Renal GFR and hepatic blood flow during voluntary diving in Weddell seals. *American Journal of Physiology*, **245**, R743–R748.

Davis, R. W., Castellini, M. A., Williams, T. M., & Kooyman, G. L. 1991. Fuel homeostasis in the harbor seal during submerged swimming. *Journal of Comparative Physiology B*, **160**, 627–635.

Davis, R. W., Fuiman, L. A., Williams, T. M., *et al.* 1999. Hunting behavior of a marine mammal beneath the Antarctic fast ice. *Science*, **283**, 993–996.

Davis, R. W. & Kanatous, S. B. 1999. Convective oxygen transport and tissue oxygen consumption in Weddell seals during aerobic dives. *Journal of Experimental Biology*, **202**, 1091–1113.

Davis, R. W. & Williams, T. M. 2012. The marine mammal dive response is exercise modulated to maximize aerobic dive duration. *Journal of Comparative Physiology A*, **198**, 583–591.

Davis, R. W., Worthy, G. A. J., Wursig, B., & Lynn, S. K. 1996. Diving behavior and at-sea movements of an Atlantic spotted dolphin in the Gulf of Mexico. *Marine Mammal Science*, **12**, 569–581.

Dawson, C., Vincent, J. F. V., Jeronimidis, G., Rice, G., & Forshaw, P. 1999. Heat transfer through penguin feathers. *Journal of Theoretical Biology*, **199**, 291–295.

De Burgh Daly, M., Elsner, R., & Angell-James, J. E. 1977. Cardiorespiratory control by carotid chemoreceptors during experimental dives in the seal. *American Journal of Physiology: Heart and Circulatory Physiology*, **232**, H508–H516.

De Leeuw, J. J. 1996. Diving costs as a component of daily energy budgets of aquatic birds and mammals: generalizing the inclusion of dive-recovery costs demonstrated in tufted ducks. *Canadian Journal of Zoology*, **74**, 2131–2142.

De Miranda, M. A., Schlater, A. E., Green, T. L., & Kanatous, S. B. 2012. In the face of hypoxia: myoglobin increases in response to hypoxic conditions and lipid supplementation in cultured Weddell seal skeletal muscle cells. *The Journal of Experimental Biology*, **215**, 806–813.

De Vries, J. & Van Eerden, M. R. 1995. Thermal conductance in aquatic birds in relation to the degree of water contact, body mass, and body fat: energetic implications of living in a strong cooling environment. *Physiological Zoology*, **68**, 1143–1163.

Dehnhardt, G., Mauck, B., & Bleckmann, H. 1998. Seal whiskers detect water movements. *Nature*, **394**, 235–236.

Dehnhardt, G., Mauck, B., Hanke, W., & Bleckmann, H. 2001. Hydrodynamic trail-following in harbor seals (*Phoca vitulina*). *Science*, **293**, 102–104.

Dejours, P. 1987. Water and air physical characteristics and their physiological consequences. In: Dejours, P., Bolis, L., Taylor, C. R. & Weibel, E. R. (eds.), *Comparative Physiology: Life in Water and On Land*. New York: Springer.

Dejours, P. 1989. From comparative physiology of respiration to several problems of environmental adaptations and to evolution. *Journal of Physiology*, **410**, 1–19.

Dempsey, J. A., Veasey, S. C., Morgan, B. J., & O'Donnell, C. P. 2010. Pathophysiology of sleep apnea. *Physiological Reviews*, **90**, 47–112.

Dendrinos, P., Karamanlidis, A. A., Androukaki, E., & Mcconnell, B. J. 2007. Diving development and behavior of a rehabilitated Mediterranean monk seal (*Monachus monachus*). *Marine Mammal Science*, **23**, 387–397.

Denison, D. M. & Kooyman, G. L. 1973. The structure and the function of the small airways in pinniped and sea otter lungs. *Respiration Physiology*, **17**, 1–10.

Denison, D. M., Warrell, D. A., & West, J. B. 1971. Airway structure and alveolar emptying in the lungs of sea lions and dogs. *Respration Physiology*, **13**, 252–260.

Dennison, S., Fahlman, A., & Moore, M. J. 2012a. The use of diagnostic imaging for identifying abnormal gas accumulations in cetaceans and pinnipeds. *Frontiers in Physiology*. DOI: 10.3389/fphys.2012.00181.

Dennison, S., Boor, M., Faquier, D., *et al.* 2011. Foramen ovale and ductus arteriosus patency in neonatal harbor seal (*Phoca vitulina*) pups in rehabilitation. *Aquatic Mammals*, **37**, 161–166.

Dennison, S., Moore, M. J., Fahlman, A., *et al.* 2012b. Bubbles in live-stranded dolphins. *Proceedings of the Royal Society B: Biological Sciences*, **279**, 1396–1404.

Deruiter, S. L., Southall, B. L., Calambokidis, J., *et al.* 2013. First direct measurements of behavioural responses by Cuvier's beaked whales to mid-frequency active sonar. *Biology Letters*, **9**. DOI: 10.1098/rsbl.2013.0223.

Dhindsa, D. S., Metcalfe, J., Hoversland, A. S., & Hartman, R. A. 1974. Comparative studies of the respiratory functions of mammalian blood X: killer whale (*Orcinus orca* Linnaeus) and beluga whale (*Delphinapterus leucas*). *Respiration Physiology*, **20**, 93–103.

Djojosugito, A. M., Folkow, B., & Yonce, L. R. 1969. Neurogenic adjustments of muscle blood flow, cutaneous A-V shunt flow and of venous tone during "diving" in ducks. *Acta Physiologica Scandinavica*, **75**, 377–386.

Dolar, M. L. L., Suarez, P., Ponganis, P. J., & Kooyman, G. L. 1999. Myoglobin in pelagic small cetaceans. *Journal of Experimental Biology*, **202**, 227–236.

Dormer, K. J., Denn, M. J., & Stone, H. L. 1977. Cerebral blood flow in the sea lion (*Zalophus californianus*) during voluntary dives. *Comparative Biochemistry and Physiology Part A: Physiology*, **58**, 11–18.

Dow, D. 1964. Diving times of wintering water birds. *The Auk*, **81**, 556–558.

Drabek, C. M. 1975. Some anatomical aspects of the cardiovascular system of Antarctic seals and their possinble functional significance in diving. *Journal of Morphology*, **145**, 85–105.

Drabek, C. M. 1977. Some anatomical and functional aspects of seal hearts and aortae. In: Harrison, J. R. (ed.), *Functional Anatomy of Marine Mammals*. San Francisco, CA: Academic Press.

Drabek, C. M. 1989. Heart and ventricle weights of Antarctic penguins. *Canadian Journal of Zoology*, **67**, 2602–2604.

Drabek, C. M. 1997. Heart and ventricle weights of the little penguin, *Eudyptula minor*. *Emu*, **97**, 258–261.

Drabek, C. M. & Burns, J. M. 2002. Heart and aorta morphology of the deep-diving hooded seal (*Cystophora cristata*). *Canadian Journal of Zoology*, **80**, 2030–2036.

Drabek, C. M. & Kooyman, G. L. 1986. Bronchial morphometry of the upper conductive zones of four odontocete cetaceans. In: Harrison, R. J. (ed.), *Research in Dolphins*. Oxford: Oxford University Press.

Drabek, C. M. & Tremblay, Y. 2000. Morphological aspects of the heart of the northern rockhopper penguin (*Eudyptes chrysocome moseley i*): possible implication in diving behavior and ecology? *Polar Biology*, **23**, 812–816.

Drouout, V., Gannier, A., & Goold, J. C. 2004. Diving and feeding behaviour of sperm whales (*Physteter macrocephalus*) in the Northwestern Mediterranean Sea. *Aquatic Mammals*, **30**, 419–426.

Drummond, P. C. & Jones, D. R. 1970. The initiation and maintenance of bradycardia in a diving mammal, the muskrat, *Ondatra zibethica*. *Journal of Physiology*, **290**, 253–271.

Du, N., Fan, J., Wu, H., Chen, S., & Liu, Y. 2007. An improved model of heat transfer through penguin feathers and down. *Journal of Theoretical Biology*, **248**, 727–735.

Duchamp, C., Barre, H., Delage, D., *et al.* 1989. Nonshivering thermogenesis and adaptation to fasting in king penguin chicks. *American Journal of Physiology: Regulatory, Integrative and Comparative Physiology*, **257**, R744–R751.

Duchamp, C., Barre, H., Rouanet, J. L., *et al.* 1991. Nonshivering thermogenesis in king penguin chicks: I. Role of skeletal muscle. *American Journal of Physiology: Regulatory, Integrative and Comparative Physiology*, **261**, R1438–R1445.

Duchamp, C., Chatonnet, J., Dittmar, A., & Barre, H. 1993. Increased role of skeletal muscle in the calorigenic response to glucagon of cold-acclimated ducklings. *American Journal of Physiology: Regulatory, Integrative and Comparative Physiology*, **265**, R1084–R1091.

Duchamp, C., Rouanet, J.-L., & Barré, H. 2002. Ontogeny of thermoregulatory mechanisms in king penguin chicks (*Aptenodytes patagonicus*). *Comparative Biochemistry and Physiology Part A: Molecular & Integrative Physiology*, **131**, 765–773.

Duffy, D. C. 1983. The foraging ecology of Peruvian seabirds. *The Auk*, **100**, 800–810.

Dumonteil, E., Barre, H., & Meissner, G. 1993. Sarcoplasmic reticulum $Ca^{(2+)}$-ATPase and ryanodine receptor in cold-acclimated ducklings and thermogenesis. *American Journal of Physiology: Cell Physiology*, **265**, C507–C513.

Duncker, H.-R. 1974. Structure of the avian respiratory tract. *Respiration Physiology*, **22**, 1–19.

Duncker, H. 1972. Structure of avian lungs. *Respiration Physiology*, **14**, 44–63.

Dunkin, R. C., McLellan, W. A., Blum, J. E., & Pabst, D. A. 2005. The ontogenetic changes in the thermal properties of blubber from Atlantic bottlenose dolphin *Tursiops truncatus*. *Journal of Experimental Biology*, **208**, 1469–1480.

Duntley, S. Q. 1963. Light in the sea. *Journal of the Optical Society of America*, **53**, 214–233.

Duran, W. N. & Renkin, E. M. 1974. Oxygen consumption and blood flow in resting mammalian skeletal muscle. *American Journal of Physiology*, **226**, 173–177.

Dutka, A. J. 2003. Long term effects on the central nervous system. In: Brubakk, A. O. & Neuman, T. S. (eds.), *Bennett and Elliott's Physiology and Medicine of Diving*. Edinburgh: Saunders.

Dutschmann, M. & Herbert, H. 1998. NMDA and $GABA_A$ receptors in the rat Kölliker-Fuse area control cardiorespiratory responses evoked by trigeminal ethmoidal nerve stimulation. *The Journal of Physiology*, **510**, 793–804.

Dyck, M. G. & Romberg, S. 2007. Observations of a wild polar bear (*Ursus maritimus*) successfully fishing Arctic charr (*Salvelinus alpinus*) and fourhorn sculpin (*Myxocephalus quadricornis*). *Polar Biology*, **30**, 1625–1628.

Dykes, R. W. 1974. Factors related to the dive reflex in harbor seals: sensory contributions from the trigeminal region. *Canadian Journal of Physiology and Pharmacology*, **52**, 259–265.

Eckenhoff, R. C., Olstad, C. S., & Carrod, G. 1990. Human dose–response relationship for decompression and endogenous bubble formation. *Journal of Applied Physiology*, **69**, 914–918.

Eckenhoff, R. G. & Knight, D. R. 1984. Cardiac arrhythmias and heart rate changes in prolonged hyperbaric exposure. *Undersea Biomedical Research*, **11**, 355–366.

Edgar, A. 1962. A note on the diving of the two New Zealand grebes. *Notornis*, **10**, 42.

Edwards, H. H. & Schnell, G. D. 2001. Body length, swimming speed, dive duration, and coloration of the dolphin *Sotalia fluviatilis* (*Tucuxi*) in Nicaragua. *Caribbean Journal of Science*, **37**, 271–298.

Eguchi, T. & Harvey, J. T. 2005. Diving behavior of the Pacific Harbor seal (*Phoca vitulina richardsi*) in Monterey Bay, California. *Marine Mammal Science*, **21**, 283–295.

El-Sayed, H., Goodall, S. R., & Hainsworth, R. 1995. Re-evaluation of Evans Blue dye dilution method of plasma volume measurements,. *Clinical & Laboratory Haematology*, **17**, 189–194.

Eliassen, E. 1960. *Cardiovascular Responses to Submersion Asphyxia in Avian Divers*. Bergen: Norwegian Universities Press .

Elliott, K. H., Davoren, G. K., & Gaston, A. G. 2007. The influence of buoyancy and drag on the dive behaviour of an Arctic seabird, the thick-billed murre. *Canadian Journal of Zoology*, **85**, 352–361.

Elliott, K. H., Shoji, A., Campbell, K. L., & Gaston, A. J. 2010. Oxygen stores and foraging behavior of two sympatric, planktivorous alcids. *Aquatic Biology*, **8**, 221–235.

Elliott, N. M., Andrews, R. D., & Jones, D. R. 2002. Pharmacological blockade of the dive response: effects on heart rate and diving behaviour in the harbour seal (*Phoca vitulina*). *Journal of Experimental Biology*, **205**, 3757–3765.

Elowson, A. M. 1984. Spread-wing postures and the water repellency of feathers: a test of Rijke's hypothesis. *The Auk*, **101**, 371–383.

Elsner, R. 1965. Heart rate response in forced versus trained experimental dives of pinnipeds. *Hvalradets Skrifter*, **48**, 24–29.

Elsner, R. 1966. Diving bradycardia in the unrestrained hippopotamus. *Nature*, **212**, 408.

Elsner, R. 1969. Cardiovascular adjustments to diving. In: Andersen, H. T. (ed.) *The Biology of Marine Mammals*. New York: Academic Press.

Elsner, R. 1988. Anaerobic contributions to metabolism in diving seals. *Canadian Journal of Zoology*, **66**, 142–143.

Elsner, R. 2015. *Diving Seals and Meditating Yogis: Strategic Metabolic Retreats*. Chicago, IL: University of Chicago Press.

Elsner, R., Angell-James, J. E., & De Burgh Daly, M. 1977. Carotid body chemoreceptor reflexes and their interactions in the seal. *American Journal of Physiology: Heart and Circulatory Physiology*, **232**, H517–H525.

Elsner, R. & De La Lande, I. S. 1998. Heterogeneous cholinergic reactions of ringed seal coronary arteries. *Comparative Biochemistry and Physiology A*, **119**, 1019–1025.

Elsner, R., Franklin, D. L., Van Citters, R. L., & Kenney, D. W. 1966a. Cardiovascular defense against asphyxia. *Science*, **153**, 941–949.

Elsner, R., Geoerge, J. C., & O'hara, T. 2004a. Vasomotor responses of isolated peripheral blood vessels from bowhead whales: thermoregulatory implications. *Marine Mammal Science*, **20**, 546–553.

Elsner, R. & Gooden, B. 1983. *Diving and Asphyxia: A Comparative Study of Animals and Man*. Cambridge: Cambridge University Press.

Elsner, R., Hammond, D. D., & Parker, H. R. 1970a. Circulatory responses to asphyxia in pregant and fetal animals: a comparative study of Weddell seals and sheep. *Yale Journal of Biology and Medicine*, **42**, 202–217.

Elsner, R., Hanafee, W. N., & Hammond, D. D. 1971. Angiography of the inferior vena cava of the harbor seal during simulated diving. *American Journal of Physiology*, **220**, 1155–1157.

Elsner, R., Kenney, D. W., & Burgess, K. 1966b. Bradycardia in the trained dolphin. *Nature*, **212**, 407.

Elsner, R. & Meiselman, H. J. 1995. Splenic oxygen storage and blood viscosity in seals. *Marine Mammal Science*, **11**, 93–96.

Elsner, R., Meiselman, H. J., & Baskurt, O. K. 2004b. Temperature–viscosity relations of bowhead whale blood: a possible mechanism for maintaing cold blood flow. *Marine Mammal Science*, **20**, 339–344.

Elsner, R., Millard, R. W., Kjekshus, J. K., *et al.* 1985. Coronary blood flow and myocardial segment dimensions during simulated dives in seals. *American Journal of Physiology*, **249**, H1119–H1126.

Elsner, R., Oyasaetr, S., Almaas, R., & Sauagstad, O. D. 1998. Diving seals, ischemia-reperfusion, and oxygen radicals. *Comparative Biochemistry and Physiology Part A: Molecular & Integrative Physiology*, **119**, 975–980.

Elsner, R., Shurley, J. T., Hammond, D. D., & Brooks, R. E. 1970b. Cerebral tolerance to hypoxemia in asphyxiated Weddell seals. *Respiration Physiology*, **9**, 287–297.

Elsner, R., Wartzok, D., Sonafrank, N. B., & Kelly, B. P. 1989. Behavioral and physiological reactions of arctic seals during under-ice pilotage. *Canadian Journal of Zoology*, **67**, 2506–2513.

Elsner, R. W., Franklin, D. L., & Van Citters, R. L. 1964a. Cardiac output during diving in an unrestrained sea lion. *Nature*, **202**, 809–810.

Elsner, R. W., Scholander, P. F., Craig, A. B., *et al.* 1964b. A venous blood oxygen reservoir in the diving elephant seal. *The Physiologist*, **7**, 124.

Elsner, R. W., Shurley, J. T., Hammond, D. D., & Brooks, R. E. 1970c. Cerebral tolerance in asphyxiated Weddell seals. *Respiration Physiology*, **9**, 287–297.

Emerling, B. M., Weinberg, F., Snyder, C., *et al.* 2009. Hypoxic activation of AMPK is dependent on mitochondrial ROS but independent of an increase in AMP/ATP ratio. *Free Radical Biology and Medicine*, **46**, 1386–1391.

English, A. W. 1976. Limb movements and locomotor function in the California sea lion (*Zalophus californianus*). *Journal of Zoology*, **178**, 341–364.

English, A. W. 1977. Structural correlates of forelimb function in fur seals and sea lions. *Journal of Morphology*, **151**, 325–352.

Enstipp, M. R., Andrews, R. D., & Jones, D. R. 1999. Cardiac responses to first ever submergence in double-crested cormorant chicks (*Phalacrocorax auritus*). *Comparative Biochemistry and Physiology Part A: Molecular & Integrative Physiology*, **124**, 523–530.

Enstipp, M. R., Andrews, R. D., & Jones, D. R. 2001. The effects of depth on the cardiac and behavioural responses of double-crested cormorants (*Phalacrocorax auritus*) during voluntary diving. *Journal of Experimental Biology*, **204**, 4081–4092.

Enstipp, M. R., Gremillet, D., & Jones, D. R. 2006. The effects of depth, temperature and food ingestion on the foraging energetics of a diving endotherm, the double-crested cormorant (*Phalacrocorax auritus*). *Journal of Experimental Biology*, **209**, 845–859.

Enstipp, M. R., Grémillet, D., & Jones, D. R. 2008. Heat increment of feeding in double-crested cormorants (*Phalacrocorax auritus*) and its potential for thermal substitution. *Journal of Experimental Biology*, **211**, 49–57.

Enstipp, M. R., Grémillet, D., & Lorentsen, S.-H. 2005. Energetic costs of diving and thermal status in European shags (*Phalacrocorax aristotelis*). *Journal of Experimental Biology*, **208**, 3451–3461.

Evans, B., Jones, D., Baldwin, J., & Gabbott, G. 1994. Diving ability of the platypus. *Journal of Zoology*, **42**, 17–27.

Evans, W. E. 1971. Orientation behavior of delphinids: radio telemetric studies. *Annals of the New York Academy of Sciences*, **188**, 142–160.

Fagius, J. & Sundolf, G. 1986. The diving response in man: effects on sympathetic activity in muscle and skin nerve fascicles. *Journal of Physiology*, **377**, 429–443.

Fago, A., Jensen, F. B., Tota, B., *et al.* 2012. Integrating nitric oxide, nitrite and hydrogen sulfide signaling in the physiological adaptations to hypoxia: a comparative approach. *Comparative Biochemistry and Physiology Part A: Molecular & Integrative Physiology*, **162**, 1–6.

Fahlman, A., Hooker, S. K., Olszowka, A., Bostrom, B. L., & Jones, D. R. 2009. Estimating the effect of lung collapse and pulmonary shunt on gas exchange during breath-hold diving: the Scholander and Kooyman legacy. *Respiration Physiology and Neurobiology*, **165**, 28–39.

Fahlman, A., Loring, S. H., Ferrigno, M., *et al.* 2011. Static inflation and deflation pressure–volume curves from excised lungs of marine mammals. *Journal of Experimental Biology*, **214**, 3822–3828.

Fahlman, A., Olszowka, A., Bostrom, B., & Jones, D. R. 2006. Deep diving mammals: dive behavior and circulatory adjustments contribute to bends avoidance. *Respiration Physiology and Neurobiology*, **153**, 66–77.

Fahlman, A., Schmidt, A., Jones, D. R., Bostrom, B. L., & Handrich, Y. 2007. To what extent might N_2 limit dive performance in king penguins? *Journal of Experimental Biology*, **210**, 3344–3355.

Fahlman, A., Svard, C., Rosen, D. A., Jones, D. R., & Trites, A. W. 2008. Metabolic costs of foraging and management of O_2 and CO_2 stores in Steller sea lions. *Journal of Experimental Biology*, **211**, 3573–3580.

Fahlman, A., Tyack, P. L., Miller, P. J., & Kvadsheim, P. H. 2014. How man-made interference might cause gas bubble emboli in deep diving whales. *Frontiers in Physiology*. DOI: 10.3389/fphys.2014.00013.

Fajardo, R. J., Hernandez, E., & O'Connor, P. M. 2007. Postcranial skeletal pneumaticity: a case study in the use of quantitative microCT to assess vertebral structure in birds. *Journal of Anatomy*, **211**, 138–147.

Falabella, V., Lewis, M., & Campagna, C. 1999. Development of cardiorespiratory patterns associated with terrestrial apneas in free-ranging southern elephant seals. *Physiological and Biochemical Zoology*, **72**, 64–70.

Falk, K., Pedersen, C. E., & Kampp, K. 2000. Measurements of diving depth in dovekies (*Alle alle*). *The Auk*, **117**, 522–525.

Falke, K. J., Busch, T., Hoffman, O., *et al.* 2008. Breathing pattern, CO_2 elimination and the absence of exhaled NO in freely diving Weddell seals. *Respiratory Physiology and Neurobiology*, **162**, 85–92.

Falke, K. J., Hill, R. D., Qvist, J., *et al.* 1985. Seal lungs collapse during free diving: evidence from arterial nitrogen tensions. *Science*, **229**, 556–558.

Faraci, F. M. 1991. Adaptations to hypoxia in birds: how to fly high. *Annual Review of Physiology*, **53**, 59–70.

Farhi, L. E. & Rahn, H. 1960. Dynamics of changes in carbon dioxide stores. *Anesthesiology*, **21**, 604–614.

Farmer, M., Weber, R. E., Bonaventura, J., Best, R. C., & Domming, D. 1979. Functional properties of hemoglobin and whole blood in an aquatic mammal, the Amazonian manatee (*Trichechus inunguis*). *Comparative Biochemistry and Physiology Part A: Comparative Physiology*, **62**, 231–238.

Fedak, M. A. 1986. Diving and exercise in seals: a benthic perspective. In: Brubakk, A. O., Kanwisher, J. W. & Sudnes, G. (eds.) *Diving in Animals and Man*. Trondheim: Tapir Publishers.

Fedak, M. A., Pullen, M. R., & Kanwisher, J. 1988. Circulatory responses of seals to periodic breathing: heart rate and breathing during exercise and diving in the laboratory and open sea. *Canadian Journal of Zoology*, **66**, 53–60.

Fedak, M. A., Rome, L., & Seeherman, H. J. 1981. One-step N2-dilution technique for calibrating open-circuit VO2 measuring systems. *Journal of Applied Physiology*, **51**, 772–776.

Feldkamp, S. D. 1987a. Forelimb propulsion in the California sea lion *Zalophus californianus*. *Journal of the Zoological Society of London*, **212**, 4333–4357.

Feldkamp, S. D. 1987b. Swimming in the California sea lion: morphometrics, drag and energetics. *Journal of Experimental Biology*, **131**, 117–135.

Feldkamp, S. D., Delong, R. L., & Antonelis, G. A. 1989. Diving patterns of California sea lions, *Zalophus californianus*. *Canadian Journal of Zoology*, **67**, 872–883.

Fernandez, A., Edwards, J., Martin, V., *et al.* 2005. "Gas and fat embolic syndrome" involving a mass stranding of beaked whales exposed to anthropogenic sonar signals. *Journal of Veterinary Pathology*, **42**, 446–457.

Ferrante, F. 1970. Oxygen conservation during submergence apnea in a diving mammal, the nutria. *American Journal of Physiology: Legacy Content*, **218**, 363–371.

Ferrante, F. L. & Opdyke, D. F. 1969. Mammalian ventricular function during submersion asphyxia. *Journal of Applied Physiology*, **26**, 561–570.

Ferretti, G., Costa, M., Ferrigno, M., *et al.* 1991. Alveolar gas composition and exchange during deep breath-hold diving and dry breath holds in elite divers. *Journal of Applied Physiology*, **70**, 794–802.

Ferrigno, M., Ferretti, G., Ellis, A., *et al.* 1997. Cardiovascular changes during deep breath-hold dives in a pressure chamber. *Journal of Applied Physiology*, **83**, 1282–1290.

Ferrigno, M. & Lundgren, C. E. 2003. Breath-hold diving. In: Brubakk, A. O. & Neuman, T. S. (eds.), *Bennett and Elliott's Physiology and Medicine of Diving*. Edinburgh: Saunders.

Field, C. L. & Tablin, F. 2012. Response of northern elephant seal platelets to pressure and temperature changes: a comparison with human platelets. *Comparative Biochemistry and Physiology Part A: Molecular & Integrative Physiology*, **162**, 289–295.

Field, C. L., Walker, N. J., & Tablin, F. 2001. Northern elephant seal platelets: analysis of shape change and response to platelet agonists. *Thrombosis Research*, **101**, 267–277.

Fischbeck, K. H. & Simon, R. P. 1981. Neurological manifestations of accidental hypothermia. *Annals of Neurology*, **10**, 384–387.

Fish, F. 1994. Influence of hydrodynamic-design and propulsive mode on mammalian swimming energetics. *Australian Journal of Zoology*, **42**, 79–101.

Fish, F. 1998. Comparative kinematics and hydrodynamics of odontocete cetaceans: morphological and ecological correlates with swimming performance. *Journal of Experimental Biology*, **201**, 2867–2877.

Fish, F. E., Howle, L. E., & Murray, M. M. 2008. Hydrodynamic flow control in marine mammals. *Integrative and Comparative Biology*, **48**, 788–800.

Fish, F. E. & Hui, C. A. 1991. Dolphin swimming: a review. *Mammal Review*, **21**, 181–195.

Fish, F. E., Innes, S., & Ronald, K. 1988. Kinematics and estimated thrust production of swimming harp and ringed seals. *Journal of Experimental Biology*, **137**, 157–173.

Fish, F. E. & Lauder, G. V. 2006. Passive and active flow control by swimming fishes and mammals. *Annual Review of Fluid Mechanics*, **38**, 193–224.

Fish, F. E., Legac, P., Williams, T. M., & Wei, T. 2014. Measurement of hydrodynamic force generation by swimming dolphins using bubble DPIV. *Journal of Experimental Biology*, **217**, 252–260.

Fish, F. E., Nicastro, A. J., & Weighs, D. 2006. Dynamics of the aerial manuvers of spinner dolphins. *Journal of Experimental Biology*, **209**, 590–598.

Flogel, U., Fago, A., & Rassaf, T. 2010. Keeping the heart in balance: the functional interactions of myoglobin with nitrogen oxides. *Journal of Experimental Biology*, **213**, 2726–2733.

Flögel, U., Fago, A., & Rassaf, T. 2010. Keeping the heart in balance: the functional interactions of myoglobin with nitrogen oxides. *Journal of Experimental Biology*, **213**, 2726–2733.

Flück, M. & Hoppeler, H. 2003. Molecular basis of skeletal muscle plasticity: from gene to form and function. *Reviews of Physiology, Biochemistry and Pharmacology*, **146**, 159–216.

Folkow, B., Fuxe, K., & Sonnenschein, R. R. 1966. Responses of skeletal musculature and it vasculature during "diving" in the duck: peculiarities of the adrenergic vasoconstrictor innervation. *Acta Physiologica Scandinavica*, **67**, 327–342.

Folkow, B., Lisander, B., & Öberg, B. 1971. Aspects of the cardiovascular nervous control in a mammalian diver (*Myocastor coypus*). *Acta Physiologica Scandinavica*, **82**, 439–446.

Folkow, B., Nilsson, N. J., & Yonce, L. R. 1967. Effects of "diving" on cardiac output in ducks. *Acta Physiologica Scandinavica*, **70**, 347–361.

Folkow, B. & Yonce, L. R. 1967. The negative inotropic effect of vagal stimulation on the heart ventricles of the duck. *Acta Physiologica Scandinavica*, **71**, 77–84.

Folkow, L. P. & Blix, A. S. 1987. Nasal heat and water exchange in gray seals. *American Journal of Physiology: Regulatory, Integrative and Comparative Physiology*, **253**, R883–R889.

Folkow, L. P. & Blix, A. S. 1989. Thermoregulatory control of expired air temperature in diving harp seals. *American Journal of Physiology: Regulatory, Integrative and Comparative Physiology*, **257**, R306–R310.

Folkow, L. P. & Blix, A. S. 1992. Metabolic rates of minke whales (*Balaenoptera acutorostrata*) in cold water. *Acta Physiologica Scandinavica*, **146**, 141–150.

Folkow, L. P. & Blix, A. S. 1999. Diving behaviour of hooded seals (*Cystophora cristata*) in the Greenland and Norwegian Seas. *Polar Biology*, **22**, 61–74.

Folkow, L. P., Nordøy, E. S., & Blix, A. S. 2004. Distribution and diving behaviour of harp seals (*Pagophilus groenlandicus*) from the Greenland Sea stock. *Polar Biology*, **27**, 281–298.

Folkow, L. P., Ramirez, J.-M., Ludvigsen, S., Ramirez, N., & Blix, A. S. 2008. Remarkable neuronal hypoxia tolerance in the deep-diving adult hooded seal (*Cystophora cristata*). *Neuroscience Letters*, **446**, 147–150.

Foot, N. J., Orgeig, S., & Daniels, C. B. 2006. The evolution of a physiological system: the pulmonary surfactant system in diving mammals. *Respiratory Physiology and Neurobiology*, **154**, 118–138.

Forbes, L. S. & Sealy, S. G. 1988. Diving behaviour of male and female western grebes. *Canadian Journal of Zoology*, **66**, 2695–2698.

Ford, T. J. & Kraus, S. D. 1992. A rete in the right whale. *Nature*, **359**, 680–680.

Ford, T. J., Werth, A. J., & George, J. C. 2013. An intraoral thermoregulatory organ in the bowhead whale (*Balaena mysticetus*), the corpus cavernosum maxillaris. *The Anatomical Record*, **296**, 701–708.

Fowler, S. L., Costa, D. P., Arnould, J. P. Y., Gales, N. J., & Burns, J. M. 2007. Ontogeny of oxygen stores and physiological diving capability in Australian sea lions. *Functional Ecology*, **21**, 922–935.

Fowler, S. L., Costa, D. P., Arnould, J. P. Y., Gales, N. J., & Kuhn, C. E. 2006. Ontogeny of diving behaviour in the Australian sea lion: trials of adolescence in a late bloomer. *Journal of Animal Ecology*, **75**, 358–367.

Francis, J., Boness, D., & Ochoa-Acuña, H. 1998. A protracted foraging and attendance cycle in female Juan Fernandez fur seals. *Marine Mammal Science*, **14**, 552–574.

Francis, T. J. R. & Mitchell, S. J. 2003. Pathophysiology of decompression sickness. In: Brubakk, A. O. & Neuman, T. S. (eds.), *Bennett and Elliott's Physiology and Medicine of Diving*. Edinburgh: Saunders.

Frankel, A. S. 2009. Sound production. In: Perrin, W. F., Wursig, B. & Thewissen, J. G. M. (eds.), *Encyclopedia of Marine Mammals*. San Diego, CA: Academic Press.

Frantzis, A. 1998. Does acoustic testing strand whales? *Nature*, **392**, 29.

Freeman, A., Nicholls, D., Wilson, K., & Bartle, J. 1997. Radio- and satellite-tracking Westland petrels *Procellaria westlandica*. *Marine Ornithology*, **25**, 31–36.

Friedlaender, A. S., Goldbogen, J. A., Nowacek, D. P., *et al.* 2014. Feeding rates and under-ice foraging strategies of the smallest lunge filter feeder, the Antarctic minke whale (*Balaenoptera bonaerensis*). *Journal of Experimental Biology*, **217**, 2851–2854.

Froget, G., Butler, P. J., Woakes, A. J., *et al.* 2004. Heart rate and energetics of free-ranging king penguins (*Aptenodytes patagonicus*). *Journal of Experimental Biology*, **207**, 3917–3926.

Frost, P. G. H., Siegfried, W. R., & Greenwood, P. J. 1975. Arterio-venous heat exchange systems in the Jackass penguin *Spheniscus demersus*. *Journal of Zoology*, **175**, 231–241.

Fujise, Y., Hidaka, H., Tatsukawa, R., & Miyazaki, N. 1985. External measurements and organ weights of five Weddell seals (*Leptonychotes weddelli*) caught near Syowa Station. *Antarctic Record (Tokyo)*, **85**, 96–101.

Furilla, R. A. & Jones, D. R. 1986. The contribution of nasal receptors to the cardiac response to diving in restrained and unrestrained redhead ducks (*Aythya americana*). *Journal of Experimental Biology*, **121**, 227–238.

Furilla, R. A. & Jones, D. R. 1987a. Cardiac responses to dabbling and diving in the mallard, *Anas platyrhynchos*. *Physiological Zoology*, **60**, 406–412.

Furilla, R. A. & Jones, D. R. 1987b. The relationship between dive and pre-dive heart rates in restrained and free dives by diving ducks. *Journal of Experimental Biology*, **127**, 333–348.

Fuson, A. L., Cowan, D. F., Kanatous, S. B., Polasek, L. K., & Davis, R. W. 2003. Adaptations to hypoxia in the heart, kidneys, and splanchnic organs of harbor seals (*Phoca vitulina*). *Journal of Experimental Biology*, **206**, 4139–4154.

Gabbott, G. R. J. & Jones, D. R. 1991. The effect of brain transection on the response to forced submergence in ducks. *Journal of the Automonic Nervous System*, **36**, 65–74.

Galantsev, V. P. 1991. Adaptational changes in the venous system of diving mammals. *Canadian Journal of Zoology*, **69**, 414–419.

Gales, N. J. & Mattlin, R. H. 1997. Summer diving behaviour of lactating New Zealand sea lions, *Phocarctos hookeri*. *Canadian Journal of Zoology*, **75**, 1695–1706.

Gales, R., Williams, C., & Ritz, D. 1990. Foraging behaviour of the little penguin, *Eudyptula minor*: initial results and assessment of instrument effect. *Journal of Zoology*, **220**, 61–85.

Galliano, R. E., Morgane, P. J., Mcfarland, W. L., Nagel, E. L., & Catherman, R. L. 1966. The anatomy of the cervicothoracic arterial system in the bottlenose dolphin (*Tursiops truncatus*) with a surgical approach suitable for guided angiography. *The Anatomical Record*, **155**, 325–337.

Gallivan, G. J. 1980. Hypoxia and hypercapnia in the respiratory control of the Amazonian manatee (*Trichechus inunguis*). *Physiological Zoology*, **53**, 254–261.

Gallivan, G. J. 1981. Ventilation and gas exchange in unrestrained harp seals. *Comparative Biochemistry and Physiology Part A: Physiology*, **69**, 809–813.

Gallivan, G. J. & Best, R. C. 1980. Metabolism and respiration of the Amazonian manatee (*Trichechus inguinis*). *Physiological Zoology*, **53**, 245–253.

Gallivan, G. J., Best, R. C., & Kanwisher, J. W. 1983. Temperature regulation in the Amazonian manatee *Trichechus inunguis*. *Physiological Zoology*, **56**, 255–262.

Gallivan, G. J., Kanwisher, J. W., & Best, R. C. 1986. Heart rates and gas exchange in the Amazonian manatee (*Trichecus manutus*) in relation to diving. *Journal of Comparative Physiology B*, **156**, 415–423.

Gallivan, G. J. & Ronald, K. 1979. Temperature regulation in freely diving harp seals (*Phoca groenlandica*). *Canadian Journal of Zoology*, **57**, 2256–2263.

Gallo-Reynoso, J. P., Figuero-Carranza, A.-L., & Leboeuf, B. J. 2008. Foraging behavior of lactating Guadalupe fur seals. In: Lorenzo, C. E. E. & Ortega, J. (eds.), *Advances en el Estudio de los Aifamijeros de Mexico*, Vol. II. Mexico: Asociacion Mexicana de Mastozoologia.

Garey, W. F. 1962. Cardiac responses of fishes in asphyxic environments. *Biological Bulletin*, **122**, 362–368.

Garry, D. J., Kanatous, S. B., & Mammen, P. P. A. 2003. Emerging roles for myoglobin in the heart. *Trends in Cardiovascular Medicine*, **13**, 111–116.

Garry, D. J., Ordway, G. A., Lorenz, J. N., *et al.* 1998. Mice without myoglobin. *Nature*, **395**, 905–908.

Garthe, S. & Furness, R. W. 2001. Frequent shallow diving by a northern fulmar feeding at Shetland. *Waterbirds*, **24**, 287–289.

Gauer, O. H. & Thron, H. L. 1965. Postural changes in circulation. In: Hamilton, W. F. & Dow, P. (eds.), *Handbook of Physiology: Circulation*. Washington, DC: American Physiological Society.

Gayeski, T. E. J. & Honig, C. 1986. O_2 gradients from sarcolemma to cell interior in red muscle at maximal V_{O2}. *American Journal of Physiology*, **251**, H1179–H1186.

Gazo, M. & Acuilar, A. 2005. Maternal attendance and diving behavior of a lactating Mediterranean monk seal. *Marine Mammal Science*, **21**, 340–345.

Geiser, F. 2004. Metabolic rate and body temperature reduction during hibernation and daily torpor. *Annual Review of Physiology*, **66**, 239–274.

Gentry, R. L. & Kooyman, G. L. (eds.) 1986. *Fur Seals: Maternal Strategies on Land and at Sea.* Princeton, NJ: Princeton University Press.

Gentry, R. L., Kooyman, G. L., & Goebel, M. E. 1986. Feeding and diving behavior of northern fur seals. In: Gentry, R. L. & Kooyman, G. L. (eds.), *Fur Seals: Maternal Strategies on Land and at Sea*. Princeton, NJ: Princeton University Press.

George, J. C. & Ronald, K. 1973. The harp seal, *Pagophilus groenlandicus*: XXV. Ultrastructure and metabolic adaptation of skeletal muscle. *Canadian Journal of Zoology*, **51**, 833–839.

Georges, J.-Y., Bonadonna, F., & Guinet, C. 2000a. Foraging habitat and diving activity of subantarctic fur seals in relation to sea-surface temperatures at Amsterdam Island. *Marine Ecology Progress Series*, **196**, 291–304.

Georges, J.-Y., Tremblay, Y., & Guinet, C. 2000b. Seasonal diving behaviour in lactating subantarctic fur seals on Amsterdam Island. *Polar Biology*, **23**, 59–69.

Gilmartin, W. G., Pierce, R. W., & Antonelis, Jr, G. A. 1974. Some physiological parameters of the blood of the California gray whale. *Marine Fisheries Review*, **36**, 28–31.

Gjertz, I., Griffiths, D., Krafft, B. A., Lydersen, C., & Wiig, O. 2001. Diving and haul out patterns of walruses *Odobenus rosmarus* on Svalbard. *Polar Biology*, **24**, 314–319.

Gjertz, I., Kovacs, K. M., Lydersen, C., & Wiig, Ø. 2000a. Movements and diving of adult ringed seals (*Phoca hispida*) in Svalbard. *Polar Biology*, **23**, 651–656.

Gjertz, I., Kovacs, K. M., Lydersen, C., & Wiig, Ø. 2000b. Movements and diving of bearded seal (*Erignathus barbatus*) mothers and pups during lactation and post-weaning. *Polar Biology*, **23**, 559–566.

Gläser, N., Wieskotten, S., Otter, C., Dehnhardt, G., & Hanke, W. 2011. Hydrodynamic trail following in a California sea lion (*Zalophus californianus*). *Journal of Comparative Physiology A: Neuroethology, Sensory, Neural, and Behavioral Physiology*, **197**, 141–151.

Glezer, I. I., Jacobs, M. S., & Morgane, P. J. 1987. Ultrastructure of the blood–brain barrier in the dolphin (*Stenella coeruleoalba*). *Brain Research*, **414**, 205–218.

Godecke, A., Flogel, U., Zanger, K., *et al.* 1999. Disruption of mice in myoglobin induced multiple compensatory mechanisms. *Proceedings of the National Academy of Sciences*, **96**, 10495–10500.

Goforth, H. W. 1986. Glycogenolytic responses and force production characteristics of a bottlenose dolphin (Tursiops truncatus) while exercising against a force transducer. PhD thesis, University of California Los Angeles.

Goldbogen, J. A., Calambokidis, J., Croll, D. A., *et al.* 2008. Foraging behavior of humpback whales: kinematic and respiratory patterns suggest a high cost for a lunge. *Journal of Experimental Biology*, **211**, 3712–3719.

Goldbogen, J. A., Calambokidis, J., Croll, D. A., *et al.* 2012. Scaling of lunge-feeding performance in rorqual whales: mass-specific energy expenditure increases with body size and progressively limits diving capacity. *Functional Ecology*, **26**, 216–226.

Goldbogen, J. A., Calambokidis, J., Oleson, E., *et al.* 2011. Mechanics, hydrodynamics and energetics of blue whale lunge feeding: efficiency dependence on krill density. *Journal of Experimental Biology*, **214**, 131–146.

Goldbogen, J. A., Calambokidis, J., Shadwick, R., *et al.* 2006. Kinematics of foraging dives and lunge-feeding in fin whales. *Journal of Experimental Biology*, **209**, 1231–1244.

Goldbogen, J. A., Pyenson, N. D., & Shadwick, R. E. 2007. Big gulps require high drag for fin whale lunge feeding. *Marine Ecology Progress Series*, **349**, 289–301.

Goldbogen, J. A., Southall, B. L., Deruiter, S. L., *et al.* 2013. Blue whales respond to simulated mid-frequency military sonar. *Proceedings of the Royal Society B: Biological Sciences*, **280**. DOI: 10.1098/rspb.2013.0657.

Goldman, S. 2010. Free energy wells for small gas bubbles in soft deformable materials. *Journal of Chemical Physics*, **132**. DOI: 10.1063/1.3394940.

Gooden, B. A. & Elsner, R. 1985. What diving mammals might tell us about blood flow regulation. *Perspectives in Biology and Medicine*, **28**, 463–474.

Gordon, K. R. 1981. Locomotor behaviour of the walrus (*Odobenus*). *Journal of Zoology*, **195**, 349–367.

Gorini, C., Jameson, H. S., & Mendelowitz, D. 2009. Serotonergic modulation of the trigeminocardiac reflex neurotransmission to cardiac vagal neurons in the nucleus ambiguus. *Journal of Neurophysiology*, **102**, 1443–1450.

Gosline, J. M. & Shadwick, R. E. 1996. The mechanical properties of fin whale arteries are explained by novel connective tissue designs. *Journal of Experimental Biology*, **199**, 985–997.

Goudappel, J. R. & Slijper, E. J. 1958. Microscopic structure of the lungs of the bottlenose whale. *Nature*, **182**, 479.

Goulet, A.-M., Hammill, M. O., & Barrette, C. 2001. Movements and diving of grey seal females (*Halichoerus grypus*) in the Gulf of St. Lawrence, Canada. *Polar Biology*, **24**, 432–439.

Graf, P., Wilson, R. Â. P., Cohen Sanchez, L., Hacklander, K., & Rosell, F. 2014. Diving behaviour and energtics of a semi-aquatic, shallow-diving species: the Eurasian beaver (*Castor fiber*). *Fifth International Bio-logging Science Symposium.* Strasbourg, France.

Green, J. A., Butler, P. J., Woakes, A. J., & Boyd, I. L. 2003. Energetics of diving in macaroni penguins. *Journal of Experimental Biology*, **206**, 43–57.

Green, K., Williams, R., & Green, M. 1998. Foraging ecology and diving behaviour of macaroni penguins *Eudyptes chrysolophus* at Heard Island. *Marine Ornithology*, **26**, 27–34.

Grémillet, D. 1995. "Wing-drying" in cormorants. *Journal of Avian Biology*, **26**, 176.

Grémillet, D., Chauvin, C., Wilson, R. P., Le Maho, Y., & Wanless, S. 2005. Unusual feather structure allows partial plumage wettability in diving great cormorants *Phalacrocorax carbo*. *Journal of Avian Biology*, **36**, 57–63.

Grémillet, D., Dell'omo, G., Ryan, P. G., *et al.* 2004. Offshore diplomacy, or how seabirds mitigate intra-specific competition: a case study based on GPS tracking of cape gannets from neighbouring colonies. *Marine Ecology Progress Series*, **268**, 265–279.

Gremillet, D., Tuschy, I., & Kierspel, M. 1998. Body temperature and insulation in diving great cormorants and European shags. *Functional Ecology*, **12**, 386–394.

Grémillet, D., Wilson, R. P., Storch, S., & Gary, Y. 1999. Three-dimensional space utilization by a marine predator. *Marine Ecology Progress Series*, **183**, 263–273.

Grinnell, S. W., Irving, L., & Scholander, P. F. 1942. Experiments on the relation between blood flow and heart rate in the living seal. *Journal of Cellular and Comparative Physiology*, **19**, 341–350.

Grocott, M. P. W., Martin, D. S., Levett, D. Z. H., *et al.* 2009. Arterial blood gases and oxygen content in climbers on Mount Everest. *New England Journal of Medicine*, **360**, 140–149.

Gros, G., Wittenberg, B. A., & Jue, T. 2010. Myoglobin's old and new clothes: from molecular structure to function in living cells. *Journal of Experimental Biology*, **213**, 2713–2725.

Groscolas, R. 1990. Metabolic adaptations to fasting in emperor and king penguins. In: Davis, L. S. & Darby, J. T. (eds.), *Penguin Biology*. San Diego, CA: Academic Press.

Groscolas, R. & Rodriquez, A. 1982. Glucose and lactate kinetics and interrelations in an Antarctic bird (emperor penguin). *American Journal of Physiology*, **242**, R458–R464.

Grubb, B. R. 1981. Blood flow and oxygen consumption in avian skeletal muscle during hypoxia. *Journal of Applied Physiology*, **50**, 450–455.

Guard, C. L. & Murrish, D. E. 1975. Effects of temperature on the viscous behavior of blood from Antarctic birds and mammals. *Comparative Biochemistry and Physiology Part A: Physiology*, **52**, 287–290.

Guillemette, M., Anthony J. Woakes, A. J., *et al.* 2004. The effect of depth on the diving behaviour of common eiders. *Canadian Journal of Zoology*, **82**, 1818–1826.

Gulland, F. M. D., Werner, L., O'Neill, S., *et al.* 1996. Baseline coagulation assay values for northern elephant seals (*Mirounga angustirostris*), and disseminated intravascular coagulation in this species. *Journal of Wildlife Diseases*, **32**, 536–540.

Gunn, A. J. & Thoresen, M. 2006. Hypothermic neuroprotection. *NeuroRX*, **3**, 154–169.

Guppy, M., Hill, R. D., Liggins, G. C., Zapol, W. M., & Hochachka, P. W. 1986. Micro-computer assisted metabolic studies of voluntary diving of Weddell seals. *American Journal of Physiology*, **250**, 175–187.

Guppy, M. & Withers, P. 1999. Metabolic depression in animals: physiological perspectives and biochemical generalizations. *Biological Reviews*, **74**, 1–40.

Gutierrez-Herrera, E., Vacas-Jaques, P., Anderson, R., Zapol, W. M., & Franco, W. 2013. Oxygen saturation in free-diving whales: optical sensor development. *Proceedings SPIE 8592 Biomedical Applcations of Light Scattering*. DOI: 10.117/12.2008547.

Gutierrez, G., Pohl, R. J., & Strong, R. 1988. Effect of flow on O_2 consumption during progressive hypoxemia. *Journal of Applied Physiology*, **65**, 601–607.

Guyton, G. P., Stanek, K. S., Schneider, R. C., *et al.* 1995. Myoglobin-saturation in free-diving Weddell seals. *Journal of Applied Physiology*, **79**, 1148–1155.

Haggblom, L., Terwilliger, R. C., & Terwilliger, N. B. 1988. Changes in myoglobin and lactate dehydrogenase in muscle tissues of a diving bird, the pigeon guillemot, during maturation. *Comparative Biochemistry and Physiology B*, **91**, 273–278.

Halasz, N. A., Elsner, R., & Garvie, R. S. 1974. Renal recovery from ischemia: a comparative study of harbor seal and dog kidneys. *American Journal of Physiology*, **227**, 1331–1335.

Hall, J. 1970. Conditioning Pacific white-striped dophins, *Lagenorhynchus obliquidens*, for open-ocean release. Naval Undersea Center Technical Paper No. 200. San Diego, California.

Hallman, M., Glumoff, V., & Rämet, M. 2001. Surfactant in respiratory distress syndrome and lung injury. *Comparative Biochemistry and Physiology Part A: Molecular & Integrative Physiology*, **129**, 287–294.

Hallman, M., Spragg, R., Harrell, J. H., Moser, K. M., & Gluck, L. 1982. Evidence of lung surfactant abnormality in respiratory failure: study of bronchoalveolar lavage phospholipids, surface activity, phospholipase activity, and plasma myoinositol. *Journal of Clinical Investigation*, **70**, 673–683.

Halsey, J. 1982. Effects of high pressure on the central nervous system. *Physiological Reviews*, **62**, 1341–1377.

Halsey, L. G., Butler, P. J., Fahlman, A., Woakes, A. J., & Handrich, Y. 2008. Behavioral and physiological significance of minimum resting metabolic rate in king penguins. *Physiological and Biochemical Zoology*, **81**, 74–86.

Hamilton, C. D., Golightly, R. T., & Takekawa, J. Y. 2005. Characteristics of diving in radio-marked Xantus's murrelet. *Marine Ornithology*, **33**, 155–159.

Hamilton, J., Moodie, D., & Levy, J. 1979. The use of the diving reflex to terminate supraventricular tachycardia in a 2-week-old infant. *American Heart Journal*, **97**, 371–374.

Hammel, H. T., Elsner, R. W., Heller, H. C., Maggert, J. A., & Bainton, C. R. 1977. Thermoregulatory responses to altering hypothalamic temperature in the harbor seal. *American Journal of Physiology: Regulatory, Integrative and Comparative Physiology*, **232**, R18–R26.

Hammond, D., Elsner, R., Simison, G., & Hubbard, R. 1969. Submersion bradycardia in the newborn elephant seal *Mirounga angustirostris*. *American Journal of Physiology* **216**, 220–222.

Hammond, J. A., Bennett, A., Waltom, M. J., & Hall, A. J. 2005. Molecular cloning and expression of leptin in gray and harbor seal blubber, bone marrow, and lung and its potential role in marine mammal respiratory physiology. *American Journal of Physiology: Regulatory, Integrative and Comparative Physiology*, **289**, R545–R553.

Hampton, I. F. G. & Whittow, G. C. 1976. Body temperature and heat exchange in the Hawaiian spinner dolphin, *Stenella longirostris*. *Comparative Biochemistry and Physiology Part A: Physiology*, **55**, 195–197.

Hance, A. J., Robin, E. D., Halter, J. B., *et al.* 1982. Hormonal changes and enforced diving in the harbor seal, *Phoca vitulina*: II. Plasma catecholamines. *American Journal of Physiology*, **242**, R528–R532.

Handrich, Y., Bevan, R. M., Charrassin, J.-B., *et al.* 1997. Hypothermia in foraging king penguins. *Nature*, **388**, 64–67.

Hanke, W., Witte, M., Miersch, L., *et al.* 2010. Harbor seal vibrissa morphology suppresses vortex-induced vibrations. *Journal of Experimental Biology*, **213**, 2665–2672.

Hansel, J., Solleder, I., Gfroerer, W., *et al.* 2009. Hypoxia and cardiac arrhythmias in breath-hold divers during voluntary immersed breath-holds. *European Journal of Applied Physiology*, **105**, 673–678.

Hanson, M. B. & Baird, R. W. 1998. Dall's porpoise reactions to tagging attempts using a remotely-deployed suction-cup tag. *Marine Technology Society Journal*, **32**, 18–23.

Harding, A., Egevang, C., Walkusz, W., *et al.* 2009. Estimating prey capture rates of a planktivorous seabird, the little auk (*Alle alle*), using diet, diving behaviour, and energy consumption. *Polar Biology*, **32**, 785–796.

Harris, R. C., Marlin, D. J., Dunnett, M., Snow, D. H., & Hultman, E. 1990. Muscle buffering capacity and dipeptide content in the thoroughbred horse, greyhound dog and man. *Comparative Biochemistry and Physiology Part A: Physiology*, **97**, 249–251.

Harrison, R. J. & Kooyman, G. L. 1968. General physiology of the pinnipedia. In: Harrison, R. J., Hubbard, R. C., Peterson, R. S., Rice, C. E., Schusterman, R. J. (ed.), *The Behavior and Physiology of Pinnipeds*. New York: Appleton-Century-Crofts.

Harrison, R. J. & Tomlinson, J. D. W. 1956. Observations on the venous system in certain Pinnipedia and Cetacea. *Proceedings of the Zoological Society of London*, **126**, 205–233.

Hartman, F. A. 1955. Heart weight in birds. *The Condor*, **57**, 221–238.

Harvey, E. N., Mcelroy, W. D., Whitely, A. H., Warren, G. H., & Peace, D. C. 1944. Bubble formation in animals: III. An analysis of gas pressure and hydrostatic tension in cats. *Journal of Cellular and Comparative Physiology*, **24**, 117–132.

Hassrick, J. L., Crocker, D. E., Teutschel, N. M., *et al.* 2010. Condition and mass impact oxygen stores and dive duration in adult female northern elephant seals. *Journal of Experimental Biology*, **213**, 585–592.

Hastie, G. D., Rosen, D. A. S., & Trites, A. W. 2007. Reductions in oxygen consumption during dive and estimated submergence limitations of Steller sea lions (*Eumetopias jubatus*). *Marine Mammal Science*, **23**, 272–286.

Hausenloy, D. J. & Yellon, D. M. 2011. The therapeutic potential of ischemic conditioning: an update. *Nature Reviews Cardiology*, **8**, 619–629.

Hawkey, C. M., Horsley, D. T., & Keymer, I. F. 1989. Haematology of wild penguins (sphenisciformes) in the Falkland Islands. *Avian Pathology*, **18**, 495–502.

Hawkins, P., Butler, P., Woakes, A., & Gabrielsen, G. 1997. Heat increment of feeding in Brunnich's guillemot. *Journal of Experimental Biology*, **200**, 1757–1763.

Hawkins, P. A., Butler, P. J., Woakes, A. J., & Speakman, J. R. 2000. Estimation of the rate of oxygen consumption of the common eider duck (*Somateria mollissima*), with some measurements of heart rate during voluntary dives. *Journal of Experimental Biology*, **203**, 2819–2832.

Heath, M. E. & Ridgway, S. H. 1999. How dolphins use their blubber to avoid heat stress during encounters with warm water. *American Journal of Physiology*, **276**, R1188–R1194.

Hedd, A., Gales, R., Brothers, N., & Robertson, G. 1997. Diving behaviour of the shy albatross *Diomedea cauta* in Tasmania: initial findings and dive recorder assessment. *Ibis*, **139**, 452–460.

Hedrick, M. S. & Duffield, D. A. 1991. Haemotological and rheological aspects of blood in seven species of marine mammals: physiological implications for diving behavior. *Journal of Zoology (London)*, **225**, 273–283.

Hedrick, M. S., Duffield, D. A., & Cornell, L. H. 1986. Blood viscosity and optimal hematocrit in a deep-diving mammal, the northern elephant seal (*Mirounga angustirostris*). *Canadian Journal of Zoology*, **64**, 2081–2085.

Heerah, K., Andrews-Goff, V., Williams, G., *et al.* 2013. Ecology of Weddell seals during winter: influence of environmental parameters on their foraging behaviour. *Deep Sea Research Part II: Topical Studies in Oceanography*, **88–89**, 23–33.

Heglund, N. C., Cavagna, G. A., & Taylor, C. R. 1982a. Energetics and mechanics of terrestrial locomotion: III. Energy changes of the centre of mass as a function of speed and body size in birds and mammals. *Journal of Experimental Biology*, **97**, 41–56.

Heglund, N. C., Fedak, M. A., Taylor, C. R., & Cavagna, G. A. 1982b. Energetics and mechanics of terrestrial locomotion: IV. Total mechanical energy changes as a function of speed and body size in birds and mammals. *Journal of Experimental Biology*, **97**, 57–66.

Heide-Jorgensen, M. P., Bloch, D., Stefansson, E., *et al.* 2002. Diving behaviour of long-finned pilot whales *Globicephalas* around the Faroe Islands. *Wildlife Biology*, **8**, 307–311.

Heide-Jorgensen, M. P., & Dietz, R. 1995. Some characteristics of narwhal, *Monodon monoceros*, diving behaviour in Baffin Bay. *Canadian Journal of Zoology*, **73**, 2120–2132.

Heintzelman, D. S. 1963. Diving times of a common goldeneye. *Wilson Bulletin*, **75**, 91.

Helbo, S. & Fago, A. 2012. Functional properties of myoglobins from five whale species with different diving capacities. *Journal of Experimental Biology*, **215**, 3403–3410.

Helbo, S., Weber, R. E., & Fago, A. 2013. Expression patterns and adaptive functional diversity of vertebrate myoglobins. *Biochimica et Biophysica Acta (BBA): Proteins and Proteomics*, **1834**, 1832–1839.

Hemmingsen, E. & Scholander, P. F. 1960. Specific transport of oxygen through hemoglobin solutions. *Science*, **132**, 1379.

Hemmingsen, E. A. 1963. Enhancement of oxygen transport by myoglobin. *Comparative Biochemistry and Physiology*, **10**, 239.

Henkel, L. A., Burkett, E. E., & Takekawa, J. Y. 2004. At-sea activity and diving behavior of a radio-tagged marbled murrelet in Central California. *Waterbirds*, **27**, 9–12.

Herrera, E. A., Reyes, R. V., Giussani, D. A., *et al.* 2008. Carbon monoxide: a novel pulmonary artery vasodilator in neonatal llamas of the Andean altiplano. *Cardiovascular Research*, **77**, 197–201.

Heyning, J. E. & Mead, J. G. 1997. Thermoregulation in the mouths of feeding gray whales. *Science*, **278**, 1138–1140.

Heyning, J. E., Mead, J. G., & Bryden, M. M. 1993. A palatal rete in the right whale? *Nature*, **361**, 24–25.

Hickey, R., Albin, M. S., Bunegin, L., & Gelineau, J. 1986. Autoregulation of spinal cord blood flow: is the cord a microcosm of the brain. *Stroke*, **17**, 1183–1189.

Hill, B. G., Dranka, B. P., Bailey, S. M., Lancaster, J. R., & Darley-Usmar, V. M. 2010. What part of NO don't you understand? Some answers to the cardinal questions in nitric oxide biology. *Journal of Biological Chemistry*, **285**, 19699–19704.

Hill, R. D. 1986. Microcomputer monitor and blood sampler for free-diving Weddell seals *Leptonychotes weddelli*. *Journal of Applied Physiology*, **61**, 1570–1576.

Hill, R. D., Schneider, R. C., Liggins, G. C., *et al.* 1987. Heart rate and body temperature during free diving of Weddell seals. *American Journal of Physiology*, **253**, R344–R351.

Hilton, J. W. & Gaskin, D. E. 1978. Comparative volumes and vascular microanatomy of the intrahepatic venous system of the harbour porpoise, *Phocoena phocoena* (L.). *Canadian Journal of Zoology*, **56**, 2292–2298.

Hindell, M. A. & Lea, M. A. 1998. Heart rate, swimming speed, and estimated oxygen consumption of a free-ranging southern elephant seal. *Physiological Zoology*, **71**, 74–84.

Hindell, M. A., Slip, D. J., & Burton, H. R. 1991. The diving behaviour of adult male and female southern elephant seals, *Mirounga leonina* (Pinnipedia: Phocidae). *Australian Journal of Zoology*, **39**, 595–619.

Hindell, M. A., Slip, D. J., Burton, H. R., & Bryden, M. M. 1992. Physiological implications of continuous, prolonged, and deep dives of southern elephant seals. *Canadian Journal of Zoology*, **70**, 370–379.

Hindle, A. G., Mellish, J.-A. E., & Horning, M. 2011. Aerobic dive limit does not decline in an aging pinniped. *Journal of Experimental Zoology Part A: Ecological Genetics and Physiology*, **315A**, 544–552.

Hindle, A. G., Senkiw, R. W., & Macarthur, R. A. 2006. Body cooling and the diving capabiities of muskrats (*Ondatra zibethicus*). *Comparative Biochemistry and Physiology Part A: Molecular and Integrative Physiology*, **144**, 232–241.

Hindle, A. G., Young, B. L., Rosen, D. A. S., Haulena, M., & Trites, A. W. 2010. Dive response differs between shallow- and deep-diving Steller sea lions. *Journal of Experimental Marine Biology and Ecology*, **394**, 141–148.

Hinds, D. S. & Calder, W. A. 1971. Tracheal dead space in the respiration of birds. *Evolution*, **25**, 429–440.

Hlastala, M. P., McKenna, H. P., Franada, R. L., & Detter, J. C. 1976. Influence of carbon monoxide on hemoglobin–oxygen binding. *Journal of Applied Physiology*, **41**, 893–899.

Hobson, R. P. & Martin, A. R. 1996. Behaviour and dive times of Arnoux's beaked whales, *Berardius arnuxii*, at narrow leads in fast ice. *Canadian Journal of Zoology*, **74**, 388–393.

Hochachka, P. W. 1986a. Balancing conflicting metabolic demands of exercise and diving. *Federation Proceedings*, **45**, 2948–2952.

Hochachka, P. W. 1986b. Defense strategies against hypoxia and hypothermia. *Science*, **231**, 234–241.

Hochachka, P. W. 1986c. Metabolic arrest. *Intensive Care Medicine*, **12**, 127–133.

Hochachka, P. W. 1988. Metabolic suppression and oxygen availability. *Canadian Journal of Zoology*, **66**, 152–158.

Hochachka, P. W. 2000. Pinniped diving response mechanism and evolution: a window on the paradigm of comparative biochemistry and physiology. *Comparative Biochemistry and Physiology Part A: Molecular and Integrative Physiology*, **126**, 435–458.

Hochachka, P. W., Buck, L. T., Doll, C. J., & Land, S. C. 1996. Unifying theory of hypoxia tolerance: molecular/metabolic defense and rescue mechanisms for surviving oxygen lack. *Proceedings of the National Academy of Sciences*, **93**, 9493–9498.

Hochachka, P. W., Castellini, J. M., Hill, R. D., *et al.* 1988. Protective metabolic mechanisms during liver ischemia: transferrable lessons from long-diving animals. *Molecular and Cellular Biochemistry*, **84**, 77–85.

Hochachka, P. W. & Foreman, R. A. 1993. Phocid and cetacean blueprints of muscle metabolism. *Canadian Journal of Zoology*, **71**, 2089–2098.

Hochachka, P. W. & Guppy, M. 1987. *Metabolic Arrest and the Control of Biological Time*. Cambridge, MA: Harvard University Press.

Hochachka, P. W., Liggins, G. C., Qvist, J., *et al.* 1977. Pulmonary metabolism for diving: conditioning blood for the brain. *Science*, **198**, 831–834.

Hochachka, P. W. & Lutz, P. L. 2001. Mechanism, origin, and evolution of hypoxia tolerance in animals. *Comparative Biochemistry and Physiology Part B: Biochemistry and Molecular Biology*, **130**, 435–459.

Hochachka, P. W., Owen, T. G., Allen, J. F., & Whittow, G. C. 1975. Multiple end products of anaerobiosis in diving vertebrates. *Comparative Biochemistry and Physiology Part B: Biochemistry and Molecular Biology*, **50**, 17.

Hochachka, P. W. & Somero, G. N. 1984. *Biochemical Adaptation*. Princeton, NJ: Princeton University Press.

Hogan, M. C., Bebout, D. E., & Wagner, P. D. 1993. Effect of blood flow reduction on maximal O_2 uptake in canine gastrocnemius muscle in situ. *Journal of Applied Physiology*, **74**, 1742–1747.

Hohimer, A. R., Hales, J. R., Rowell, L. B., & Smith, O. A. 1983. Regional distribution of blood flow during mild dynamic leg exercise in the baboon. *Journal of Applied Physiology*, **55**, 1173–1177.

Hol, R., Blix, A. S., & Myhre, H. O. 1975. Selective redistribution of the blood volume in the diving seal (*Pagophilus groenlandicus*). *Rapports et Proces-Verbaux del Reúnions Conseil permanent international pour l'exploration de la mer*, **169**, 423–432.

Hong, S. K., Ashwell-Erickson, S., Gigliotti, P., & Elsner, R. 1982. Effects of anoxia and low pH on organic ion transport and electrolyte distribution in harbor seal (*Phoca vitulina*) kidney slices. *Journal of Comparative Physiology*, **149**, 19–24.

Hong, S. K., Song, S. H., Kim, P. K., & Suh, C. S. 1967. Seasonal observations on the cardiac rhythm during diving in the Korean ama. *Journal of Applied Physiology*, **23**, 18–22.

Hooker, S. K. & Baird, R. W. 1999. Deep-diving behaviour of the northern bottlenose whale, *Hyperoodon ampullatus* (Cetacea: Ziphiidae). *Proceedings of the Royal Society London B*, **266**, 671–676.

Hooker, S. K. & Baird, R. W. 2001. Diving and ranging behaviour of odontocetes: a methodological review and critique. *Mammal Review*, **31**, 81–105.

Hooker, S. K., Baird, R. W., & Fahlman, A. 2009. Could beaked whales get the bends? Effect of diving behaviour and physiology on modelled gas exchange for three species: *Ziphius cavirostris*, *Mesoplodon densirostris* and *Hyperoodon ampullatus*. *Respiratory Physiology & Neurobiology*, **167**, 235–246.

Hooker, S. K., Fahlman, A., Moore, M. J., *et al.* 2012. Deadly diving? Physiological and behavioural management of decompression stress in diving mammals. *Proceedings of the Royal Society B: Biological Sciences*, **279**, 1041–1050.

Hooker, S. K., Miller, P. J. O., Johnson, M. P., Cox, O. P., & Boyd, I. L. 2005. Ascent exhalations of Antarctic fur seals: a behavioural adaptation for breath-hold diving? *Proceedings of the Royal Society London B*, **272**, 355–363.

Horning, M. 2012. Constraint lines and performance envelopes in behavioral physiology: the case of the aerobic dive limit. *Frontiers in Physiology,* **3**. DOI: 10.3389/fphys.2012.00381.

Horning, M. & Trillmich, F. 1997. Ontogeny of diving behaviour in the Galápagos fur seal. *Behaviour*, **134**, 1211–1257.

Horsburgh, J. M., Morrice, M., Lea, M.-A., & Hindell, M. A. 2008. Determining feeding events and prey encounter rates in a southern elephant seal: a method using swim speed and stomach temperature. *Marine Mammal Science*, **24**, 207–217.

Houser, D. S., Dankiewicz-Talmadge, L. A., Stockard, T. K., & Ponganis, P. J. 2010. Investigation of the potential for vascular bubble formation in a repetitively diving dolphin. *Journal of Experimental Biology*, **213**, 52–62.

Houser, D. S., Howard, R., & Ridgway, S. 2001. Can diving-induced tissue nitrogen supersaturation increase the chance of acoustically driven bubble growth in marine mammals. *Journal of Theoretical Biology*, **213**, 183–195.

Howell, A. B. 1930. *Aquatic Mammals*. New York: Dover Publications.

Hudson, C. C., Liu, M., Chiang, G. G., *et al.* 2002. Regulation of hypoxia-inducible factor 1α expression and function by the mammalian target of rapamycin. *Molecular and Cellular Biology*, **22**, 7004–7014.

Hudson, D. M. & Jones, D. R. 1986. The influence of body mass on the endurance to restrained submersion in the Pekin duck. *Journal of Experimental Biology*, **120**, 351–367.

Hui, C. 1975. Thoracic collapse as affected by the retia thoracica in the dolphin. *Respiration*, **25**, 63–70.

Hui, C. A. 1988a. Penguin swimming: I. Hydrodynamics. *Physiological Zoology*, **61**, 333–343.

Hui, C. A. 1988b. Penguin swimming: II. Energetics and behavior. *Physiological Zoology*, **61**, 344–350.

Huin, N. 1994. Diving depths of white-chinned petrels. *The Condor*, **96**, 1111–1113.

Huin, N. & Prince, P. A. 1997. Diving behaviour of the grey-headed albatross. *Antarctic Science*, **9**, 243–249.

Hull, C. L. 2000. Comparative diving behaviour and segregation of the marine habitat by breeding royal penguins, *Eudyptes schlegeli*, and eastern rockhopper penguins, *Eudyptes chrysocome filholi*, at Macquarie Island. *Canadian Journal of Zoology*, **78**, 333–345.

Hurford, W. E., Hochachka, P. W., Schneider, R. C., *et al.* 1996. Splenic contraction, catecholamine release, and blood volume redistribution during diving in the Weddell seal. *Journal of Applied Physiology*, **80**, 298–306.

Hurford, W. E., Hong, S. K., Park, Y. S., *et al.* 1990. Splenic contraction during breath-hold diving in the Korean ama. *Journal of Applied Physiology*, **69**, 932–936.

Hurley, J. A. & Costa, D. P. 2001. Standard metabolic rate at the surface and during trained submersions in adult California sea lions (*Zalophus californianus*). *Journal of Experimental Biology*, **204**, 3273–3281.

Hyatt, R. E., Schilder, D. P., & Fry, D. L. 1958. Relationship between maximum expiratory flow and degree of lung inflation. *Journal of Applied Physiology*, **13**, 331–336.

Hyman, A. L. & Kadowitz, P. J. 1975. Effects of alveolar and perfusion hypoxia and hypercapnia on pulmonary vascular resistance in the lamb. *American Journal of Physiology*, **228**, 397–403.

Hyvärinen, H. 1989. Diving in darkness: whiskers as sense organs of the ringed seal (*Phoca hispida saimensis*). *Journal of Zoology*, **218**, 663–678.

Ignarro, L. J., Buga, G. M., Wood, K. S., Byrns, R. E., & Chaudhuri, G. 1987. Endothelium-derived relaxing factor produced and released from artery and vein is nitric oxide. *Proceedings of the National Academy of Sciences*, **84**, 9265–9269.

Ikels, K. G. 1970. Production of gas bubbles in fluids by tribonucleation. *Journal of Applied Physiology*, **28**, 524–527.

Irvine, A. B. 1983. Manatee metabolism and its influence on distribution in Florida. *Biological Conservation*, **25**, 315–334.

Irving, L. 1934. On the ability of warm-blooded animals to survive without breathing. *The Scientific Monthly*, **38**, 422–428.

Irving, L. 1938. Changes in the blood flow through the brain and muscles during the arrest of breathing. *American Journal of Physiology*, **122**, 207–214.

Irving, L. 1939. Respiration in diving mammals. *Physiological Reviews*, **19**, 112–134.

Irving, L. 1963. Bradycardia in human divers. *Journal of Applied Physiology*, **18**, 489–491.

Irving, L. & Hart, J. S. 1957. The metabolism and insulation of seals as bare-skinned mammals in cold water. *Canadian Journal of Zoology*, **35**, 497–511.

Irving, L. & Orr, M. D. 1935. The diving habits of the beaver. *Science*, **82**, 569.

Irving, L., Peyton, L. J., Bahn, C. H., & Peterson, R. S. 1962. Regulation of temperature in fur seals. *Physiological Zoology*, **35**, 275–284.

Irving, L., Peyton, L. J., Bahn, C. H., & Peterson, R. S. 1963. Action of the heart and breathing during the development of fur seals (*Callorhinus ursinus*). *Physiological Zoology*, **36**, 1–20.

Irving, L., Scholander, P. F., & Grinnell, S. W. 1940. Respiratory metabolism of the porpoise. *Science*, **91**, 455.

Irving, L., Scholander, P. F., & Grinnell, S. W. 1941a. The respiration of the porpoise, *Tursiops truncatus*. *Journal of Cellular and Comparative Physiology*, **17**, 145–168.

Irving, L., Scholander, P. F., & Grinnell, S. W. 1941b. Significance of the heart rate to the diving ability of seals. *Journal of Cellular and Comparative Physiology*, **18**, 283–297.

Irving, L., Scholander, P. F., & Grinnell, S. W. 1942a. Experimental studies of the respiration of sloths. *Journal of Cellular and Comparative Physiology*, **20**, 189–210.

Irving, L., Scholander, P. F., & Grinnell, S. W. 1942b. The regulation of arterial blood pressure in the seal during diving. *American Journal of Physiology*, **135**, 557–566.

Irving, L., Solandt, O. M., & Solandt, D. Y. 1935a. The respiratory metabolism of the seal and its adjustment to diving. *Journal of Cellular and Comparative Physiology*, **7**, 137–151.

Irving, L., Solandt, O. M., Solandt, D. Y., & Fisher, K. C. 1935b. Respiratory characteristics of the blood of the seal. *Journal of Cellular and Comparative Physiology*, **6**, 393–403.

Iyer, N. V., Kotch, L. E., Agani, F., *et al.* 1998. Cellular and developmental control of O2 homeostasis by hypoxia-inducible factor 1α. *Genes & Development*, **12**, 149–162.

Janes, D. N. & Chappell, M. A. 1995. The effect of ration size and body size on specific dynamic action in Adélie penguin chicks, *Pygoscelis adeliae*. *Physiological Zoology*, **68**, 1029–1044.

Jaquet, N., Dawson, S., & Slooten, E. 2000. Seasonal distribution and diving behaviour of male sperm whales off Kaikoura: foraging implications. *Canadian Journal of Zoology*, **78**, 407–419.

Jarman, M. 1973. Experiments on the emperor penguin, *Aptendoytes forsteri*, in various thermal environments. *British Antarctic Survey Bulletin*, **33–34**, 57–63.

Jay, C. V., Farley, S. D., & Garner, G. W. 2001. Summer diving behavior of male walruses in Bristol Bay, Alaska. *Marine Mammal Science*, **17**, 617–631.

Jefferson, T. A. 1987. A study of the behavior of Dall's porpoise (*Phocoenoides dalli*) in the Johnstone Strait, British Columbia. *Canadian Journal of Zoology*, **65**, 736–744.

Jenni, D. A. & Gambs, R. D. 1974. Diving times of grebes and masked ducks. *The Auk*, **91**, 415–417.

Jensen, F. B., Hansen, M. N., Montesanti, G., & Wang, T. 2014. Nitric oxide metabolites during anoxia and reoxygenation in the anoxia-tolerant vertebrate *Trachemys scripta*. *Journal of Experimental Biology*, **217**, 423–431.

Jepson, P. D., Arbelo, M., Deaville, R., *et al.* 2003. Gas-bubble lesions in stranded cetaceans. *Nature*, **425**, 575–576.

Jepson, P. D., Deaville, R., Patterson, I. A. P., *et al.* 2005. Acute and chronic gas bubble lesions in cetaceans stranded in the United Kingdom. *Veterinary Pathology Online*, **42**, 291–305.

Jobe, A. H. 1993. Pulmonary surfactant therapy. *New England Journal of Medicine*, **328**, 861–868.

Jobsis, P. D., Ponganis, P. J., & Kooyman, G. L. 2001. Effects of training on forced submersion responses in harbor seals. *Journal of Experimental Biology*, **204**, 3877–3885.

Jodice, P. G. & Collopy, M. W. 1999. Diving and foraging patterns of marbled murrelets (*Brachyramphus inarmoratus*): testing predictions from optimal-breathing models. *Canadian Journal of Zoology*, **77**, 1409–1418.

Johansen, K. 1959. Heart activity during experimental diving of snakes. *American Journal of Physiology: Legacy Content*, **197**, 604–606.

Johansen, K. 1964. Regional distribution of circulating blood during submersion asphyxia in the duck. *Acta Physiologica Scandinavica*, **62**, 1–9.

Johansen, K. & Aakhus, T. 1963. Central cardiovascular responses to submersion asphyxia in the duck. *American Journal of Physiology: Legacy Content*, **205**, 1167–1171.

Johansen, K., Berger, M., Bicudo, J. E. P. W., Ruschi, A., & De Alameida, P. J. 1987. Respiratory properties of blood and myoglobin in hummingbirds. *Physiological Zoology*, **60**, 269–278.

Johansen, K. & Krog, J. 1959. Peripheral circulatory response to submersion asphyxia in the duck. *Acta Physiologica Scandinavica*, **46**, 194–200.

Johansson, L. C. & Aldrin, B. S. W. 2002. Kinematics of diving Atlantic puffins (*Fratercula arctica* L.): evidence for an active upstroke. *Journal of Experimental Biology*, **205**, 371–378.

Johansson, L. C. & Lindhe Norberg, U. M. 2001. Lift-based paddling in diving grebe. *Journal of Experimental Biology*, **204**, 1687–1696.

Johansson, L. C. & Norberg, R. A. 2003. Delta-wing function of webbed feet gives hydrodynamic lift for swimming propulsion in birds. *Nature*, **424**, 65–68.

Johnson, M., Madsen, P. T., Zimmer, W. M. X., De Soto, N. A., & Tyack, P. L. 2004. Beaked whales echolocate on prey. *Proceedings of the Royal Society of London B: Biological Sciences*, **271**, S383–S386.

Johnson, M., Madsen, P. T., Zimmer, W. M. X., De Soto, N. A., & Tyack, P. L. 2006. Foraging Blainville's beaked whales (*Mesoplodon densirostris*) produce distinct click types matched to different phases of echolocation. *Journal of Experimental Biology*, **209**, 5038–5050.

Johnson, M. P. & Tyack, P. L. 2003. A digital acoustic recording tag for measuring the response of wild marine mammals to sound. *IEEE Journal of Oceanic Engineering*, **28**, 3–12.

Jones, D. R., Bryan Jr., R. M., West, N. H., Lord, R. H., & Clark, B. 1979. Regional distribution of blood flow during diving in the duck (*Anas platyrhynchos*). *Canadian Journal of Zoology*, **57**, 995–1002.

Jones, D. R., Fisher, H. D., Mctaggart, S., & West, N. H. 1973. Heart rate during breath-holding and diving in the unrestrained harbor seal (*Phoca vitulina richardi*). *Canadian Journal of Zoology*, **51**, 671–680.

Jones, D. R. & Furilla, R. A. 1987. The anatomical, physiological, behavioral, and metabolic consequences of voluntary and forced diving. In: Seller, T. J. (ed.), *Bird Respiration*. Boca Raton, FL: CRC Press.

Jones, D. R., Furilla, R. A., Heieis, M. R. A., Gabbott, G. R. J., & Smith, F. M. 1988. Forced and voluntary diving in ducks: cardiovascular adjustments and their control. *Canadian Journal of Zoology*, **66**, 75–83.

Jones, D. R. & Holeton, G. F. 1972. Cardiac output of ducks during diving. *Comparative Biochemistry and Physiology*, **41**, 639–645.

Jones, D. R., Milsom, W. K., & Gabbott, G. R. J. 1982. Role of central and peripheral chemoreceptors in diving responses of ducks. *American Journal of Physiology*, **243**, R537–R545.

Jones, D. R., Milsom, W. K., Smith, F. M., West, N. H., & Bamford, O. S. 1983. Diving responses in ducks after acute barodenervation. *American Journal of Physiology: Regulatory, Integrative and Comparative Physiology*, **245**, R222–R229.

Jones, D. R. & Purves, M. J. 1970. The carotid body in the duck and the consequences of its denervation upon the cardiac responses to immersion. *Journal of Physiology*, **211**, 279–294.

Jones, J. H., Effmann, E. L., & Schmidt-Nielsen, K. 1985. Lung volume changes during respiration in ducks. *Respiration Physiology*, **59**, 15–25.

Joulia, F., Lemaitre, F., Fontanari, P., Mille, M. L., & Barthelemy, P. 2009. Circulatory effects of apnoea in elite breath-hold divers. *Acta Physiologica*, **197**, 75–82.

Joyner, M. J. & Wilkins, B. W. 2007. Exercise hyperaemia: is anything obligatory but the hyperaemia? *Journal of Physiology*, **583**, 855–860.

Jurgens, K. D., Papadopoulos, S., Peters, T., & Gros, G. 2000. Myoglobin: just an oxygen store or also an oxygen transporter? *News in Physiological Sciences*, **15**, 269–274.

Jurgens, K. D., Peters, T., & Gros, G. 1994. Diffusivity of myoglobin in intact skeletal muscle cells. *Proceedings of the National Academy of Sciences*, **91**, 3829–3833.

Jury, J. 1986. Razorbill swimming at depth of 140 m. *British Birds*, **79**, 339.

Kadowaki, A., Matsukawa, K., Wakasugi, R., Nakamoto, T., & Liang, N. 2011. Central command does not decrease cardiac parasympathetic efferent nerve activity during spontaneous fictive motor activity in decerebrate cats. *American Journal of Physiology: Heart and Circulatory Physiology*, **300**, H1373–H1385.

Kaiser, M. J., Galanidi, M., Showler, D., *et al.* 2006. Distribution and behaviour of common scoter *Melanitta nigra* relative to prey resources and environmental parameters. *Ibis*, **148**, 110–128.

Kajimura, M., Fukuda, R., Bateman, R. M., Yamamoto, T., & Suematsu, M. 2010. Interactions of multiple gas-transducing systems: hallmarks and uncertainties of CO, NO, and H_2S gas biology. *Antioxidants & Redox Signaling*, **13**, 157–192.

Kanatous, S. B., Davis, R. W., Watson, R., *et al.* 2002. Aerobic capacities in the skeletal muscles of Weddell seals: key to longer dive durations? *Journal of Experimental Biology*, **205**, 3601–3608.

Kanatous, S. B., Dimichele, L. V., Cowan, D. F., & Davis, R. W. 1999. High aerobic capacities in the skeletal muscles of pinnipeds: adaptations to diving hypoxia. *Journal of Applied Physiology*, **86**, 1247–1256.

Kanatous, S. B., Elsner, R., & Mathieu-Costello, O. 2001. Muscle capillary supply in harbor seals. *Journal of Applied Physiology*, **90**, 1919–1926.

Kanatous, S. B., Hawke, T. J., Trumble, S. J., *et al.* 2008. The ontogeny of aerobic and diving capacity in the skeletal muscles of Weddell seals. *Journal of Experimental Biology*, **211**, 2559–2565.

Kanatous, S. B., Mammen, P. P. A., Rosenberg, P. B., *et al.* 2009. Hypoxia reprograms calcium signaling and regulates myoglobin expression. *American Journal of Physiology: Cell Physiology*, **296**, C393–C402.

Kanwisher, J. W., Gabrielsen, G., & Kanwisher, N. 1981. Free and forced diving in birds. *Science*, **211**, 717–719.

Kaseloo, P. & Lovvorn, J. 2006. Substitution of heat from exercise and digestion by ducks diving for mussels at varying depths and temperatures. *Journal of Comparative Physiology B: Biochemical, Systemic, and Environmental Physiology*, **176**, 265–275.

Kaseloo, P. A. & Lovvorn, J. R. 2003. Heat increment of feeding and thermal substitution in mallard ducks feeding voluntarily on grain. *Journal of Comparative Physiology B: Biochemical, Systemic, and Environmental Physiology*, **173**, 207–213.

Kasting, N. W., Adderley, S. A. L., Safford, T., & Hewlett, K. G. 1989. Thermoregulation in beluga (*Delphinapterus leucas*) and killer (*Orcinus orca*) whales. *Physiological Zoology*, **62**, 687–701.

Kato, A., Croxall, J. P., Watanuki, Y., & Naito, Y. 1992. Diving patterns and performance in male and female blue-eyed cormorants *Phalacrocorax atriceps* at South Georgia. *Marine Ornithology*, **19**, 117–129.

Kato, A., Ropert-Coudert, Y., Gremillet, D., & Cannell, B. 2006. Locomotion and foraging strategy in foot-propelled and wing-propelled shallow-diving seabirds. *Marine Ecology Progress Series*, **308**, 293–301.

Kato, A., Watanuki, Y., & Naito, Y. 2003. Foraging behaviour of chick-rearing rhinoceros auklets *Cerorhinca monocerata* at Teuri Island, Japan, determined by acceleration-depth recording micro data loggers. *Journal of Avian Biology*, **34**, 282–287.

Kato, A., Watanuki, Y., Nishumi, I., *et al.* 2000. Variation in foraging and parental behavior of king cormorants. *The Auk*, **117**, 718–730.

Kaufman, M. P. & Forster, H. V. 1996. Reflexes controlling circulatory, ventilatory and airway responses to exercise. In: Rowell, L. B. & Shepherd, J. T. (eds.), *Handbook of Physiology*. Bethesda, MD: American Physiological Society.

Kawakami, Y., Natelson, B. H., & Dubois, A. R. 1967. Cardiovascular effects of face immersion and factors affecting diving reflex in man. *Journal of Applied Physiology*, **23**, 964–970.

Keijer, E. & Butler, P. J. 1982. Volumes of the respiratory and circulatory systems in tufted and mallard ducks. *Journal of Experimental Biology*, **101**, 213–220.

Keitt, B. S., Croll, D. A., & Tershy, B. R. 2001. Dive depth and diet of the black-vented shearwater (*Puffinus opisthomelas*). *The Auk*, **117**, 507–510.

Kelly, B. P. & Wartzok, D. 1996. Ringed seal diving behavior in the breeding season. *Canadian Journal of Zoology*, **74**, 1547–1555.

Kempainen, R. R. & Brunette, D. D. 2004. The evaluation and management of accidental hypothermia. *Respiratory Care*, **49**, 192–205.

Kenny, R. 1979. Breathing and heart rates of the southern elephant seal, *Mirounga leonina* L. *Papers and Proceedings of the Royal Society of Tasmania*, **113**, 21–27.

Kenyon, K. 1969. *The Sea Otter in the Eastern Pacific Ocean*. Washington, DC: Government Printing Office.

Kenyon, K. W. 1975. *The Sea Otter in the Eastern North Pacific*. New York: Dover Publications.

Kerem, D. & Elsner, R. 1973. Cerebral tolerance to asphyxial hypoxia in the harbor seal. *Respiration Physiology*, **19**, 188–200.

Kerem, D. H., Hammond, D. D., & Elsner, R. 1973. Tissue glycogen levels in the Weddell seals (*Leptonychotes weddelli*): a possible adaptation to asphyxial hypoxia. *Comparative Biochemistry and Physiology Part A: Comparative Physiology*, **45**, 731–736.

Kerem, D. H., Kylstra, J. A., & Saltzman, H. A. 1975. Respiratory flow rates in the sea lion. *Undersea Biomedical Research*, **2**, 20–27.

Kielhorn, C. E., Dillaman, R. M., Kinsey, S. T., *et al.* 2013. Locomotor muscle profile of a deep (*Kogia breviceps*) versus shallow (*Tursiops truncatus*) diving cetacean. *Journal of Morphology*, **274**, 663–675.

King, A. S. & Cowie, A. F. 1969. The functional anatomy of the bronchial muscle of the bird. *Journal of Anatomy*, **105**, 323–336.

King, J. E. 1983. *Seals of the World*. Ithaca, NY: Cornell University Press.

Kipps, E. K., McLellan, W. A., Rommel, S. A., & Pabst, D. A. 2002. Skin density and its influence on buoyancy in the manatee (*Trichechus manatus latirostris*), harbor porpoise (*Phocoena phocoena*), and bottlenose dolphin (*Tursiops truncatus*). *Marine Mammal Science*, **18**, 765–778.

Kirac, C. O., Savas, Y., Güclüsoy, H., & Veryeri, N. O. 2002. Observations on diving behaviour of free-ranging Mediterranean monk seals *Monachus monachus* on Turkish coasts. *The Monachus Guardian*, **5**.

Kirkwood, R. & Robertson, G. 1997. The foraging ecology of female emperor penguins in winter. *Ecological Monographs*, **67**, 155–176.

Kjekshus, K. J., Blix, A. S., Elsner, R., Hol, R., & Amundsen, E. 1982. Myocardial blood flow in the diving seal. *American Journal of Physiology*, **242**, R97–R104.

Klabunde, R. E. 2011. *Cardiovascular Physiology Concepts*. Philadelphia, PA: Lippincott Williams & Wilkins.

Klatsky, L. J., Wells, R. S., & Sweeney, J. C. 2007. Offshore bottlenose dolphins (*Tursiops truncatus*): movement and dive behavior near the Bermuda Pedestal. *Journal of Mammalogy*, **88**, 59–66.

Kleiber, M. 1975. *The Fire of Life*. New York: Wiley and Sons.

Knieriem, A. & Hartmann, M. G. 2001. Comparative histopathology of lungs from by-caught Atlantic white-sided dolphins (*Leucopleurus acutus*). *Aquatic Mammals*, **27**, 73–81.

Koehler, R. C., Traystman, R. J., Zeger, S., Rogers, M. C., & Jones, M. D. 1984. Comparison of cerebrovascular response to hypoxic and carbon monoxide hypoxia in newborn and adult sheep. *Journal of Cerebral Blood Flow & Metabolism*, **4**, 115–122.

Kohin, S., Williams, T. M., & Ortiz, C. L. 1999. Effects of hypoxia and hypercapnia on aerobic metabolic processes in northern elephant seals. *Respiration Physiology*, **117**, 59–72.

Kojeszewski, T. & Fish, F. E. 2007. Swimming kinematics of the Florida manatee (*Trichechus manatus latirostris*): hydrodynamic analysis of an undulatory mammalian swimmer. *Journal of Experimental Biology*, **210**, 2411–2418.

Kolluru, G. K., Shen, X., Bir, S. C., & Kevil, C. G. 2013. Hydrogen sulfide chemical biology: pathophysiological roles and detection. *Nitric Oxide*, **35**, 5–20.

Koopman, H. N. & Westgate, A. J. 2012. Solubility of nitrogen in marine mammal blubber depends on its lipid composition. *Journal of Experimental Biology*, **215**, 3856–3863.

Kooyman, G. L. 1965. Maximum diving capacities of the Weddell seal (*Leptonychotes weddelli*). *Science*, **151**, 1553–1554.

Kooyman, G. L. 1968. An analysis of some behavioral and physiological characteristics related to diving in the Weddell seal. *Antarctic Research Series*, **11**, 227–261.

Kooyman, G. L. 1972. Deep diving behaviour and effects of pressure in reptiles, birds, and mammals. *Symposia of the Society for Experimental Biology*, **26**, 295–311.

Kooyman, G. L. 1973. Respiratory adaptations in marine mammals. *American Zoologist*, **13**, 457–468.

Kooyman, G. L. 1981. *Weddell Seal: Consummate Diver*. Cambridge: Cambridge University Press.

Kooyman, G. L. 1985. Physiology without restraint in diving mammals. *Marine Mammal Science*, **1**, 166–178.

Kooyman, G. L. 1989. *Diverse Divers: Physiology and Behavior*. Berlin: Springer-Verlag.

Kooyman, G. L. 2006. Mysteries of adaptation to hypoxia and pressure in marine mammals: the Kenneth S. Norris Lifetime Achievement Award Lecture. *Marine Mammal Science*, **22**, 507–526.

Kooyman, G. L. & Andersen, H. T. 1968. Deep diving. In: Andersen, H. T. (ed.), *The Biology of Marine Mammals*. New York: Academic Press.

Kooyman, G. L. & Campbell, W. B. 1972. Heart rates in freely diving Weddell seals, *Leptonychotes weddelli. Comparative Biochemistry and Physiology Part A: Physiology*, **43**, 31–36.

Kooyman, G. L., Castellini, M. A., Davis, R. W., & Maue, R. A. 1983. Aerobic dive limits in immature Weddell seals. *Journal of Comparative Physiology*, **151**, 171–174.

Kooyman, G. L., Cherel, Y., Le Maho, Y., *et al.* 1992a. Diving behavior and energetics during foraging cycles in king penguins. *Ecological Monographs*, **62**, 143–163.

Kooyman, G. L. & Cornell, L. H. 1981. Flow properties of expiration and inspiration in a trained bottlenose porpoise. *Physiological Zoology*, **54**, 55–61.

Kooyman, G. L., Davis, R. W., & Croxall, J. P. 1986. Diving behavior of Antarctic fur seals. In: Gentry, R. L. & Kooyman, G. L. (eds.), *Fur Seals: Maternal Strategies on Land and at Sea*. Princeton, NJ: Princeton University Press.

Kooyman, G. L., Drabek, C. M., Elsner, R., & Campbell, W. B. 1971a. Diving behavior of the emperor penguin, *Aptenodytes forsteri. The Auk*, **88**, 775–795.

Kooyman, G. L., Gentry, R. L., Bergman, W. P., & Hammel, H. T. 1976a. Heat loss in penguins during immersion and compression. *Comparative Biochemistry and Physiology*, **54A**, 75–80.

Kooyman, G. L., Gentry, R. L., & Urquhart, D. L. 1976b. Northern fur seal diving behavior: a new approach to its study. *Science*, **193**, 411–412.

Kooyman, G. L., Hammond, D. D., & Schroeder, J. P. 1970. Bronchograms and tracheograms of seals under pressure. *Science*, **169**, 82–84.

Kooyman, G. L., Kerem, D. H., Campbell, W. B., & Wright, J. J. 1971b. Pulmonary function in freely diving Weddell seals, *Leptonychotes weddelli*. *Respiration Physiology*, **12**, 271–282.

Kooyman, G. L., Kerem, D. H., Campbell, W. B., & Wright, J. J. 1973a. Pulmonary gas exchange in freely diving Weddell seals (*Leptonychotes weddelli*). *Respiration Physiology*, **17**, 283–290.

Kooyman, G. L. & Kooyman, T. G. 1995. Diving behavior of emperor penguins nurturing chicks at Coulman Island, Antarctica. *Condor*, **97**, 536–549.

Kooyman, G. L., Kooyman, T. G., Horning, M., & Kooyman, C. A. 1996. Penguin dispersal after fledging. *Nature*, **383**, 397.

Kooyman, G. L., Norris, K. S., & Gentry, R. L. 1975. Spout of the gray whale: its physical characteristics. *Science*, **190**, 908–910.

Kooyman, G. L. & Ponganis, P. J. 1990. Behavior and physiology of diving in emperor and king penguins. In: Davis, L. S. & Darby, J. T. (eds.), *Penguin Biology*. San Diego, CA: Academic Press.

Kooyman, G. L. & Ponganis, P. J. 1994. Emperor penguin oxygen consumption, heart rate and plasma lactate levels during graded swimming exercise. *Journal of Experimental Biology*, **195**, 199–209.

Kooyman, G. L. & Ponganis, P. J. 1998. The physiological basis of diving to depth: birds and mammals. *Annual Review of Physiology*, **60**, 19–32.

Kooyman, G. L. & Ponganis, P. J. 2008. The initial journey of juvenile emperor penguins. *Aquatic Conservation: Marine and Freshwater Ecosystems*, **17**, 537–543.

Kooyman, G. L., Ponganis, P. J., Castellini, M. A., *et al.* 1992b. Heart rates and swim speeds of emperor penguins diving under sea ice. *Journal of Experimental Biology*, **165**, 161–180.

Kooyman, G. L., Ponganis, P. J., & Howard, R. S. 1999. Diving animals. In: Lundgren, C. & Miller, J. (eds.), *The Lung at Depth*. New York: Marcel Dekker.

Kooyman, G. L., Schroeder, J. P., Denison, D. M., *et al.* 1973b. Blood N_2 tensions of seals during simulated deep dives. *American Journal of Physiology*, **223**, 1016–1020.

Kooyman, G. L., Schroeder, J. P., Greene, D. G., & Smith, V. A. 1973c. Gas exchange in penguins during simulated dives to 30 and 68 m. *American Journal of Physiology*, **225**, 1467–1471.

Kooyman, G. L. & Sinnett, E. E. 1979. Mechanical properties of the harbor porpoise lung. *Respiration Physiology*, **36**, 287–300.

Kooyman, G. L. & Sinnett, E. E. 1982. Pulmonary shunts in harbor seals and sea lions during simulated dives to depth. *Physiological Zoology*, **55**, 105–111.

Kooyman, G. L. & Trillmich, F. 1986. Diving behavior of Galapagos fur seals. In: Gentry, R. L. & Kooyman, G. L. (eds.), *Fur Seals: Maternal Strategies on Land and at Sea*. Princeton, NJ: Princeton University Press.

Kooyman, G. L., Wahrenbrock, E. A., Castellini, M. A., Davis, R. W., & Sinnett, E. E. 1980. Aerobic and anaerobic metabolism during diving in Weddell seals: evidence of preferred pathways from blood chemistry and behavior. *Journal of Comparative Physiology*, **138**, 335–346.

Kooyman, M. M. & Kooyman, G. L. 2009. Historical perspectives. *Aquatic Mammals*, **35**, 523–556.

Koschier, F. J., Elsner, R. W., Holleman, D. F., & Hong, S. K. 1978. Organic anion transport by renal cortical slices of harbor seals (*Phoca vitulina*). *Comparative Biochemistry and Physiology Part A: Physiology*, **60**, 289–292.

Kovacs, C. E. & Meyers, R. E. 2000. Anatomy and histochemistry of flight muscles in a wing-propelled diving bird, the Atlantic puffin, *Fratercula artcica*. *Journal of Morphology*, **244**, 109–125.

Krafft, B. A., Lydersen, C., Kovacs, K. M., Gjertz, I., & Haug, T. 2000. Diving behaviour of lactating bearded seals (*Erignathus barbatus*) in the Svalbard area. *Canadian Journal of Zoology*, **78**, 1408–1418.

Kramer, D. L. 1988. The behavioral ecology of air breathing by aquatic mammals. *Canadian Journal of Zoology*, **66**, 89–94.

Krogh, A. 1934. The physiology of the blue whale. *Nature*, **133**, 635–637.

Krutzikowsky, G. K. & Mate, B. R. 2000. Dive and surfacing characteristics of bowheadwhales (*Balaena mysticetus*) in the Beaufort and Chukchi seas. *Canadian Journal of Zoology*, **78**, 1182–1198.

Kuhn, C. E. & Costa, D. P. 2006. Identifying and quantifying prey consumption using stomach temperature change in pinnipeds. *Journal of Experimental Biology*, **209**, 4524–4532.

Kuhn, C. E., Crocker, D. E., Tremblay, Y., & Costa, D. P. 2009. Time to eat: measurements of feeding behaviour in a large marine predator, the northern elephant seal *Mirounga angustirostris*. *Journal of Animal Ecology*, **78**, 512–523.

Kuhn, C. E., McDonald, B. I., Shaffer, S. A., *et al.* 2006. Diving physiology and winter foraging activity of a juvenile leopard seal (*Hydrurga leptonyx*). *Polar Biology*, **29**, 303–307.

Kuhn, R. A., Ansorge, H., Godynicki, S., & Meyer, W. 2010. Hair density in the Eurasian otter *Lutra lutra* and the sea otter *Enhycra lutris*. *Acta Theriologica*, **55**, 211–222.

Kuroki, M., Kato, A., Watanuki, Y., Takahashi, A., & Naito, Y. 2003. Diving behavior of an epipelagically feeding alcid, the rhinoceros auklet (*Cerorhinca monocerata*). *Canadian Journal of Zoology*, **81**, 1249–1256.

Kvadsheim, P. H. & Aarseth, J. J. 2002. Thermal function of phocid seal fur. *Marine Mammal Science*, **18**, 952–962.

Kvadsheim, P. H. & Folkow, L. P. 1997. Blubber and flipper heat transfer in harp seals. *Acta Physiologica Scandinavica*, **161**, 385–395.

Kvadsheim, P. H., Folkow, L. P., & Blix, A. S. 1996. Thermal conductivity of minke whale blubber. *Journal of Thermal Biology*, **21**, 123–128.

Kvadsheim, P. H., Folkow, L. P., & Blix, A. S. 2005. Inhibition of shivering in hypothermic seals during diving. *American Journal of Physiology*, **289**, R326–R331.

Kvadsheim, P. H., Miller, P. J. O., Tyack, P. L., *et al.* 2012. Estimated tissue and blood N2 levels and risk of decompression sickness in deep-, intermediate-, and shallow-diving toothed whales during exposure to naval sonar. *Frontiers in Physiology*. DOI: 10.3389/fphys.2012.00125.

Kvietys, P. R. & Granger, D. N. 1982. Relation between intestinal blood flow and oxygen uptake. *American Journal of Physiology*, **242**, G202–G208.

Lacombe, A. M. & Jones, D. R. 1990. The source of circulating catecholamines in forced dived ducks. *General and Comparative Endocrinology*, **80**, 41–47.

Lacombe, A. M. & Jones, D. R. 1991a. Neural and humoral effects on hindlimb vascular resistance of ducks during forced submergence. *American Journal of Physiology: Regulatory, Integrative and Comparative Physiology*, **261**, R1579–R1586.

Lacombe, A. M. & Jones, D. R. 1991b. Role of adrenal catecholamines during forced submergence in ducks. *American Journal of Physiology: Regulatory, Integrative and Comparative Physiology*, **261**, R1364–R1372.

Ladhams, D. 1968. Diving times of grebes. *British Birds*, **61**, 27–30.

Lafortuna, C., Jahoda, M., Azzellino, A., Saibene, F., & Colombini, A. 2003. Locomotor behaviours and respiratory pattern of the Mediterranean fin whale (*Balaenoptera physalus*). *European Journal of Applied Physiology*, **90**, 387–395.

Laidre, K. L., Heide-Jørgensen, M., & Dietz, R. 2002. Diving behaviour of narwhals (*Monodon monoceros*) at two coastal localities in the Canadian High Arctic. *Canadian Journal of Zoology*, **80**, 624–635.

Laidre, K. L., Heide-Jørgensen, M., & Nielsen, T. G. 2007. Role of the bowhead whale as a predator in West Greenland. *Marine Ecology Progress Series*, **346**, 285–297.

Lancel, S., Hassoun, S. M., Favory, R., *et al.* 2009. Carbon monoxide rescues mice from lethal sepsis by supporting mitochondrial energetic metabolism and activating mitochondrial biogenesis. *Journal of Pharmacology and Experimental Therapeutics*, **329**, 641–648.

Lander, M. E., Gulland, F. M., & Delong, R. L. 2000. Satellite tracking a rehabilitated Guadalupe fur seal (*Arctocephalus townsendi*). *Aquatic Mammals*, **26**, 137–142.

Lang, T. G. & Pryor, K. 1966. Hydrodynamic perfomance of porpoises (*Stenella attenuata*). *Science*, **152**, 531–533.

Lanphier, E. H. & Rahn, H. 1963. Alveolar gas exchange during breath-hold diving. *Journal of Applied Physiology*, **18**, 471–477.

Lasiewski, R. C. & Calder, W. A. 1971. A preliminary allometric analysis of respiratory variables in resting birds. *Respiration Physiology*, **11**, 152–166.

Laughlin, M. H. & Armstrong, R. B. 1987. Adrenoreceptor effects on rat muscle blood flow during treadmill exercise. *Journal of Applied Physiology*, **62**, 1465–1472.

Laughlin, M. H., Korthuis, R. J., Duncker, D. J., & Bache, R. J. 2010. Control of blood flow to cardiac and skeletal muscle during exercise. In: Terjung, R. (ed.), *Comprehensive Physiology*. Bethesda, MD: John Wiley & Sons, Inc.

Lavigne, D. M., Innes, S., Worthy, G. A. J., *et al.* 1986. Metabolic rates of seals and whales. *Canadian Journal of Zoology*, **64**, 279–284.

Lawrie, R. 1953. The activity of the cytochrome system in muscle and its relation to myoglobin. *Biochemical Journal*, **55**, 298–305.

Le Boeuf, B. J., Costa, D. P., Huntley, A. C., & Feldkamp, S. D. 1988. Continuous, deep diving in female Northern seals, *Mirounga angustirostris*. *Canadian Journal of Zoology*, **66**, 446–458.

Le Boeuf, B. J., Crocker, D. E., Blackwell, S. B., Morris, P. A., & Thorson, P. H. 1993. Sex differences in diving and foraging behaviour of northern elephant seals. *Symposia Zoological Society London*, **66**, 149–178.

Lecorre, M. 1997. Diving depths of two tropical pelecaniformes: the red-tailed tropicbird and the red-footed booby. *The Condor*, **99**, 1004–1007.

Lei, J., Vodovotz, Y., Tzeng, E., & Billiar, T. R. 2013. Nitric oxide, a protective molecule in the cardiovascular system. *Nitric Oxide*, **35**, 175–185.

Leith, D. 1976. Comparative mammalian respiratory mechanics. *The Physiologist*, **19**, 485–510.

Leith, D., Lowe, R., & Gillespie, J. 1972. Mechanics of baleen whale lungs. *Federation Proceedings*, **31**, 335.

Leith, D. E. 1989. Adaptations to deep breath-hold diving: respiratory and circulatory mechanics. *Undersea Biomedical Research*, **16**, 345–353.

Lenfant, C. 1969. Physiological properties of blood of marine mammals. In: Anderson, H. T. (ed.), *The Biology of Marine Mammals*. New York: Academic Press.

Lenfant, C., Johansen, K., & Torrance, J. D. 1970. Gas transport and oxygen storage capacity in some pinnipeds and the sea otter. *Respiration Physiology*, **9**, 277–286.

Lenfant, C., Kenney, D. W., & Aucutt, C. 1968. Respiratory function in the killer whale (*Orcinus orca* Linnaeus). *American Journal of Physiology*, 215.

Lenfant, C., Kooyman, G. L., Elsner, R., & Drabek, C. M. 1969a. Respiratory function of blood of the adult and fetus Weddell seal (*Leptonychotes weddelli*). *American Journal of Physiology*, **216**, 1595–1597.

Lenfant, C., Kooyman, G. L., Elsner, R., & Drabek, C. M. 1969b. Respiratory function of the blood of the Adelie penguins (*Pygoscelis adleiae*). *American Journal of Physiology*, **216**, 1598–1600.

Lesage, V., Hammill, M. O., & Kovacs, K. M. 1999. Functional classification of harbor seal (*Phoca vitulina*) dives using depth profiles, swimming velocity, and an index of foraging success. *Canadian Journal of Zoology*, **77**, 74–87.

Lescroel, A. & Bost, C.-A. 2005. Foraging under contrasting oceanographic conditions: the gentoo penguin at Kerguelen Archipelago. *Marine Ecology Progress Series*, **302**, 245–261.

Lestyk, K., Folkow, L., Blix, A., Hammill, M., & Burns, J. 2009. Development of myoglobin concentration and acid buffering capacity in harp (*Pagophilus groenlandicus*) and hooded (*Cystophora cristata*) seals from birth to maturity. *Journal of Comparative Physiology B: Biochemical, Systemic, and Environmental Physiology*, **179**, 985–996.

Levenson, D. H. & Dizon, A. 2003. Genetic evidence for the ancestral loss of short-wavelength-sensitive cone pigments in mysticete and odontocete cetaceans. *Proceedings of the Royal Society of London Series B: Biological Sciences*, **270**, 673–679.

Levenson, D. H., Ponganis, P. J., Crognale, M. A., *et al.* 2006. Visual pigments of marine carnivores: pinnipeds, polar bears, and sea otters. *Journal of Comparative Physiology A: Neuroethology, Sensory, Neural, and Behavioral Physiology*, **192**, 833–843.

Levenson, D. H. & Schusterman, R. J. 1997. Pupillometry in seals and sea lions: ecological implications. *Canadian Journal of Zoology*, **75**, 2050–2057.

Levenson, D. H. & Schusterman, R. J. 1999. Dark adaptation and visual sensitivity in shallow- and deep-diving pinnipeds. *Marine Mammal Science*, **15**, 1303–1313.

Levy, M. N. 1971. Brief reviews: sympathetic–parasympathetic interactions in the heart. *Circulation Research*, **29**, 437–445.

Lewis, S., Benvenuti, S., Dall'antonia, L., *et al.* 2002. Sex-specific foraging behaviour in a monomorphic seabird. *Proceedings of the Royal Society London, Series B*, **269**, 1687–1693.

Liebsch, N., Wilson, R. P., Bornemann, H., Adelung, D., & Plötz, J. 2007. Mouthing off about fish capture: jaw movement in pinnipeds reveals the real secrets of ingestion. *Deep Sea Research Part II: Topical Studies in Oceanography*, **54**, 256–269.

Liggins, G. C., France, J. T., Knox, B. S., & Zapol, W. M. 1979. High corticosteroid levels in plasma of adult and foetal Weddell seals (*Leptonychotes weddelli*). *Acta Endocrinologica Copenhagen*, **90**, 718–726.

Liggins, G. C., France, J. T., Schneider, R. C., Knox, B. S., & Zapol, W. M. 1993. Concentrations, metabolic clearance rates, production rates and plasma binding of cortisol in Antarctic phocid seals. *Acta Endocrinologica*, **129**, 356–359.

Liggins, G. C., Qvist, J., Hochachka, P. W., *et al.* 1980. Fetal cardiovascular and metabolic responses to simulated diving in the Weddell seal. *Journal of Applied Physiology*, **49**, 424–430.

Lillie, M. A., Piscitelli, M. A., Vogl, A. W., Gosline, J. M., & Shadwick, R. E. 2013. Cardiovascular design in fin whales: high-stiffness arteries protect against adverse pressure gradients at depth. *Journal of Experimental Biology*, **216**, 2548–2563.

Lin, P.-C., Kreutzer, U., & Jue, T. 2007. Myoglobin translational diffusion in rat myocardium and its implication on intracellular oxygen transport. *Journal of Physiology*, **578**, 595–603.

Lin, Z., Demello, D. E., Wallot, M., & Floros, J. 1998. An SP-B gene mutation responsible for SP-B deficiency in fatal congenital alveolar proteinosis: evidence for a mutation hotspot in exon 4. *Molecular Genetics and Metabolism*, **64**, 25–35.

Lindholm, P., Ekborn, A., Oberg, D., & Gennser, M. 2008. Pulmonary edema and hemoptypsis after breath-hold diving at residual volume. *Journal of Applied Physiology*, **102**, 912–917.

Lindholm, P. & Lundgren, C. E. G. 1996. Alveolar gas composition before and after maximal breath-holds in competitive divers. *Undersea and Hyperbaric Medicine*, **33**, 463–467.

Lindholm, P. & Lundgren, C. E. G. 2009. The physiology and pathophysiology of human breath-hold diving. *Journal of Applied Physiology*, **106**, 284–292.

Liner, M. H. & Andersson, J. P. A. 2008. Pulmlonary edema after competitive breath-hold diving. *Journal of Applied Physiology*, **102**, 986–990.

Liner, M. H. & Andersson, J. P. A. 2009. Hypoxic syncope in a competitive breath-hold diver with elevation of the brain damage marker S100B. *Aviation Space and Environmental Medicine*, **80**, 1066–1068.

Liu, L., Cash, T. P., Jones, R. G., *et al.* 2006. Hypoxia-induced energy stress regulates mRNA translation and cell growth. *Molecular Cell*, **21**, 521–531.

Liwanag, H. E. M., Berta, A., Costa, D. P., Abney, M., & Williams, T. M. 2012. Morphological and thermal properties of mammalian insulation: the evolution of fur for aquatic living. *Biological Journal of the Linnean Society*, **106**, 926–939.

Liwanag, H. E. M., Williams, T. M., Costa, D. P., *et al.* 2009. The effects of water temperature on the energetic costs of juvenile and adult California sea lions (*Zalophus californianus*): the importance of skeletal muscle thermogenesis for thermal balance. *Journal of Experimental Biology*, **212**, 3977–3984.

Ljubkovic, M., Dujic, Z., Mollerlokken, A., *et al.* 2011. Venous and arterial bubbles at rest after no-decompression air dives. *Medicine & Science in Sports & Exercise*, **43**, 990–995.

Lockyer, C. 1976. Body weights of some species of large whales. *Journal du Conseil/Conseil Permanent International pour l'Exploration de la Mer*, **36**, 259–273.

Lockyer, C. & Waters, T. 1986. Weights and anatomical measurements of northeastern Atlantic fin (*Balaenoptera physalus*, Linnaeus) and sei (*B. borealis*, Lesson) whales. *Marine Mammal Science*, **2**, 169–185.

Lohman, S., Folkow, L. P., Osterud, B., & Sager, G. 1998. Changes in fibrinolytic activity in diving grey seals. *Comparative Biochemistry and Physiology A*, **120**, 693–698.

London, J., Johnson, D., Boveng, P. L., & Cameron, M. F. 2014. Dive behavior and spatial variability of bearded, ribbon, and spotted seals in the Bering and Chukchi Seas. *Fifth International Bio-logging Science Symposium*. Strasbourg, France.

Longo, L. D., Koos, B. J., & Power, G. G. 1973. Fetal myoglobin: quantitative determination and importance for oxygenation. *American Journal of Physiology*, **224**, 1032–1036.

Lovick, T. A. 1987. Differential control of cardiac and vasomotor activity by neurones in nucleus paragigantocellularis lateralis in the cat. *Journal of Physiology*, **389**, 23–35.

Lovvorn, J. R. 2001. Upstroke thrust, drag effects, and stroke–glide cycles in wing-propelled swimming by birds. *American Zoologist*, **41**, 154–165.

Lovvorn, J. R., Croll, D. A., & Liggins, G. A. 1999. Mechanical versus physiological determinants of swimming speeds in diving Brunnich's guillemots. *Journal of Experimental Biology*, **202**, 1741–1752.

Lovvorn, J. R. & Jones, D. R. 1991. Effects of body size, body fat, and change in pressure with depth on buoyancy and costs of diving in ducks (*Aythya* spp.). *Canadian Journal of Zoology*, **69**, 2879–2887.

Lovvorn, J. R., Jones, D. R., & Blake, R. W. 1991. Mechanics of underwater locomotion in diving ducks: drag, buoyancy and acceleration in a size gradient of species. *Journal of Experimental Biology*, **159**, 89–108.

Lovvorn, J. R. & Liggins, G. A. 2002. Interactions of body shape, body size and stroke-acceleration patterns in costs of underwater swimming by birds. *Functional Ecology*, **16**, 106–112.

Lovvorn, J. R., Liggins, G. A., Borstad, M. H., Calisal, S. M., & Mikkelsen, J. 2001. Hydrodynamic drag of diving birds: effects of body size, body shape and feathers at steady speeds. *Journal of Experimental Biology*, **204**, 1547–1557.

Lovvorn, J. R., Watanuki, Y., Kato, A., Naito, Y., & Liggins, G. A. 2004. Stroke patterns and regulation of swim speed and energy cost in free-ranging Brünnich's guillemots. *Journal of Experimental Biology*, **207**, 4679–4695.

Lowry, L. F., Frost, K., Davis, R., Suydam, R. S., & Demaster, D. 1994. Movements and behavior of satellite-tagged spotted seals (*Phoca largha*) in the Bering and Chukchi Seas. Springfield, VA: US Department of Commerce.

Ludynia, K., Garthe, S., & Luna-Jorquera, G. 2010. Distribution and foraging behaviour of the Peruvian booby (*Sula variegata*) off northern Chile. *Journal of Ornithology*, **151**, 103–111.

Lukton, A. & Olcott, H. S. 1958. Content of free imidazole compounds in the muscle tissue of aquatic animals. *Journal of Food Science*, **23**, 611–618.

Luna-Jorquera, G., & Culik, B. M. 1999. Diving behaviour of Humboldt penguins *Spheniscus humboldti* in northern Chile. *Marine Ornithology*, **27**, 67–76.

Lutz, J., Henrich, H., & Bauereisen, E. 1975. Oxygen supply and uptake in the liver and the intestine. *Pflugers Archiv*, **360**, 7–15.

Lyamin, O. I., Manger, P. R., Mukhametov, L. M., Siegel, J. M., & Shpak, O. V. 2000. Rest and activity states in a gray whale. *Journal of Sleep Research*, **9**, 261–267.

Lyamin, O. I., Manger, P. R., Ridgway, S. H., Mukhametov, L. M., & Siegel, J. M. 2008. Cetacean sleep: an unusual form of mammalian sleep. *Neuroscience & Biobehavioral Reviews*, **32**, 1451–1484.

Lydersen, C. & Kovacs, K. M. 1993. Diving behaviour of lactating harp seal, *Phoca groenlandica*, females from the Gulf of St. Lawrence, Canada. *Animal Behviour*, **46**, 1213–1221.

Lydersen, C., Ryg, M. S., Hammill, M. O., & O'Brien, P. J. 1992. Oxygen stores and aerobic dive limit of ringed seals (*Phoca hispida*). *Canadian Journal of Zoology*, **70**, 458–461.

Lydersen, C. L., Kovacs, K. K., Ries, S. R., & Knauth, M. K. 2002. Precocial diving and patent foramen ovale in bearded seal (*Erignathus barbatus*) pups. *Journal of Comparative Physiology B: Biochemical, Systemic, and Environmental Physiology*, **172**, 713–717.

Lythgoe, J. N. & Dartnall, H. J. A. 1970. A "deep sea rhodopsin" in a mammal. *Nature*, **227**, 955–956.

Macarthur, R. A. 1986. Brown fat and aquatic temperature regulation in muskrats, *Ondatra zibethicus*. *Physiological Zoology*, **59**, 306–317.

Macarthur, R. A., Humphries, M. M., Fines, G. A., & Campbell, K. L. 2001. Body oxygen stores, aerobic dive limits, and the diving abilities of juvenile and adult muskrats (*Ondatra zibethicus*). *Physiological and Biochemical Zoology*, **74**, 178–190.

Macarthur, R. A. & Karpan, C. M. 1989. Heart rates of muskrats diving under simulated field conditions: persistence of the bradycardia response and factors modifying its expression. *Canadian Journal of Zoology*, **67**, 1783–1792.

Macklem, P. T., Bouverot, P., & Scheid, P. 1979. Measurement of the distensibility of the parabronchi in duck lungs. *Respiration Physiology*, **38**, 23–35.

Madsen, P. T., Johnson, M. P., Aguilar De Soto, N., Zimmer, W. M. X., & Tyack, P. L. 2005. Biosonar performance of foraging beaked whales (*Mesoplodon densirostris*). *Journal of Experimental Biology*, **208**, 181–194.

Madsen, P. T., Lammers, M., Wisniewska, D., & Beedholm, K. 2013. Nasal sound production in echolocating delphinids (*Tursiops truncatus* and *Pseudorca crassidens*) is dynamic, but unilateral: clicking on the right side and whistling on the left side. *Journal of Experimental Biology*, **216**, 4091–4102.

Madsen, P. T., Payne, R., Kristiansen, N. U., *et al.* 2002. Sperm whale sound production studied with ultrasound time/depth-recording tags. *Journal of Experimental Biology*, **205**, 1899–1906.

Madsen, P. T., Wilson, M., Johnson, M., *et al.* 2007. Clicking for calamari: toothed whales can echolocate squid *Loligo pealeii*. *Aquatic Biology*, **1**, 141–150.

Maina, J. A. & King, A. S. 1987. A morphometric study of the lung of the Humboldt penguin (*Spheniscus humboldti*). *Zentralblat of Veterinary Medicine, C: Anatomy, Histology and Embryology*, **16**, C293–C297.

Maina, J. M. 2006. Development, structure, and function of a novel respiratory organ, the lung–air sac system of birds: to go where no other vertebrate has gone. *Biological Review*, **81**, 5455–5479.

Maina, J. N. & Nathaniel, C. 2001. A qualitative and quantitative study of the lung of an ostrich, *Struthio camelus*. *Journal of Experimental Biology*, **204**, 2313–2330.

Maina, J. N., West, J. B., Orgeig, S., *et al.* 2010. Recent advances into understanding some aspects of the structure and function of mammalian and avian lungs. *Physiological and Biochemical Zoology*, **83**, 792–807.

Malan, A. 2014. The evolution of mammalian hibernation: lessons from comparative acid–base physiology. *Integrative and Comparative Biology*, **54**, 484–496.

Mallet, M. L. 2002. Pathophysiology of accidental hypothermia. *Quarterly Journal of Medicine*, **95**, 775–785.

Manohar, M., Goetz, T. E., & Hassan, A. S. 2001. Effect of prior high-intensity exercise on exercise-induced arterial hypoxemia in thoroughbred horses. *Journal of Applied Physiology*, **90**, 2371–2377.

Maresh, J. L., Simmons, S. E., Crocker, D. E., *et al.* 2014. Free-swimming northern elephant seals have low field metabolic rates that are sensitive to an increased cost of transport. *Journal of Experimental Biology*, **217**, 1485–1495.

Markussen, N. H., Ryg, M., & Oritsland, N. A. 1994. The effect of feeding on the metabolic rate in harbour seals (*Phoca vitulina*). *Journal of Comparative Physiology B*, **164**, 89–93.

Marsh, H., Spain, A. V., & Heinsohn, G. E. 1978. Physiology of the dugong. *Comparative Biochemistry and Physiology Part A: Physiology*, **61**, 159–168.

Marshall, C. D. 2009. Feeding morphology. In: Perrin, W. F., Wursig, B. & Thewissen, J. G. M. (eds.), *Encyclopedia of Marine Mammals*. San Diego, CA: Academic Press.

Marshall, C. D., Amin, H., Kovacs, K. M., & Lydersen, C. 2006. Microstructure and innervation of the mystacial vibrissal follicle–sinus complex in bearded seals, *Erignathus barbatus* (Pinnipedia: Phocidae). *The Anatomical Record Part A: Discoveries in Molecular, Cellular, and Evolutionary Biology*, **288A**, 13–25.

Marshall, C. D., Huth, G. D., Edmonds, V. M., Halin, D. L., & Reep, R. L. 1998. Prehensile use of perioral bristles during feeding and associated behaviors of the Florida manatee (*Trichechus manatus latirostris*). *Marine Mammal Science*, **14**, 274–289.

Martin, A. R. & Smith, T. G. 1992. Deep diving in wild, free-ranging beluga whales, *Delphinapterus leucas*. *Canadian Journal of Fisheries and Aquatic Sciences*, **49**, 462–466.

Martin, A. R. & Smith, T. G. 1999. Strategy and capability of wild belugas, *Delphinapterus leucas*, during deep, benthic diving. *Canadian Journal of Zoology*, **77**, 1783–1793.

Martin, G. R. 1998. Eye structure and amphibious foraging in albatrosses. *Proceedings of the Royal Society B*, **265**, 665–671.

Martin, G. R. 1999. Eye structure and foraging in king penguins *Aptenodytes patagonicus*. *Ibis*, **141**, 444–450.

Martin, G. R. & Brooke, M. D. L. 1991. The eye of a procellariiform seabird the Manx shearwater *Puffinus-puffinus* visual fields and optical structure. *Brain Behavior and Evolution*, **37**, 65–78.

Mason, R. J., Murray, J. F., Broaddus, V. C., & Nadel, T. A. 2005. *Murray and Nadel's Textbook of Respiratory Medicine*. Philadelphia, PA: Saunders.

Mass, A. M. & Supin, A. Y. 2009. Vision. In: Perrin, W. F., Wursig, B. & Thewissen, J. G. M. (eds.), *Encyclopedia of Marine Mammals*. San Diego, CA: Academic Press.

Masuda, K., Kent, T., Ping-Chang, L., *et al.* 2008. Determination of myoglobin concentration in blood perfused tissue. *European Journal of Applied Physiology*, **104**, 41–48.

Mate, B., Mesecar, R., & Lagerquist, B. 2007. The evolution of satellite-monitored radio tags for large whales: one laboratory's experience. *Deep Sea Research Part II: Topical Studies in Oceanography*, **54**, 224–247.

Mate, B. R., Rossbach, K. A., Nieukirk, S. L., *et al.* 1995. Satellite-monitored movements and dive behavior of a bottlenose dolphin (*Tursiops truncatus*) in Tampa Bay, Florida. *Marine Mammal Science*, **11**, 452–463.

Mate, B. R., Stafford, K. M., Nawojchik, R., & Dunn, J. L. 1994. Movements and dive behavior of a satellite-monitored Atlantic white-sided dolphin (*Lagenorhynchus acutus*) in the Gulf of Maine. *Marine Mammal Science*, **10**, 116–121.

Matsukawa, K. 2012. Central command: control of cardiac sympathetic and vagal efferent nerve activity and the arterial baroreflex during spontaneous motor behaviour in animals. *Experimental Physiology*, **97**, 20–28.

Matsukawa, K., Mitchell, J. H., Wall, P. T., & Wilson, L. B. 1991. The effect of static exercise on renal sympathetic nerve activity in conscious cats. *Journal of Physiology*, **434**, 453–467.

Matsukawa, K., Wall, P. T., Wilson, L. B., & Mitchell, J. H. 1990. Reflex responses of renal nerve activity during isometric muscle contraction in cats. *American Journal of Physiology: Heart and Circulatory Physiology*, **259**, H1380–H1388.

Matsukawa, K., Wall, P. T., Wilson, L. B., & Mitchell, J. H. 1992. Neurally mediated renal vasoconstriction during isometric muscle contraction in cats. *American Journal of Physiology: Heart and Circulatory Physiology*, **262**, H833–H838.

Mattlin, R. H., Gales, N. J., & Costa, D. P. 1998. Seasonal dive behaviour of lactating New Zealand fur seals (*Arctocephalus forsteri*). *Canadian Journal of Zoology*, **76**, 350–360.

Matyukhin, V. A., Neshumova, T. V., & Cherepanova, V. A. 1988. *Energetics of Muscular Activity in Diving Animals*. Novosibirsk: Siberian Academy of Sciences Science Publishing.

McAnulty, J. F. 2010. Hypothermic organ preservation by static storage methods: current status and a view to the future. *Cryobiology*, **60**, S13–S19.

McClelland, S. J., Gay, M., Pabst, D. A., *et al.* 2012. Microvascular patterns in the blubber of shallow and deep diving odontocetes. *Journal of Morphology*, **273**, 932–942.

McCulloch, P. F. 2012. Animal models for investigating central control of the mammalian diving reflex. *Frontiers in Physiology*, **3**: 169. DOI: 10.3389/fphys.2012.00169.

McCulloch, P. F., Paterson, I. A., & West, N. H. 1995. An intact glutamatergic trigeminal pathway is essential for the cardiac response to simulated diving. *American Journal of Physiology: Regulatory, Integrative and Comparative Physiology*, **269**, R669–R677.

McDonald, B. I. & Ponganis, P. J. 2012. Lung collapse in the diving sea lion: hold the nitrogen and save the oxygen. *Biology Letters*, **8**, 1047–1049.

McDonald, B. I. & Ponganis, P. J. 2013. Insights from venous oxygen profiles: oxygen utilization and management in diving California sea lions. *Journal of Experimental Biology*, **216**, 3332–3341.

McDonald, B. I. & Ponganis, P. J. 2014. Deep-diving sea lions exhibit extreme bradycardia in long-duration dives. *Journal of Experimental Biology*, **217**, 1525–1534.

McFarland, W. L., Jacobs, M. S., & Morgane, P. J. 1979. Blood supply to the brain of the dolphin, *Tursiops truncatus*, with comparative observations on special aspects of the cerebrovascular supply of other vertebrates. *Neuroscience and Biobehavior Research*, **3**, 193.

McGinnis, S. M., Whittow, G. C., Ohata, C. A., & Huber, H. 1972. Body heat dissipation and conservation in two species of dolphins. *Comparative Biochemistry and Physiology Part A: Physiology*, **43**, 417–423.

McIntyre, I. W., Campbell, K. L., & MacArthur, R. A. 2002. Body oxygen stores, aerobic dive limits and diving behaviour of the star-nosed mole (*Condylura cristata*) and comparisons with non-aquatic talpids. *Journal of Experimental Biology*, **205**, 45–54.

McKean, T. 1982. Cardiovascular adjustments to laboratory diving in beavers and nutria. *American Journal of Physiology: Regulatory, Integrative and Comparative Physiology*, **242**, R434–R440.

Meagher, E. M., McLellan, W. A., Westgate, A. J., *et al.* 2002. The relationship between heat flow and vasculature in the dorsal fin of wild bottlenose dolphins *Tursiops truncatus*. *Journal of Experimental Biology*, **205**, 3475–3486.

Medway, W. & Moldovan, F. 1966. Blood studies on the North Atlantic pilot (pothead) whale, *Globicephala melaena* (Traill, 1809). *Physiological Zoology*, **39**, 110–116.

Meir, J. U., Champagne, C. D., Costa, D. P., Williams, C. L., & Ponganis, P. J. 2009. Extreme hypoxemic tolerance and blood oxygen depletion in diving elephant seals. *American Journal of Physiology: Regulatory, Integrative and Comparative Physiology*, **297**, R927–R939.

Meir, J. U. & Ponganis, P. J. 2009. High-affinity hemoglobin and blood oxygen saturation in diving emperor penguins. *Journal of Experimental Biology*, **212**, 3330–3338.

Meir, J. U. & Ponganis, P. J. 2010. Blood temperature profiles of diving elephant seals. *Physiological and Biochemical Zoology*, **83**, 531–540.

Meir, J. U., Robinson, P. W., Vilchis, L. I., *et al.* 2013. Blood oxygen depletion is independent of dive function in a deep diving vertebrate, the northern elephant seal. *PloS ONE*, **8**, e83248.

Meir, J. U., Stockard, T. K., Williams, C. L., Ponganis, K. V., & Ponganis, P. J. 2008. Heart rate regulation and extreme bradycardia in diving emperor penguins. *Journal of Experimental Biology*, **211**, 1169–1179.

Meis, L., Arruda, A. P., & Carvalho, D. P. 2005. Role of sarco/endoplasmic reticulum Ca^{2+}-ATPase in thermogenesis. *Bioscience Reports*, **25**, 181–190.

Meiselman, H. J., Castellini, M. A., & Elsner, R. 1992. Hemorheological behavior of seal blood. *Clinical Hemorheology*, **12**, 657–675.

Mekjavic, I. G., Tipton, M. J., & Eiken, O. 2003. Thermal considerations in diving. In: Brubakk, A. O. & Neuman, T. S. (eds.), *Bennett and Elliott's Physiology and Medicine of Diving*. Edinburgh: Saunders.

Melin, S. R., Delong, R. L., & Siniff, D. B. 2008. The effects of El Niño on the foraging behavior of lactating California sea lions (*Zalophus californianus californianus*) during the nonbreeding season. *Canadian Journal of Zoology*, **86**, 192–206.

Mellish, J.-A., Thomton, J., & Horning, M. 2007. Physiological and behavioral response to intra-abdominal transmitter implantation in Steller sea lions. *Journal of Experimental Marine Biology and Ecology*, **351**, 283–293.

Melnikov, V. V. 1997. The arterial system of the sperm whale (*Physeter macrocephalus*). *Journal of Morphology*, **234**, 37–50.

Melrose, W. D. & Nicol, S. C. 1992. Haematology, red cell metabolism and blood chemistry of the black-faced cormorant *Leucocarbo fuscescens*. *Comparative Biochemistry and Physiology A*, **102**, 67–70.

Merrick, R. L. & Loughlin, T. R. 1997. Foraging behavior of adult female and young-of-the-year Steller sea lions in Alaskan waters. *Canadian Journal of Zoology*, **75**, 776–786.

Meyers, R. A. & Stakebake, E. F. 2005. Anatomy and histochemistry of spread-wing posture in birds: 3. Immunohistochemistry of flight muscles and the "shoulder lock" in albatrosses. *Journal of Morphology*, **263**, 12–29.

Midtgård, U. 1980a. Arteriovenous anastomoses and vascularity in the feet of eiders and gulls (*Aves*). *Zoomorphology*, **96**, 263–270.

Midtgård, U. 1980b. Heat loss from the feet of mallards *Anas platyrhynchos* and arterio-venous heat exhcange in the rete tibiotarsale. *Ibis*, **122**, 354–359.

Midtgård, U. 1981. The rete tibiotarsale and arterio-venous association in the hind limb of birds: a comparative morphological study on counter-current heat exchange systems. *Acta Zoologica*, **62**, 67–87.

Midtgård, U. 1983. Scaling of the brain and the eye cooling system in birds: a morphometric analysis of the rete ophthalmicum. *Journal of Experimental Zoology*, **225**, 197–207.

Midtgård, U. 1984. Density of arteriovenous anastomoses in some skin areas of the domestic fowl (*Gallus domesticus*). *The Anatomical Record*, **209**, 455–459.

Midtgard, U., Sejrsen, P., & Johansen, K. 1985. Blood flow in the brood patch of Bantam hens: evidence of cold vasodilatation. *Journal of Comparative Physiology B*, **155**, 703–709.

Milde, L. N. 1992. Clinical use of mild hypothermia for brain protection: a dream revisited. *Journal of Neurosurgical Anesthesia*, **4**, 211–215.

Mill, G. K. & Baldwin, J. 1983. Biochemical correlates of swimming and diving behavior in the little blue penguin *Eudyptula minor*. *Physiological Zoology*, **56**, 242–254.

Millard, R. W., Johansen, K., & Milsom, W. K. 1973. Radiotelemetry of cardiovascular responses to exercise and diving in penguins. *Journal of Comparative Biochemistry and Physiology A*, **46**, 227–240.

Miller, A. K. & Trivelpiece, W. Z. 2008. Chinstrap penguins alter foraging and diving behavior in response to the size of their principle prey, Antarctic krill. *Marine Biology*, **154**, 201–208.

Miller, N. J., Daniels, C. B., Costa, D. P., & Orgeig, S. 2004a. Control of surfactant secretion in adult California sea lions. *Biochemical and Biophysical Research Communications*, **313**, 727–732.

Miller, N. J., Daniels, C. B., Schurch, S., Schoel, W. M., & Orgeig, S. 2006a. The surface activity of pulmonary surfactant from diving mammals. *Respiratory Physiology and Neurobiology*, **150**, 220–232.

Miller, N. J., Postle, A. D., Orgeig, S., Coster, G., & Daniels, C. B. 2006b. The composition of pulmonary surfactant from diving mammals. *Respiratory Physiology and Neurobiology*, **152**, 152–168.

Miller, N. J., Postle, A. D., Scurch, S., *et al.* 2005. The development of the pulmonary surfactant system in California sea lions. *Comparative Biochemistry and Physiology A*, **141**, 191–199.

Miller, P. J., Johnson, M. P., Tyack, P. L., & Terray, E. A. 2004b. Swimming gait, passive drag and buoyancy of diving sperm whales. *Journal of Experimental Biology*, **207**, 1953–1967.

Miller, P. J. O., Johnson, M. P., & Tyack, P. L. 2004c. Sperm whale behaviour indicates the use of echolocation click buzzes "creaks" in prey capture. *Proceedings of the Royal Society of London Series B: Biological Sciences*, **271**, 2239–2247.

Millikan, G. A. 1939. Muscle hemoglobin. *Physiological Reviews*, **19**, 503–523.

Mills, K. 2000. Diving behaviour of two Galapagos penguins *Spheniscus mendiculus*. *Marine Ornithology*, **28**, 75–79.

Milsom, W., Castellini, M., Harris, M., *et al.* 1996. Effects of hypoxia and hypercapnia on patterns of sleep-associated apnea in elephant seal pups. *American Journal of Physiology*, **271**, R1017–R1024.

Milsom, W. K., Johansen, K., & Millard, R. W. 1973. Blood respiratory properties in some Antarctic birds. *The Condor*, **75**, 472–474.

Milsom, W. K., Jones, D. R., & Gabbott, G. R. 1981. On chemoreceptor control of ventilatory responses to CO2 in unanesthetized ducks. *Journal of Applied Physiology*, **50**, 1121–1128.

Minamikawa, S., Iwasaki, T., & Kishiro, T. 2007. Diving behaviour of a Baird's beaked whale, *Berardius bairdii*, in the slope water region of the western North Pacific: first dive records using a data logger. *Fisheries Oceanography*, **16**, 573–577.

Minamikawa, S., Watanabe, H., & Iwasaki, T. 2013. Diving behavior of a false killer whale, *Pseudorca crassidens*, in the Kuroshio–Oyashio transition region and the Kuroshio front region of the western North Pacific. *Marine Mammal Science*, **29**, 177–185.

Mirceta, S., Signore, A. V., Burns, J. M., *et al.* 2013. Evolution of mammalian diving capacity traced by myoglobin net surface charge. *Science*, **340**, 1324–1327.

Mitani, Y., Andrews, R. D., Sato, K., *et al.* 2010. Three-dimensional resting behaviour of northern elephant seals: drifting like a falling leaf. *Biology Letters*, **6**, 163–166.

Mitchell, J. H., Kaufman, M. P., & Iwamoto, G. A. 1983. The exercise pressor reflex: its cardiovascular effects, afferent mechanisms, and central pathways. *Annual Review of Physiology*, **45**, 229–242.

Mitz, S. A., Reuss, S., Folkow, L. P., *et al.* 2009. When the brain goes diving: glial oxidative metabolism may confer hypoxia tolerance to the seal brain. *Neuroscience*, **163**, 552–560.

Miyazaki, N. 2001. A review of studies on ringed, Caspian, and Baikal seals. *Otsuchi Marine Science*, **26**, 1–6.

Mizuno, M., Kimura, Y., Iwakawa, T., *et al.* 2003. Regional differences in blood flow and oxygen consumption in resting muscle and their relationship during recovery from exhaustive exercise. *Journal of Applied Physiology*, **95**, 2204–2210.

Molyneux, G. S. & Bryden, M. M. 1975. Arteriovenous anastomoses in the skin of the Weddell seal (*Leptonychotes weddelli*). *Science*, **189**, 1100–1102.

Molyneux, G. S. & Bryden, M. M. 1978. Arteriovenous anastomoses in the skin of seals I: the Weddell seal (*Leptonychotes weddelli*) and the elephant seal (*Mirounga leonina*) (Pinnipedia: Phocidae). *Anatomical Record*, **191**, 239–252.

Montague, T. 1985. A maximum dive recorder for little penguins. *Emu*, **85**, 264–267.

Moore, C., Crocker, D. E., Fahlman, A., *et al.* 2014. Ontogenetic changes in skeletal muscle fiber type, fiber diameter and myoglobin concentration in the Northern elephant seal (*Mirounga angustirostris*). *Frontiers in Physiology*, **5**. DOI: 10.3389/fphys.2014.00217.

Moore, M. J., Bogomolni, A. L., Dennison, S. E., *et al.* 2009. Gas bubbles in seals, dolphins, and porpoises entangled and drowned at depth in gillnets. *Veterinary Pathology*, **46**, 536–547.

Moore, M. J. & Early, G. A. 2004. Cumulative sperm whale bone damage and the bends. *Science*, **306**, 2215.

Moore, M. J., Hammar, T., Arruda, J., *et al.* 2011. Hyperbaric computed tomographic measurement of lung compression in seals and dolphins. *Journal of Experimental Biology*, **214**, 2390–2397.

Morrison, P. 1962. Body temperatures in some Australian mammals: III. Cetacea (Megaptera). *The Biological Bulletin*, **123**, 154–169.

Morrison, P., Rosenmann, M., & Estes, J. A. 1974. Metabolism and thermoregulation in the sea otter. *Physiological Zoology*, **47**, 218–229.

Morrison, S. F. 2001. Differential control of sympathetic outflow. *American Journal of Physiology: Regulatory, Integrative and Comparative Physiology*, **281**, R683–R698.

Moser, E. & Andersen, P. 1994. Conserved spatial learning in cooled rats in spite of slowing of dentate field potentials. *Journal of Neuroscience*, **14**, 4458–4466.

Motterlini, R. & Otterbein, L. E. 2010. The therapeutic potential of carbon monoxide. *Nature Reviews Drug Discovery*, **9**, 728–743.

Moudgil, R., Michelakis, E. D., & Archer, S. L. 2005. Hypoxic pulmonary vasoconstriction. *Journal of Applied Physiology*, **98**, 390–403.

Mougin, J.-L. & Mougin, M.-C. 2000. Maximum diving depths for feeding attained by Bulwer's petrels (*Bulweria bulwerii*) during the incubation period. *Journal of Zoology London*, **250**, 75–77.

Moylan, T. J. & Sidell, B. D. 2000. Concentrations of myoglobin and myoglobin mRNA in heart ventricles from Antarctic fish. *Journal of Experimental Biology*, **203**, 1277–1286.

Mozo, J., Emre, Y., Bouillaud, F., Ricquier, D., & Criscuolo, F. 2005. Thermoregulation: what role for UCPs in mammals and birds? *Bioscience Reports*, **25**, 227–249.

Murdaugh, H. V. & Jackson, J. E. 1962. Heart rate and blood lactic acid concentration during experimental diving of water snakes. *American Journal of Physiology: Legacy Content*, **202**, 1163–1165.

Murdaugh, Jr., H. V., Cross, C. E., Millen, J. E., Gee, J. B. L., & Robin, E. D. 1968. Dissociation of bradycardia and arterial constriction during diving in the seal *Phoca vitulina*. *Science*, **162**, 364–365.

Murdaugh, H. V., Schmidt-Nielsen, B., Wood, J. W., & Mitchell, W. L. 1961. Cessation of renal function during diving in the trained seal (*Phoca vitulina*). *Journal of Cellular and Comparative Physiology*, **58**, 261–265.

Murphy, E. & Steenbergen, C. 2008. Mechanisms underlying acute protection from cardiac ischemia-reperfusion injury. *Physiological Reviews*, **88**, 581–609.

Murray, J., Escobar, E., & Rapaport, E. 1969. Effects of blood viscosity on hemodynamic responses in acute normovolemic anemia. *American Journal of Physiology: Legacy Content*, **216**, 638–642.

Murrish, D. E. 1970. Responses to diving in the dipper, *Cinclus mexicanus*. *Comparative Biochemistry and Physiology*, **34**, 853–858.

Murrish, D. E. 1982. Acid–base balance in three species of Antarctic penguins exposed to thermal stress. *Physiological Zoology*, **55**, 137–143.

Murry, C. E., Jennings, R. B., & Reimer, K. A. 1986. Preconditioning with ischemia: a delay of lethal cell injury in ischemic myocardium. *Circulation*, **74**, 1124–1136.

Murry, C. E., Richard, V. J., Reimer, K. A., & Jennings, R. B. 1990. Ischemic preconditioning slows energy metabolism and delays ultrastructural damage during a sustained ischemic episode. *Circulation Research*, **66**, 913–931.

Musch, T. I., Friedman, D. B., Pitetti, K. H., *et al.* 1987a. Regional distribution of blood flow of dogs during graded dynamic exercise. *Journal of Applied Physiology*, **63**, 2269–2277.

Musch, T. I., Haidet, G. C., Ordway, G. A., Longhurst, J. C., & Mitchell, J. H. 1987b. Training effects on regional blood flow response to maximal exercise in foxhounds. *Journal of Applied Physiology*, **62**, 1724–1732.

Nagy, K. A. 1980. CO_2 production in animals: analysis of potential errors in the doubly labeled water method. *American Journal of Physiology: Regulatory, Integrative and Comparative Physiology*, **238**, R466–R473.

Nagy, K. A., Girard, I. A., & Brown, T. K. 1999. Energetics of free-ranging mammals, reptiles, and birds. *Annual Review of Nutrition*, **19**, 247–277.

Nagy, K. A., Kooyman, G. L., & Ponganis, P. J. 2001. Energetic cost of foraging in free-diving emperor penguins. *Physiological and Biochemical Zoology*, **74**, 541–547.

Naito, Y., Costa, D. P., Adachi, T., *et al.* 2013. Unravelling the mysteries of a mesopelagic diet: a large apex predator specializes on small prey. *Functional Ecology*, **27**, 710–717.

Nakaoa, A., Netoa, J. S., Kannob, S., *et al.* 2005. Protection against ischemia/reperfusion injury in cardicac and renal transplantation with carbon monoxide, bilverdin, and both. *American Journal of Transplantation*, **5**, 282–291.

Nawojchik, R., St. Aubin, D. J., & Johnson, A. 2003. Movements and dive behavior of two stranded, rehabilitated long-finned pilot whales (*Globicephala melas*) in the Northwest Atlantic. *Marine Mammal Science*, **19**, 232–239.

Neshumova, T. V., Cherapanova, V. A., & Petrov, E. A. 1983. Myoglobin concentration in muscles of the seal *Pusa sibirica. Journal of Evolutionary Biochemistry and Physiology*, **19**, 93–95.

Neshumova, T. V., Cherapanova, V. A., & Zhuzhgin, S. M. 1986. Electrical activity in muscles of the seal *Pusa sibirica* during locomotion in water and on land. *Journal of Evolutionary Biochemistry and Physiology*, **22**, 323–326.

Neshumova, T. V. & Cherepanova, V. A. 1984. Blood supply and myoglobin stocks in muscles of the seal *Pusa siberica* and muskrat *Ondatra zibethica. Journal of Evolutionary Biochemistry and Physiology*, **20**, 282–287.

Neves, V., Bried, J., Gonzalez-Solis, J., Roscales, J., & Clarke, M. 2012. Feeding ecology and movements of the Barolo shearwater *Puffinus baroli baroli* in the Azores, NE Atlantic. *Marine Ecology Progress Series*, **452**, 269–285.

Nevitt, G. A., Veit, R. R., & Kareiva, P. 1995. Dimethyl sulphide as a foraging cue for Antarctic Procellariiform seabirds *Nature*, **376**, 680–682.

Nichols, J. W. & Weber, L. J. 1989. Comparative oxygen affinity of fish and mammalian myoglobins. *Journal of Comparative Physiology B*, **159**, 205–210.

Nicol, S. 1991. Respiratory properties of the blood of the little blue penguin *Eudyptula minor. Comparative Biochemistry and Physiology A*, **98**, 17–22.

Niizuma, Y., Gabrielsen, G. W., Sato, K., Watanuki, Y., & Naito, Y. 2007. Brunnich's guillemots (*Uria lomvia*) maintain high temperature in the body core during dives. *Comparative Biochemistry and Physiology A*, **147**, 438–444.

Ninomiya, H., Imamura, E., & Inomata, T. 2014. Comparative anatomy of the ophthalmic rete and its relationship to ocular blood flow in three species of marine mammal. *Veterinary Ophthalmology*, **17**, 100–105.

Ninomiya, H., Inomata, T., Shirouzu, H., & Katsumata, E. 2005. Microanatomy of the terminal air spaces of Baird's beaked whale (*Berardius bairdii*) lungs. *Journal of Veterinary Medical Science*, **67**, 473–479.

NOAA. 2002. *Report of the Workshop on Acoustic Resonance as a Source of Tissue Trauma in Cetaceans, 24 and 25 April 2002*. Silver Spring, MD: US Department of Commerce, National Oceanic and Atmospheric Administration, National Marine Fisheries Service. [Available at www.nmfs.noaa.gov/pro_tres/readingrm/MMSURTASS/Res_Wkshp_Rpt_Fin.pdf.

Nocera, J. J. & Burgess, N. M. 2002. Diving schedules of common loons in varying environments. *Canadian Journal of Zoology*, **80**, 1643.

Nolet, B. A., Wansink, D. E. H., & Kruuk, H. 1993. Diving of otters (*Lutra lutra*) in a marine habitat: use of depths by a single-prey loader. *Journal of Animal Ecology*, **62**, 22–32.

Nordgarden, U., Folkow, L. P., Walloe, L., & Blix, A. S. 2000. On the direction and velocity of blood flow in the extradural vein of harp seals (*Phoca groenlandica*) during simulated diving. *Acta Physiologica Scandinavica*, **168**, 271–276.

Nordoy, E. S. & Blix, A. S. 2009. Movements and dive behaviour of two leopard seals (*Hydrurga leptonyx*) off Queen Maud Land, Antarctica. *Polar Biology*, **32**, 263–270.

Noren, D. P., Kendall, T., Cuccurullo, V., & Williams, T. M. 2012a. The dive response redefined: underwater behavior influences cardiac variability in freely diving dolphins. *Journal of Experimental Biology*, **215**, 2735–2741.

Noren, D. P., Williams, T. M., Berry, P., & Butler, E. 1999. Thermoregulation during swimming and diving in bottlenose dolphins (*Tursiops truncatus*). *Journal of Comparative Physiology B*, **169**, 93–99.

Noren, S. R. 2004. Buffering capacity of the locomotor muscle in cetaceans: correlates with postpartum development, dive duration, and swim performance. *Marine Mammal Science*, **20**, 808–822.

Noren, S. R., Cuccurullo, V., & Williams, T. M. 2004. The development of diving bradycardia in bottlenose dolphins (*Tursiops truncatus*). *Journal of Comparative Physiology B: Biochemical, Systemic, and Environmental Physiology*, **174**, 139–147.

Noren, S. R., Iverson, S. J., & Boness, D. J. 2005. Development of the blood and muscle oxygen stores in gray seals (*Halichoerus grypus*): implications for juvenile diving capacity and the necessity of a terrestrial postweaning fast. *Physiological and Biochemical Zoology*, **78**, 482–490.

Noren, S. R. & Williams, T. M. 2000. Body size and skeletal muscle myoglobin of cetaceans: adaptations for maximizing dive duration. *Comparative Biochemistry and Physiology A*, **126**, 181–191.

Noren, S. R., Williams, T. M., Pabst, D. A., McLellan, W. A., & Dearolf, J. L. 2001. Development of diving in marine endotherms: preparing the skeletal muscles of dolphins, penguins, and seals for activity during submergence. *Journal of Comparative Physiology B*, **171**, 127–134.

Noren, S., Williams, T., Ramirez, K., *et al.* 2012b. Changes in partial pressures of respiratory gases during submerged voluntary breath hold across odontocetes: is body mass important? *Journal of Comparative Physiology B: Biochemical, Systemic, and Environmental Physiology*, **182**, 299–309.

Norman, F. & Ward, S. 1993. Foraging group size and dive duration of Adélie penguins *Pygoscelis adeliae* at sea off Hop Island, Rauer group, East Antarctica. *Marine Ornithology*, **21**, 37–47.

Norris, K. S. 1968. The evolution of acoustic mechanisms in odontocete cetaceans. In: Drake, E. T. (ed.) *Evolution and Environment*. New Haven, CT: Yale University Press.

Norris, K. S. & Dohl, T. P. 1980. Behavior of the Hawaiian spinner dolphin, *Stenella longirostris*. *Fishery Bulletin*, **77**, 821–849.

Norris, K. S. & Harvey, G. W. 1972. A theory of the function of the spermaceti organ of the sperm whale (*Physeter catodon* L.). In: Galler, S. F., Schmidt-Koenig, K., Jacobs, G. J. & Belleville, R. E. (eds.), *Animal Orientation and Navigation*. Washington, DC: NASA.

Norton, J. M. 2001. Toward consistent definitions for preload and afterload. *Advances in Physiology Education*, **25**, 53–61.

Nowacek, D. P., Johnson, M. P., Tyack, P. L., Shorter, K. A., Mclellan, J., & Pabst, D. A. 2001. Buoyant Balaenids: the ups and downs of buoyancy in right whales. *Proceedings of the Royal Society London B*, **268**, 1811–1816.

Nowicki, S. N., Stirling, I., & Sjare, B. 1997. Duration of stereotyped underwater vocal displays by male Atlantic walruses in relation to aerobic dive limit. *Marine Mammal Science*, **13**, 566–575.

Nummela, S. 2009. Hearing. In: Perrin, W. F., Wursig, B. & Thewissen, J. G. M. (eds.), *Encyclopedia of Marine Mammals*. San Diego, CA: Academic Press.

Nummela, S., Thewissen, J. G. M., Bajpai, S., Hussain, T., & Kumar, K. 2007. Sound transmission in archaic and modern whales: anatomical adaptations for underwater hearing. *The Anatomical Record: Advances in Integrative Anatomy and Evolutionary Biology*, **290**, 716–733.

Nunn, J. F. 1977. *Applied Respiratory Physiology*. London: Butterworths.

O'Connor, P. M. 2004. Pulmonary pneumaticity in the postcranial skeleton of extant *Aves*: a case study examining Anseriformes. *Journal of Morphology*, **261**, 141–161.

O'Connor, P. M. 2009. Evolution of archosaurian body plans: skeletal adaptations of an air-sac-based breathing apparatus in birds and other archosaurs. *Journal of Experimental Zoology Part A: Ecological Genetics and Physiology*, **311A**, 629–646.

O'Leary, D. S. 1993. Autonomic mechanisms of muscle metaboreflex control of heart rate. *Journal of Applied Physiology*, **74**, 1748–1754.

Ochrymowych, C. & Lambertsen, R. H. 1984. Anatomy and vasculature of a minke whale heart. *American Journal of Anatomy*, **169**, 165–175.

Odden, A., Folkow, L. P., Caputa, M., Hotvedt, R., & Blix, A. S. 1999. Brain cooling in seals. *Acta Physiologica Scandinavica*, **166**, 77–78.

Odend'hal, S. & Poulter, T. C. 1966. Pressure regulation in the middle ear cavity of sea lions: a possible mechanism. *Science*, **153**, 768–769.

Oelschlager, H. H. A. & Oeschlager, J. S. 2008. Brain. In: Perrin, W. F., Wursig, B. & Thewissen, J. G. M. (eds.), *Encyclopedia of Marine Mammals*. 2nd edn. Amsterdam: Elsevier.

Oka, N. 1994. Underwater feeding of three shearwaters: pale-footed (*Puffinus carneipes*), sooty (*Puffinus griseus*) and streaked (*Calonectris leucomelas*) shearwaters. *Journal of the Yamashina Institute for Ornithology*, **26**, 81–84.

Okuma, E. & Abe, H. 1992. Major buffering constituents in animal muscle. *Comparative Biochemistry and Physiology Part A: Physiology*, **102**, 37–41.

Oliver, G. W., Morris, P. A., Thorson, P. H., & Le Boeuf, B. J. 1998. Homing behavior of juvenile northern elephant seals. *Marine Mammal Science*, **14**, 245–256.

Olsen, C. R., Elsner, R., Hale, F. C., & Kenney, D. W. 1969a. "Blow" of the pilot whale. *Science*, **163**, 953–955.

Olsen, C. R., Hale, F. C., & Elsner, R. 1969b. Mechanics of ventilation in the pilot whale. *Respiration Physiology*, **7**, 137–149.

Olson, K. R., Dombkowski, R. A., Russell, M. J., *et al.* 2006. Hydrogen sulfide as an oxygen sensor/transducer in vertebrate hypoxic vasoconstriction and hypoxic vasodilation. *Journal of Experimental Biology*, **209**, 4011–4023.

Olson, K. R., Whitfield, N. L., Bearden, S. E., *et al.* 2010. Hypoxic pulmonary vasodilation: a paradigm shift with a hydrogen sulfide mechanism. *American Journal of Physiology: Regulatory, Integrative and Comparative Physiology*, **298**, R51–R60.

Ordway, G. A. & Garry, D. J. 2004. Myoglobin: an essential hemoprotein in striated muscle. *Journal of Experimental Biology*, **207**, 3441–3446.

Orgeig, S., Bernhard, W., Biswas, S. C., *et al.* 2007. The anatomy, physics, and physiology of gas exchange surfaces: is there a universal function for pulmonary surfactant in animal respiratory structures? *Integrative and Comparative Biology*, **47**, 610–627.

Otani, S., Naito, Y., Kawamura, A., *et al.* 1998. Diving behavior and performance of harbor porpoises, *Phocoena phocoena*, in Funka Bay, Hokkaido, Japan. *Marine Mammal Science*, **14**, 209–220.

Overgaard, K., Friis, S., Pedersen, B. K., & Lykkeboe, G. 2006. Influence of lung volume, glossopharyngeal inhalation, and P_{ET} O_2 and P_{ET} CO_2 on apnea performance in trained breath-hold divers. *European Journal of Applied Physiology*, **97**, 158–164.

Pabst, D. A. 1990. Axial muscles and connective tissues of the bottlenose dolphin. In: Leatherwood, S. & Reeves, R. R. (eds.) *The Bottlenose Dolphin*. San Diego, CA: Academic Press.

Pabst, D. A. 1993. Intramuscular morphology and tendon geometry of the epaxial swimming muscles of dolphins. *Journal of Zoology*, **230**, 159–176.

Pabst, D. A., Rommel, S. A., & McLellan, W. A. 1999. The functional morphology of marine mammals. In: Reynolds, J. E., & Rommel, S. A. (eds.), *Biology of Marine Mammals*. Washington, DC: Smithsonian University Press.

Pabst, D. A., Rommel, S. A., McLellan, W. A., Williams, T. M., & Rowles, T. K. 1995. Thermoregulation of the intra-abdominal testes of the bottlenose dolphin (*Tursiops truncatus*) during exercise. *Journal of Experimental Biology*, **198**, 221–226.

Panigada, S., Zanardelli, M., Canese, S., & Jahoda, M. 1999. How deep can baleen whales dive? *Marine Ecology Progress Series*, **187**, 309–311.

Panneton, W. M. 2013. The mammalian diving response: an enigmatic reflex to preserve life? *Physiology*, **28**, 284–297.

Panneton, W. M., Gan, Q., & Juric, R. 2010. The rat: a laboratory model for studies of the diving response. *Journal of Applied Physiology*, **108**, 811–820.

Panneton, W. M. & Yavari, P. 1995. A medullary dorsal horn relay for the cardiorespiratory responses evoked by stimulation of the nasal mucosa in the muskrat *Ondatra zibethicus*: evidence for excitatory amino acid transmission. *Brain Research*, **691**, 37–45.

Papadopoulos, S., Jurgens, K. D., & Gros, G. 1995. Diffusion of myoglobin in skeletal muscle cells: dependence on fibre type, contraction and temperature. *Pflugers Archiv*, **430**, 519–525.

Papaioannou, T. G. & Stefanadis, C. 2005. Basic principles of the intraaortic balloon pump and mechanisms affecting its performance. *ASAIO Journal*, **51**, 296–300.

Papastavrou, V., Smith, S. C., & Whitehead, H. 1989. Diving behaviour of the sperm whale, *Physeter macrocephalus*, off the Galapagos Islands. *Canadian Journal of Zoology*, **67**, 839–846.

Park, K. H., Rubin, L. E., Gross, S. S., & Roberto, L. 1992. Nitric oxide is a mediator of hypoxic coronary vasodilation: regulation to adenosine and cyclooxygenase-derived metabolites. *Circulation Research*, **71**, 992–1001.

Parkos, C. A. & Wahrenbrock, E. A. 1987. Acute effects of hypercapnia and hypoxia on minute ventilation in unrestrained Weddell seals. *Respiration Physiology*, **67**, 197–207.

Parrish, F. A., Abernathy, K., Marshall, G. J., & Buhleier, B. M. 2002. Hawaiian monk seals (*Monachus schauinslandi*) foraging in deep-water coral beds. *Marine Mammal Science*, **18**, 244–258.

Påsche, A. 1976a. The effect of hypercapnia on respiratory characteristics and diving behaviour of freely diving seals. *Respiration Physiology*, **26**, 183–193.

Påsche, A. 1976b. Hypoxia in freely diving hooded seal, *Cystophora cristata*. *Comparative Biochemistry and Physiology Part A: Physiology*, **55**, 319–322.

Paton, J. F. R., Boscan, P., Pickering, A. E., & Nalivaiko, E. 2005. The Yin and Yang of cardiac autonomic control: vago-sympathetic interactions revisited. *Brain Research Reviews*, **49**, 555–565.

Paton, J. F. R., Nalivaiko, E., Boscan, P., & Pickering, A. E. 2006. Reflexively evoked coactivation of cardiac vagal and sympathetic motor outflows: observations and functional implications. *Clinical and Experimental Pharmacology and Physiology*, **33**, 1245–1250.

Patterson, W. R., Dalton, L. M., & McGlasson, D. L. 1998. A comparison of human and killer whale platelet fatty acid composition. *Comparative Biochemistry and Physiology Part B: Biochemistry and Molecular Biology*, **120**, 247–252.

Patterson, W. R., Dalton, L. M., McGlasson, D. L., & Cissik, J. H. 1993. Aggregation of killer whale platelets. *Thrombosis Research*, **70**, 225–231.

Paulev, P. 1965. Decompression sickness following repeated breath-hold dives. *Journal of Applied Physiology*, **20**, 1028–1031.

Peichl, L., Behrmann, G., & Kröger, R. H. H. 2001. For whales and seals the ocean is not blue: a visual pigment loss in marine mammals. *European Journal of Neuroscience*, **13**, 1520–1528.

Perko, M. J., Nielsen, H. B., Skak, C., *et al.* 1998. Mesenteric, coeliac and splanchnic blood flow in humans during exercise. *Journal of Physiology*, **513**, 907–913.

Perrin, W. F., Wursig, B., & Thewissen, J. G. M. (eds.) 2009. *Encyclopedia of Marine Mammals*. San Diego, CA: Academic Press.

Persson, S. G. B., Ekman, L., Lyden, G., & Tufvesson, G. 1973. Circulation effects of splenectomy in the horse I–IV. II. Effect on plasma volume and total and circulating red cell volume. *Zentrabl. Veterinaermed*, **20**, 456–468.

Peters, G., Wilson, R. P., Scolaro, J. A., *et al.* 1998. The diving behavior of Magellanic penguins at Punta Norte, Peninsula Valdés, Argentina. *Colonial Waterbirds*, **21**, 1–10.

Petersen, S. L., Ryan, P. G., & Gremillet, D. 2006. Is food availability limiting African penguins *Spheniscus demersus* at Boulders? A comparison of foraging effort at mainland and island colonies. *Ibis*, **148**, 14–26.

Petrov, E. A. 1985. Myoglobin concentration and distribution in the tissue of foetal Baikalian seal (*Phoca sibirica*). *Journal of Evolutionary Biochemistry and Physiology*, **21**, 83–86.

Petrov, E. A. & Shoshenko, K. A. 1987. Total store of oxygen and duration of diving of the Nerpa. In: Galazii, G. I. (ed.), *Morphology and Ecology of Fish*. Novbosibirsk: Academy of Sciences of Russia, Siberian Division.

Petschow, D., Wardinger, I., Baumann, R., *et al.* 1977. Causes of high blood O_2 affinity in animals living at high altitude. *Journal of Applied Physiology*, **42**, 139–143.

Piantadosi, C. A. 2008. Carbon monoxide, reactive oxygen signaling, and oxidative stress. *Free Radical Biology and Medicine*, **45**, 562–569.

Piantadosi, C. A., Carraway, M. S., Babiker, A., & Suliman, H. B. 2008. Heme Oxygenase-1 regulates cardiac mitochondrial biogenesis via Nrf2-mediated transcriptional control of nuclear respiratory factor-1. *Circulation Research*, **103**, 1232–1240.

Piatt, J. F. & Nettleship, D. N. 1985. Diving depths of four alcids. *The Auk*, **102**, 293–297.

Pierard, J. 1971. Osteology and myology of the Weddell seal (*Leptonychotes weddelli*)(Lesson, 1826). *Antarctic Research Series*, **18**, 53–108.

Piscitelli, M., McLellan, W., Rommel, S., Blum, J., Barco, S., & Pabst, D. A. 2010. Lung size and thoracic morphology in shallow (*Tursiops truncatus*) and deep (*Kogia* spp.) diving cetaceans. *Journal of Morphology*, **271**, 654–673.

Piscitelli, M. A., Raverty, S. A., Lillie, M. A., & Shadwick, R. E. 2013. A review of cetacean lung morphology and mechanics. *Journal of Morphology*, **274**, 1425–1440.

Pitcher, K. W., Rehberg, M. J., Pendleton, G. W., *et al.* 2004. Ontogeny of dive performance in pup and juvenile Steller sea lions in Alaska. *Canadian Journal of Zoology*, **83**, 1214–1231.

Polasek, L. & Davis, R. W. 2001. Heterogeneity of myoglobin distribution in the locomotory muscles of five cetacean species. *Journal of Experimental Biology*, **204**, 209–215.

Polasek, L., Dickson, K. A., & Davis, R. W. 2006. Metabolic indicators in the skeletal muscles of harbor seals (*Phoca vitulina*). *American Journal of Physiology*, **290**, R1720–R1727.

Ponganis, P. J. 2007. Bio-logging of physiological parameters in higher marine vertebrates. *Deep-Sea Research II*, **54**, 183–192.

Ponganis, P. J. 2011. Diving mammals. In: Terjung, R. (ed.) *Comprehensive Physiology*. Bethesda, MD: John Wiley & Sons, Inc.

Ponganis, P. J., Costello, M. L., Starke, L. N., Mathieu-Costello, O., & Kooyman, G. L. 1997a. Structural and biochemical characteristics of locomotory muscles of emperor penguins, *Aptenodytes forsteri*. *Respiration Physiology*, **109**, 73–80.

Ponganis, P. J., Gentry, R. L., Ponganis, E. P., & Ponganis, K. V. 1992a. Analysis of swim velocities during deep and shallow dives of two northern fur seals, *Callorhinus ursinus*. *Marine Mammal Science*, **8**, 69–75.

Ponganis, P. J. & Kooyman, G. L. 1999. Heart rate and electrocardiogram characteristics of a young California gray whale (*Eschrictius robustus*). *Marine Mammal Science*, **15**, 1198–1207.

Ponganis, P. J., Kooyman, G. L., Baronov, E. A., Thorson, P. H., & Stewart, B. S. 1997b. The aerobic submersion limit of Baikal seals, *Phoca sibirica*. *Canadian Journal of Zoology*, **75**, 1323–1327.

Ponganis, P. J., Kooyman, G. L., & Castellini, M. A. 1993a. Determinants of the aerobic dive limit of Weddell seals: analysis of diving metabolic rates, post-dive end tidal P_{O2}'s, and blood and muscle oxygen stores. *Physiological Zoology*, **66**, 732–749.

Ponganis, P. J., Kooyman, G. L., & Castellini, M. A. 1995. Multiple sightings of Arnoux's beaked whales along the Victoria Land coast. *Marine Mammal Science*, **11**, 247–250.

Ponganis, P. J., Kooyman, G. L., Castellini, M. A., Ponganis, E. P., & Ponganis, K. V. 1993b. Muscle temperature and swim velocity profiles during diving in a Weddell seal, *Leptonychotes weddellii*. *Journal of Experimental Biology*, **183**, 341–348.

Ponganis, P. J., Kooyman, G. L., & Ridgway, S. H. 2003a. Comparative diving physiology. In: Brubakk, A. O. & Neuman, T. S. (eds.), *Bennett and Elliott's Physiology and Medicine of Diving*. Edinburgh: Saunders.

Ponganis, P. J., Kooyman, G. L., Sartoris, D., & Jobsis, P. F. 1992b. Pinniped splenic volumes. *American Journal of Physiology*, **262**, R322–R325.

Ponganis, P. J., Kooyman, G. L., Starke, L. N., Kooyman, C. A., & Kooyman, T. G. 1997c. Post-dive blood lactate concentrations in emperor penguins, *Aptenodytes forsteri*. *Journal of Experimental Biology*, **200**, 1623–1626.

Ponganis, P. J., Kooyman, G. L., Van Dam, R., & Le Maho, Y. 1999a. Physiological responses of king penguins during simulated diving to 136 m depth. *Journal of Experimental Biology*, **202**, 2819–2822.

Ponganis, P. J., Kooyman, G. L., Winter, L. M., & Starke, L. N. 1997d. Heart rate and plasma lactate responses during submerged swimming and diving in California sea lions (*Zalophus californianus*). *Journal of Comparative Physiology B*, **167**, 9–16.

Ponganis, P. J., Kooyman, G. L., & Zornow, M. H. 1991. Cardiac output in swimming California sea lions, *Zalophus californianus*. *Physiological Zoology*, **64**, 1296–1306.

Ponganis, P. J., Kooyman, G. L., Zornow, M. H., Castellini, M. A., & Croll, D. A. 1990. Cardiac output and stroke volume in swimming harbor seals. *Journal of Comparative Physiology*, **160B**, 473–482.

Ponganis, P. J., Kreutzer, U., Stockard, T. K., *et al.* 2008. Blood flow and metabolic regulation in seal muscle during apnea. *Journal of Experimental Biology*, **211**, 3323–3332.

Ponganis, P. J., Meir, J. U., & Williams, C. L. 2010a. Oxygen store depletion and the aerobic dive limit in emperor penguins. *Aquatic Biology*, **8**, 237–245.

Ponganis, P. J., Meir, J. U., & Williams, C. L. 2011. In pursuit of Irving and Scholander: a review of oxygen store management in seals and penguins. *Journal of Experimental Biology*, **214**, 3325–3339.

Ponganis, P. J. & Pierce, R. W. 1978. Muscle metabolic profiles and fiber-type composition in some marine mammals. *Comparative Biochemistry and Physiology B*, **59**, 99–102.

Ponganis, P. J., St. Leger, J., & Scadeng, M. 2015. Penguin lungs and air sacs: implications for baroprotection, oxygen stores and buoyancy. *Journal of Experimental Biology*, **218**, 720–730.

Ponganis, P. J., Starke, L. N., Horning, M., & Kooyman, G. L. 1999b. Development of diving capacity of emperor penguins. *Journal of Experimental Biology*, **202**, 781–786.

Ponganis, P. J., Stockard, T., Levenson, D. H., Berg, L., & Baranov, E. A. 2006a. Intravascular pressure profiles in elephant seals: a hypothesis on the extradural vein and venous return to the heart. *Comparative Biochemistry and Physiology A*, **145**, 123–130.

Ponganis, P. J., Stockard, T., Levenson, D. H., Berg, L., & Barnov, E. A. 2006b. Cardiac output and muscle blood flow during rest-associated apneas in elephant seals. *Comparative Biochemistry and Physiology A*, **144**, 105–111.

Ponganis, P. J., Stockard, T. K., Meir, J. U., *et al.* 2009. O_2 store management in diving emperor penguins. *Journal of Experimental Biology*, **212**, 217–224.

Ponganis, P. J., Stockard, T. K., Meir, J. U., *et al.* 2007. Returning on empty: extreme blood O_2 depletion underlies dive capacity of emperor penguins. *Journal of Experimental Biology*, **210**, 4279–4285.

Ponganis, P. J., Van Dam, R. P., Knower, T., & Levenson, D. H. 2001. Temperature regulation in emperor penguins foraging under sea ice. *Comparative Biochemistry and Physiology A*, **129**, 811–820.

Ponganis, P. J., Van Dam, R. P., Knower, T., Levenson, D. H., & Ponganis, K. V. 2004. Deep dives and aortic temperatures of emperor penguins: new directions for bio-logging at the isolated dive hole. *Memoirs of the National Institute of Polar Research, Special Issue*, **58**, 155–161.

Ponganis, P. J., Van Dam, R. P., Levenson, D. H., *et al.* 2003b. Regional heterothermy and conservation of core temperature in emperor penguins diving under sea ice. *Comparative Biochemistry and Physiology A*, **135**, 477–487.

Ponganis, P. J., Van Dam, R. P., Marshall, G., Knower, T., & Levenson, D. H. 2000. Sub-ice foraging behavior of emperor penguins. *Journal of Experimental Biology*, **203**, 3275–3278.

Ponganis, P. J., Welch, T. J., Welch, L. S., & Stockard, T. K. 2010b. Myoglobin production in emperor penguins. *Journal of Experimental Biology*, **213**, 1901–1906.

Postle, A. D., Heeley, E. L., & Wilton, D. C. 2001. A comparison of the molecular species compositions of mammalian lung surfactant phospholipids. *Comparative Biochemistry and Physiology Part A: Molecular & Integrative Physiology*, **129**, 65–73.

Potvin, J., Goldbogen, J. A., & Shadwick, R. E. 2009. Passive versus active engulfment: verdict from trajectory simulations of lunge-feeding fin whales *Balaenoptera physalus*. *Journal of the Royal Society Interface*, **6**, 1005–1025.

Powell, F. L. 2000. Respiration. In: Whittow, G. C. (ed.), *Sturkie's Avian Physiology*. San Diego, CA: Academic Press.

Powers, S. K. & Jackson, M. J. 2008. Exercise-induced oxidative stress: cellular mechanisms and impact on muscle force production. *Physiologcial Reviews*, **88**, 1243–1276.

Prince, P. A., Huin, N., & Weimerskirch, H. 1994. Diving depths of albatrosses. *Antarctic Science*, **6**, 353–354.

Prince, P. A. & Jones, M. 1992. Maximum dive depths attained by South Georgia diving petrel *Pelecanoides georgicus* at Bird Island, South Georgia. *Antarctic Science*, **4**, 433–434.

Puerta, M. L., Del Campo, A. L. G., Huecas, V., & Abelenda, M. 1991. Hematology and blood chemistry of the white pelican (*Pelecanus onocrotalus*). *Comparative Biochemistry and Physiology Part A: Physiology*, **98**, 393–394.

Pugh, L. G. C. E. 1959. Carbon monoxide content of the blood and other observations on Weddell seals. *Nature*, **183**, 74–76.

Pütz, K. & Cherel, Y. 2005. The diving behaviour of brooding king penguins (*Aptenodytes patagonicus*) from the Falkland Islands: variation in dive profiles and synchronous underwater swimming provide new insights into their foraging strategies. *Marine Biology*, **147**, 281–290.

Pütz, K., Wilson, R. P., Charrassin, J. B., *et al.* 1998. Foraging strategy of king penguins (*Aptnodytes patagonicus*) during summer at the Crozet Islands. *Ecology*, **79**, 1905–1921.

Quinones, Q. J., Ma, Q., Zhang, Z. L., Barnes, B. M., & Podgoreanu, M. V. 2014. Organ protective mechanisms common to extremes of physiology: a window through hibernating biology. *Integrative and Comparative Biology*, **54**, 497–515.

Qvist, J., Hill, R. D., Schneider, R. C., *et al.* 1986. Hemoglobin concentrations and blood gas tensions of free-diving Weddell seals. *Journal of Applied Physiology*, **61**, 1560–1569.

Qvist, J., Hurford, W. E., Park, Y. S., *et al.* 1993. Arterial blood gas tensions during breath-hold diving in the Korean ama. *Journal of Applied Physiology*, **75**, 285–293.

Qvist, J., Weber, R. E., & Zapol, W. M. 1981. Oxygen equilibrium properties of blood and hemoglobin of fetal and adult Weddell seals. *Journal of Applied Physiology*, **50**, 999–1005.

Radermacher, P., Falke, K. J., Park, Y. S., *et al.* 1992. Nitrogen tensions in brachial vein blood of Korean ama divers. *Journal of Applied Physiology*, **73**, 2592–2595.

Ralls, K., Hatfield, B. B., & Siniff, D. B. 1995. Foraging patterns of California sea otters as indicated by telemetry. *Canadian Journal of Zoology*, **73**, 523–531.

Ramirez, J.-M., Folkow, L. P., & Blix, A. S. 2007. Hypoxia tolerance in mammals and birds: from the wilderness to the clinic. *Annual Review of Physiology*, **69**, 113–143.

Rayner, M. J., Taylor, G. A., Tompson, D. R., Torres, L. G., Sagar, P. M., & Shaffer, S. A. 2011. Migration and diving activity in three non-breeding flesh-footed shearwaters *Puffinus carneipes*. *Journal of Avian Biology*, **42**, 266–270.

Reed, J. Z., Butler, P. J., & Fedak, M. A. 1994a. The metabolic characteristics of the locomotory muscles of grey seals (*Halichoerus grypus*), harbour seals (*Phoca vitulina*), and Antarctic fur seals (*Arctocephalus gazella*). *Journal of Experimental Biology*, **194**, 33–46.

Reed, J. Z., Chambers, C., Fedak, M. A., & Butler, P. J. 1994b. Gas exchange of captive freely diving grey seals (*Halichoerus grypus*). *Journal of Experimental Biology*, **191**, 1–18.

Reed, J. Z., Chambers, C., Hunter, C. J., *et al.* 2000. Gas exchange and heart rate in the harbour porpoise, *Phocoena phocoena*. *Journal of Comparative Physiology B*, **170**, 1–10.

Reeder, B. J. 2010. The redox activity of hemoglobins: from physiologic functions to pathologic mechanisms. *Antioxidants & Redox Signaling*, **13**, 1087–1123.

Reidenberg, J. S. & Laitman, J. T. 2007. Discovery of a low frequency sound source in Mysticeti (baleen whales): anatomical establishment of a vocal fold homolog. *The Anatomical Record: Advances in Integrative Anatomy and Evolutionary Biology*, **290**, 745–759.

Renjun, L., Gewalt, W., Neurohr, B., & Winkler, A. 1994. Comparative studies on the behaviour of *Inia geoffrensis* and *Lipotes vexillifer* in artificial environments. *Aquatic Mammals*, **20**, 39–45.

Revsbech, I. G., Shen, X., Chakravarti, R., *et al.* 2014. Hydrogen sulfide and nitric oxide metabolites in the blood of free-ranging brown bears and their potential roles in hibernation. *Free Radical Biology and Medicine*, **73**, 349–357.

Reynafarje, B. 1963. Simplified method for the determination of myoglobin. *Journal of Laboratory and Clinical Medicine*, **61**, 139–145.

Reynolds III, J. E. 1981. Behavior patterns in the West Indian manatee, with emphasis of feeding and diving (Florida). *Florida Scientist*, **44**, 232–242.

Reynolds III, J. E. & Odell, D. K. 1991. *Manatees and Dugongs*. New York: Facts on File, Inc.

Reynolds III, J. E. & Rommel, S. A. (eds.) 1999. *Biology of Marine Mammals*. Washington, DC: Smithsonian Institution Press.

Rhode, E. A., Elsner, R., Peterson, T. M., Campbell, K. B., & Spangler, W. 1986. Pressure–volume characteristics of aortas of harbor and Weddell seals. *American Journal of Physiology*, **251**, R174–R180.

Rhodes, M. A., Carraway, M. S., Piantadosi, C. A., *et al.* 2009. Carbon monoxide, skeletal muscle oxidative stress, and mitochondrial biogenesis in humans. *American Journal of Physiology: Heart and Circulatory Physiology*, **297**, H392–H399.

Ribak, G., Weihs, D., & Arad, Z. 2004. How do cormorants counter buoyancy during submerged swimming? *Journal of Experimental Biology*, **207**, 2101–2114.

Ribak, G., Weihs, D., & Arad, Z. 2005a. Submerged swimming of the great cormorant *Phalacrocorax carbo sinensis* is a variant of the burst-and-glide gait. *Journal of Experimental Biology*, **208**, 3835–3849.

Ribak, G., Weihs, D., & Arad, Z. 2005b. Water retention in the plumage of diving great cormorants *Phalacrocorax carbo sinensis*. *Journal of Avian Biology*, **36**, 89–95.

Richardson, R. S., Noyszewski, E. A., Kendrick, K. F., Leigh, J. J., & Wagner, P. D. 1995. Myoglobin O_2 dissociation during exercise: evidence of limited O_2 transport. *Journal of Clinical Investigation*, **96**, 1916–1926.

Richmond, J. P., Burns, J. M., & Rea, L. D. 2006. Ontogeny of total body oxygen stores and aerobic dive potential in the Steller sea lion (*Eumetopias jubatus*). *Journal of Comparative Physiology B*, **176**, 535–545.

Ridgway, S. H. 1972. *Mammals of the Sea: Biology and Medicine*. Springfield, IL: Charles C. Thomas.

Ridgway, S. H. 1986. Diving by Cetaceans. In: Brubakk, A. O., Kanwisher, J. W. & Sundnes, G. (eds.), *Diving in Animals and Man*. Trondheim: Royal Norwegian Society of Science and Letters.

Ridgway, S. H., Bowers, C. A., Miller, D., *et al.* 1984. Diving and blood oxygen in the white whale. *Canadian Journal of Zoology*, **62**, 2349–2351.

Ridgway, S. H., Carder, D. A., & Clark, W. 1975a. Conditioned bradycardia in the sea lion *Zalophus californianus*. *Nature*, **256**, 37–38.

Ridgway, S. H., Harrison, R. J., & Joyce, P. L. 1975b. Sleep and cardiac rhythm in the gray seal. *Science*, **187**, 553–555.

Ridgway, S. H. & Harrison, R. J. 1981–1998. *Handbook of Marine Mammals*. Vols 1–6. New York: Academic Press.

Ridgway, S. H. & Howard, R. 1979. Dolphin lung collapse and intramuscular circulation during free diving: evidence from nitrogen washout. *Science*, **206**, 1182–1183.

Ridgway, S. H. & Johnston, D. G. 1966. Blood oxygen and ecology of porpoises of three genera. *Science*, **151**, 456–458.

Ridgway, S. H. & Kohin, S. 1995. The relationship between heart mass and body mass for three cetacean genera: narrow allometry demonstrates interspecific differences. *Marine Mammal Science*, **11**, 72–80.

Ridgway, S. H., Scronce, B. L., & Kanwisher, J. 1969. Respiration and deep diving in the bottlenose porpoise. *Science*, **166**, 1651–1654.

Rijke, A. M. 1968. The water repellency and feather structure of cormorants, Phalacrocoracidae. *Journal of Experimental Biology*, **48**, 185–189.

Rivero, J.-L. L., Talmadge, R. J., & Edgerton, V. R. 1999. Interrelationships of myofibrillar ATPase activity and metabolic properties of myosin heavy chain-based fibre types in rat skeletal muscle. *Histochemistry and Cell Biology*, **111**, 277–287.

Riveros-Moreno, V. & Wittenberg, J. B. 1972. The self-diffusion coefficients of myoglobin and hemoglobin in concentrated solution. *Journal of Biological Chemistry*, **247**, 895–901.

Robinson, A. J., Kropatkin, M., & Aggeler, P. M. 1969. Hageman Factor (Factor XII) deficiency in marine mammals. *Science*, **166**, 1420–1422.

Robinson, P. W., Costa, D. P., Crocker, D. E., *et al.* 2012. Foraging behavior and success of a mesopelagic predator in the northeast Pacific Ocean: insights from a data-rich species, the northern elephant seal. *PloS ONE*, **7**, e36728.

Robinson, P. W., Simmons, S. E., Crocker, D. E., & Costa, D. P. 2010. Measurements of foraging success in a highly pelagic marine predator, the northern elephant seal. *Journal of Animal Ecology*, **79**, 1146–1156.

Robinson, S. A. & Hindell, M. A. 1996. Foraging ecology of gentoo penguins *Pygoscelis papua* at Macquarie Island during the period of chick care. *Ibis*, **138**, 722–731.

Roca, J., Agusti, A. G. N., Alonso, A., *et al.* 1992. Effects of training on muscle O_2 transport at VO_{2MAX}. *Journal of Applied Physiology*, **73**, 1067–1076.

Rodary, D., Bonneau, W., Le Maho, Y., & Bost, C. A. 2000. Benthic diving in male emperor penguins foraging in winter. *Marine Ecology Progress Series*, **207**, 171–181.

Rommel, S. A. & Caplan, H. 2003. Vascular adaptations for heat conservation in the tail of Florida manatees (*Trichechus manatus latirostris*). *Journal of Anatomy*, **202**, 343–353.

Rommel, S. A., Costidis, A. M., Fernandez, A., *et al.* 2006. Elements of beaked whale anatomy and diving physiology and some hypothetical causes of sonar-related stranding. *Journal of Cetacean Research Management*, **7**, 189–209.

Rommel, S. A., Early, G. A., Matassa, K. A., Pabst, D. A., & McLellan, W. A. 1995. Venous structures associated with thermoregulation of phocid seal reproductive organs. *The Anatomical Record*, **243**, 390–402.

Rommell, S. A., Pabst, D. A., McLellan, W. A., Mead, J. A., & Potter, C. W. 1992. Anatomical evidence for a countercurrent heat exchanger associated with dolphin testes. *Anatomical Record*, **232**, 150–156.

Rommell, S. A., Pabst, D. A., McLellan, W. A., Williams, T. M., & Friedl, W. A. 1994. Temperature regulation of the testes of the bottlenose dolphin (*Tursiops truncatus*): evidence from colonic temperatures. *Journal of Comparative Physiology B*, **164**, 130–134.

Ron, Y. & Guillemette, M. 1991. Diving and foraging in the common eider. *Ornis Scandinavica*, **22**, 349–352.

Ronald, K., McCarter, R., & Selley, L. J. 1977. Venous circulation of the harp seal (*Pagophilus groenlandicus*). In: Harrison, R. J. (ed.), *Functional Anatomy of Marine Mammals*. New York: Academic Press.

Ropert-Coudert, Y., Chiaradia, A., & Kato, A. 2006. An exceptionally deep dive by a little penguin *Eudyptula minor*. *Marine Ornithology*, **34**, 71–74.

Ropert-Coudert, Y., Daunt, F., Kato, A., *et al.* 2009. Underwater wingbeats extend depth and duration of plunge dives in northern gannets *Morus bassanus*. *Journal of Avian Biology*, **40**, 380–387.

Ropert-Coudert, Y., Grémillet, D., Kato, A., *et al.* 2004a. A fine-scale time budget of Cape gannets provides insights into their foraging strategies. *Animal Behaviour*, **67**, 985–992.

Ropert-Coudert, Y., Grémillet, D., Ryan, P. G., *et al.* 2004b. Between air and water: the plunge dive of the Cape gannet *Morus capensis*. *Ibis*, **146**, 281–290.

Ropert-Coudert, Y. & Kato, A. 2009. Diving activity of hoary-headed (*Poliocephalus poliocephalus*) and Australasian little (*Tachybaptus novaehollandiae*) grebes. *Waterbirds*, **32**, 157–161.

Ropert-Coudert, Y., Kato, A., Naito, Y., & Cannell, B. L. 2003. Individual diving strategies in the little penguin. *Waterbirds*, **26**, 403–408.

Rosen, D. A. & Trites, A. W. 1997. Heat increment of feeding in Steller sea lions, *Eumetopias jubatus*. *Comparative Biochemistry and Physiology A*, **118**, 877–881.

Rosen, D. A. S., Winship, A. J., & Hoopes, L. A. 2007. Thermal and digestive constraints to foraging behaviour in marine mammals. *Philosophical Transactions of the Royal Society B: Biological Sciences*, **362**, 2151–2168.

Rosser, B. W. C., Waldbillig, D. M., Wick, M., & Bandman, E. 1994. Muscle fiber types in the pectoralis of the white pelican, a soaring bird. *Acta Zoologica*, **75**, 329–336.

Roughton, F. J. W. & Darling, R. C. 1944. The effect of carbon monoxide on the oxyhemoglobin dissociation curve. *American Journal of Physiology*, **141**, 17–31.

Rowell, L. B. & O'Leary, D. S. 1990. Reflex control of the circulation during exercise: chemoreflexes and mechanoreflexes. *Journal of Applied Physiology*, **69**, 407–418.

Rowlatt, U. 1981. The cardiac ventricles of a baleen whale (*Balaenoptera acutorostrata*: minke whale) and a toothed whale (*Hyperoodon ampullatus*: bottlenose whale). *Journal of Morphology*, **168**, 85–96.

Rowlatt, U. & Gaskin, D. E. 1975. Functional anatomy of the heart of the harbor porpoise, *Phocaena phocaena*. *Journal of Morphology*, **146**, 479–493.

Ryan, P. G. 2007. Diving in shallow water: the foraging ecology of darters (Aves: Anhingidae). *Journal of Avian Biology*, **38**, 507–514.

Ryan, P. G. & Nel, D. C. 1999. Foraging behaviour of diving petrels *Pelecanoides*. *Emu*, **99**, 72–74.

Ryan, P. G., Pichegru, L., Ropert-Coudert, Y., Grémillet, D., & Kato, A. 2010. On a wing and a prayer: the foraging ecology of breeding Cape cormorants. *Journal of Zoology*, **280**, 25–32.

Ryg, M., Lydersen, C., Knutsen, L. Ø., *et al.* 1993. Scaling of insulation in seals and whales. *Journal of Zoology*, **230**, 193–206.

Sadé, J., Handrich, Y., Bernheim, J., & Cohen, D. 2008. Pressure equilibration in the penguin middle ear. *Acta Oto-laryngologica*, **128**, 18–21.

Saito, H., Poon, M.-C., Goldsmith, G. H., Ratnoff, O. D., & Árnason, Ú. 1976. Studies on the blood clotting and fibrinolytic system in the plasma from a sei (baleen) whale. *Experimental Biology and Medicine*, **152**, 503–507.

Sakamoto, K. Q., Takahashi, A., Iwata, T., & Trathan, P. N. 2009. From the eye of the albatrosses: a bird-borne camera shows an association between albatrosses and a killer whale in the Southern Ocean. *PLoS ONE*, **4**, e7322.

Sakamoto, T. & Monafo, W. W. 1989. Regional blood flow in the brain and spinal cord of hypothermic rats. *American Journal of Physiology: Heart and Circulatory Physiology*, **257**, H785–H790.

Saltin, B. 2007. Exercise hyperaemia: magnitude and aspects on regulation in humans. *Journal of Physiology*, **583**, 819–823.

Saltin, B., Gagge, A. P., & Stolwijk, J. A. J. 1968. Muscle temperature during submaximal exercise in man. *Journal of Applied Physiology*, **25**, 679–688.

Sanders, T. M., Werner, R. A., & Bloor, C. M. 1976. Visceral blood flow distribution during exercise to exhaustion in conscious dogs. *Journal of Applied Physiology*, **40**, 927–931.

Sato, K., Daunt, F., Watanuki, Y., Takahashi, A., & Wanless, S. 2008. A new method to quantify prey acquisition in diving seabirds using wing stroke frequency. *Journal of Experimental Biology*, **211**, 58–65.

Sato, K., Mitani, Y., Camerson, M. F., Siniff, D. B., & Naito, Y. 2003. Factors affecting stroking patterns and body angle in diving Weddell seals under natural conditions. *Journal of Experimental Biology*, **206**, 1461–1470.

Sato, K., Naito, Y., Kato, A., *et al.* 2002. Buoyancy and maximal diving depth in penguins: do they control inhaling air volume? *Journal of Experimental Biology*, **205**, 1189–1197.

Sato, K., Shiomi, K., Marshall, G., Kooyman, G. L., & Ponganis, P. J. 2011. Stroke rates and diving air volumes of emperor penguins: implications for dive performance. *Journal of Experimental Biology*, **214**, 2854–2863.

Sato, K., Shiomi, K., Watanabe, Y., *et al.* 2009. Scaling of swim speed and stroke frequency in geometrically similar penguins: they swim optimally to minimize cost of transport. *Proceedings of the Royal Society B*, **277**, 707–714.

Sato, K., Watanuki, Y., & Naito, Y. 2006. The minimum air volume kept in diving Adelie penguins: evidence for regulation of air volume in the respiratory system. *Coastal Marine Science*, **30**, 439–442.

Sato, K., Watanuki, Y., Takahashi, A., *et al.* 2007. Stroke frequency, but not swimming speed, is related to body size in free-ranging seabirds, pinnipeds and cetaceans. *Proceedings of the Royal Society B: Biological Sciences*, **274**, 471–477.

Schaeffer, K. E., Allison, R. D., Dougherty, Jr, J. H., *et al.* 1968. Pulmonary and circulatory adjustments determining the limits of depths in breathhold diving. *Science*, **162**, 1020–1023.

Schagatay, E. 2012. Size matters: spleen and lung volumes predict performance in human apneic diving. *Frontiers in Physiology*. DOI: 10.3389/fphys.2012.00173.

Scheffer, V. B. 1964. Hair patterns in seals (Pinnipedia). *Journal of Morphology*, **115**, 291–303.

Scheid, P. 1979. Mechanisms of gas exchange in bird lungs. *Reviews in Physiology, Biochemistry, and Physiology*, **86**, 137–186.

Scheid, P. & Piiper, J. 1987. Gas exchange and transport. In: Seller, T. J. (ed.) *Bird Respiration*. Boca Raton, FL: CRC Press.

Scheid, P., Slama, H., & Willmer, H. 1974. Volume and ventilation of air sacs in ducks studied by inert gas wash-out. *Respiration Physiology*, **21**, 19–36.

Schenkman, K. A., Marble, D. A., Burns, D. H., & Feigl, E. O. 1997. Myoglobin oxygen dissociation by multiwavelength spectroscopy. *Journal of Applied Physiology*, **82**, 86–92.

Schiavani, A. & Rey, A. R. 2004. Long days, long trips: foraging ecology of female rockhopper penguins *Eudyptes chrysocome chrysocome* at Tierra del Fuego. *Marine Ecology Progress Series*, **275**, 231–262.

Schmid, D., Grémillet, D. J. H., & Culik, B. M. 1995. Energetics of underwater swimming in the great cormorant (*Phalacrocorax carbo sinensis*). *Marine Biology*, **123**, 875–881.

Schmidt-Nielsen, B., Murdaugh, H. V., O'Dell, R., & Bacsanyi, J. 1959. Urea excretion and diving in the seal (*Phoca vitulina* L.). *Journal of Cellular and Comparative Physiology*, **53**, 393–411.

Schmidt-Nielsen, K. 1984. *Scaling: Why is Animal Size so Important?* Cambridge: Cambridge University Press.

Schmidt, A., Alard, F., & Handrich, Y. 2006. Changes in body temperature in king penguins at sea: the result of fine adjustments in peripheral heat loss. *American Journal of Physiology*, **291**, R608–R618.

Schneuer, M., Flachsbarth, S., Czech-Damal, N. U., *et al.* 2012. Neuroglobin of seals and whales: evidence for a divergent role in the diving brain. *Neuroscience*, **223**, 35–44.

Scholander, P. F. 1940. Experimental investigations on the respiratory function in diving mammals and birds. *Hvalradets Skrifter*, **22**, 1–131.

Scholander, P. F. 1955. Counter current exchange: a principle in biology. *Hvalradets Skrifter*, **44**, 1–24.

Scholander, P. F. 1960. Oxygen transport through hemoglobin solutions. *Science*, **131**, 585–590.

Scholander, P. F. 1963. The master switch of life. *Scientific American*, **209**, 92–106.

Scholander, P. F., Hammel, H. T., Lemessurier, H., Hemmingsen, E., & Garey, W. 1962. Circulatory adjustment in pearl divers. *Journal of Applied Physiology*, **17**, 184–190.

Scholander, P. F., Hock, R., Walters, V., & Irving, L. 1950a. Adaptation to cold in Arctic and tropical mammals and birds in relation to body temperature, insulation, and basal metabolic rate. *The Biological Bulletin*, **99**, 259–271.

Scholander, P. F., Hock, R., Walters, V., Johnson, F., & Irving, L. 1950b. Heat regulation in some arctic and tropical mammals and birds. *The Biological Bulletin*, **99**, 237–258.

Scholander, P. F. & Irving, L. 1941. Experimental investigations on the respiration and diving of the Florida manatee. *Journal of Cellular and Comparative Physiology*, **17**, 169–191.

Scholander, P. F., Irving, L., & Grinnell, S. W. 1942a. Aerobic and anaerobic changes in seal muscle during diving. *Journal of Biological Chemistry*, **142**, 431–440.

Scholander, P. F., Irving, L., & Grinnell, S. W. 1942b. On the temperature and metabolism of the seal during diving. *Journal of Cellular and Comparative Physiology*, **19**, 67–78.

Scholander, P. F. & Schevill, W. E. 1955. Counter-current vascular heat exchange in the fins of whales. *Journal of Applied Physiology*, **8**, 279–282.

Scholander, P. F., Walters, V., Hock, R., & Irving, L. 1950c. Body insulation of some artic and tropical mammals and birds. *Biological Bulletin*, **99**, 225–236.

Schorger, A. 1947. The deep diving of the loon and old-squaw and its mechanism. *The Wilson Bulletin*, **59**, 151–159.

Schorr, G. S., Falcone, E. A., Moretti, D. J., & Andrews, R. D. 2014. First long-term behavioral records from Cuvier's beaked whales (*Ziphius cavirostris*) reveal record-breaking dives. *PLoS ONE*. DOI: 10.1371/journal.pone.0092633.

Schreer, J. F. & Testa, J. W. 1996. Classification of Weddell seal diving behavior. *Marine Mammal Science*, **12**, 227–250.

Schreiber, E. A. & Burger, J. (eds.) 2002. *Biology of Marine Birds*. Boca Raton, FL: CRC Press.

Schulte-Pelkum, N., Wieskotten, S., Hanke, W., Dehnhardt, G., & Mauck, B. 2007. Tracking of biogenic hydrodynamic trails in harbour seals (*Phoca vitulina*). *Journal of Experimental Biology*, **210**, 781–787.

Schumacher, U. & Welsch, U. 1987. Histological, histochemical, and fine structural observations on the spleen of seals. *American Journal of Anatomy*, **179**, 356–368.

Scott, M. S. & Chivers, S. J. 2009. Movements and diving behavior of pelagic spotted dolphins. *Marine Mammal Science*, **25**, 137–160.

Sebert, P. (ed.) 2010. *Comparative High Pressure Biology*. Enfield: Science Publishers.

Seddon, P. J. & Vanheezik, Y. 1990. Diving depths of the yellow-eyed penguin *Megadyptes antipodes*. *Emu*, **90**, 53–57.

Semenza, G. L. 2004. O2-regulated gene expression: transcriptional control of cardiorespiratory physiology by HIF-1. *Journal of Applied Physiology*, **96**, 1173–1177.

Semenza, G. L. 2007. Oxygen-dependent regulation of mitochondrial respiration by hypoxia-inducible factor 1. *Biochemical Journal*, **405**, 1–9.

Sendroy, J., Dillon, R. T., & Van Slyke, D. D. 1934. Studies of gas and electrolyte equilibria in blood: XIX. The solubility and physical state of uncombined oxygen in blood. *Journal of Biological Chemistry*, **105**, 597–632.

Sessler, D. I. 2000. Perioperative heat balance. *Anesthesiology*, **92**, 578–596.

Sessler, D. I. M. D. 2009. Thermoregulatory defense mechanisms. *Critical Care Medicine Therapeutic Temperature Management: State of the Art in the Critically Ill*, **37**, S203–S210.

Shadwick, R. E. & Gosline, J. M. 1994. Arterial mechanics in the fin whale suggest a unique hemodynamic design. *American Journal of Physiology*, **267**, R805–R818.

Shaffer, S. A. 2011. A review of seabird energetics using the doubly labeled water method. *Comparative Biochemistry and Physiology Part A: Molecular & Integrative Physiology*, **158**, 315–322.

Shaffer, S. A., Costa, D. P., Williams, T. M., & Ridgway, S. H. 1997. Diving and swimming performance of white whales, *Delphinapterus leucas*: an assessment of plasma lactate and blood gas levels and respiratory rates. *Journal of Experimental Biology*, **200**, 3091–3099.

Shaffer, S. A., Tremblay, Y., Weimerskirch, H., *et al.* 2006. Migratory shearwaters integrate oceanic resources across the Pacific Ocean in an endless summer. *Proceedings of the National Academy of Sciences*, **103**, 12799–12802.

Shaffer, S. A., Weimerskirch, H., Scott, D., *et al.* 2009. Spatiotemporal habitat use by breeding sooty shearwaters *Puffinus griseus*. *Marine Ecology Progress Series*, **391**, 209–220.

Shankaran, S., Laptook, A. R., Ehrenkranz, R. A., *et al.* 2005. Whole-body hypothermia for neonates with hypoxic–ischemic encephalopathy. *New England Journal of Medicine*, **353**, 1574–1584.

Shelton, G. & Jones, D. R. 1965. Central blood pressure and heart output in surfaced and submerged frogs. *Journal of Experimental Biology*, **42**, 339–357.

Shiomi, K., Sato, K., Mitamura, H., *et al.* 2008. Effect of ocean current on the dead-reckoning estimation of 3D dive paths of emperor penguins. *Aquatic Biology*, **3**, 265–270.

Shiomi, K., Sato, K., & Ponganis, P. J. 2012. Point of no return in diving emperor penguins: is the timing of the decision to return limited by the number of strokes? *Journal of Experimental Biology*, **215**, 135–140.

Shiva, S., Huang, Z., Grubina, R., *et al.* 2007. Deoxymyoglobin is a nitrite reductase that generates nitric oxide and regulates mitochondrial respiration. *Circulation Research*, **100**, 654–661.

Shrestha, R., Mckinley, C., Showalter, R., *et al.* 1997. Quantitative liver function tests define the functional severity of liver disease in early-stage cirrhosis. *Liver Transplantation and Surgery*, **3**, 166–173.

Sibley, C. G. & Monroe, Jr., B. L. 1990. *Distribution and Taxonomy of Birds of the World*. New Haven, CT: Yale University Press.

Signore, P. E. & Jones, D. R. 1995. Effect of pharmacological blockade on cardiovascular responses to voluntary and forced diving in muskrats. *Journal of Experimental Biology*, **198**, 2307–2315.

Simmins, K. 1969. Duration of dives in the red-necked grebe. *British Birds*, **64**, 300–302.

Simmonds, M. P. & Lopez-Jurado, L. F. 1991. Whales and the military. *Nature*, **351**, 448.

Simon, L., Robin, E., Elsner, R., Van Kessel, A., & Theodore, J. 1974. A biochemical basis for differences in maximal diving time in aquatic mammals. *Comparative Biochemistry and Physiology B*, **47**, 209–215.

Simon, M., Johnson, M., & Madsen, P. T. 2012. Keeping momentum with a mouthful of water: behavior and kinematics of humpback whale lunge feeding. *Journal of Experimental Biology*, **215**, 3786–3798.

Simpson, J. G. & Gardner, M. B. 1972. Comparative microscopic anatomy of selected marine mammals. In: Ridgway, S. H. (ed.), *Mammals of the Sea: Biology and Medicine*. Springfield, IL: C.C. Thomas.

Simpson, J. G., Gilmartin, W. G., & Ridgway, S. H. 1970. Blood volume and other hematologic values in young elephant seals (*Mirounga angustirostris*). *American Journal of Veterinary Research*, **31**, 1449–1452.

Sinnett, E. E., Kooyman, G. L., & Wahrenbrock, E. A. 1978. Pulmonary circulation of the harbor seal. *Journal of Applied Physiology*, **45**, 718–727.

Sjare, B. L. & Smith, T. G. 1986. The relationship between behavioral activity and underwater vocalizations of the white whale, *Delphinapterus leucas*. *Canadian Journal of Zoology*, **64**, 2824–2831.

Skinner, L. A. & Milsom, W. K. 2004. Respiratory chemosensitivity during wake and sleep in harbour seal pups (*Phoca vitulina richardsii*). *Physiological and Biochemical Zoology*, **77**, 847–863.

Skinner, M. R. & Marshall, J. M. 1996. Studies on the role of ATP, adenosine, and nitric oxide in mediating muscle vasodilatation induced in the rat by acute systemic hypoxia. *Journal of Physiology*, **495**.2, 553–560.

Skrovan, R. C., Williams, T. M., Berry, P. S., *et al.* 1999. The diving physiology of bottlenose dolphins (*Tursiops truncatus*): II. Biomechanics and changes in buoyancy at depth. *Journal of Experimental Biology*, **202**, 2749–2761.

Sleet, R. B., Sumich, J. L., & Weber, L. J. 1981. Estimates of total blood volume and total body weight of sperm whale (*Physeter catodon*). *Canadian Journal of Zoology*, **59**, 567–570.

Slijper, E. J. 1961. Foramen ovale and ductus arteriosus botalli in aquatic mammals. *Mammalia*, **25**, 528–570.

Slijper, E. J. 1962. *Whales*. London: Hutchinson and Co.

Smith, N. D. 2012. Body mass and foraging ecology predict evolutionary patterns of skeletal pneumaticity in the diverse "waterbird" clade. *Evolution*, **66**, 1059–1078.

Smith, N. P., Barclay, C. J., & Loiselle, D. S. 2005. The efficiency of muscle contraction. *Progress in Biophysics and Molecular Biology*, **88**, 1–58.

Smith, S. A., Mitchell, J. H., & Garry, M. G. 2006. The mammalian exercise pressor reflex in health and disease. *Experimental Physiology*, **91**, 89–102.

Smodlaka, H., Henry, R. W., Schumacher, J., & Reed, R. B. 2008. Macroscopic anatomy of the heart of the ringed seal (*Phoca hispida*). *Anatomia, Histologia, Embryologia*, **37**, 30–35.

Snow, D. H., Harris, R. C., & Gash, S. P. 1985. Metabolic responses of equine muscle to intermittent maximal exercise. *Journal of Applied Physiology*, **58**, 1689–1697.

Soegaard, L. B., Hansen, M. N., Van Elk, C., Brahm, J., & Jensen, F. B. 2012. Respiratory properties of blood in the harbor porpoise, *Phocoena phocoena*. *Journal of Experimental Biology*, **215**, 1938–1943.

Sommer, N., Dietrich, A., Schermuly, R. T., *et al.* 2008. Regulation of hypoxic pulmonary vasoconstriction: basic mechanisms. *European Respiratory Journal*, **32**, 1639–1651.

Sommer, L. S., Mcfarland, W. L., Galliano, R. E., Nagel, E. L., & Morgane, P. J. 1968. Hemodynamic and coronary angiographic studies in the bottlenose dolphin (*Tursiops truncatus*). *American Journal of Physiology*, **215**, 1498–1505.

Sparling, C. E. & Fedak, M. A. 2004. Metabolic rates of captive grey seals during voluntary diving. *Journal of Experimental Biology*, **207**, 1615–1624.

Sparling, C. E., Fedak, M. A., & Thompson, D. 2007. Eat now, pay later? Evidence of deferred food processing costs in diving seals. *Biology Letters*, **3**, 94–98.

Sparling, C. E., Thompson, D., Fedak, M. A., Gallon, S. L., & Speakman, J. R. 2008. Estimating field metabolic rates of pinnipeds: doubly labelled water gets the seal of approval. *Functional Ecology*, **22**, 245–254.

Speakman, J. R. 1998. The history and theory of the doubly labeled water technique. *The American Journal of Clinical Nutrition*, **68**, 932S–938S.

Spragg, R., Ponganis, P. J., Marsh, J. J., Rau, G. A., & Bernhard, W. 2004. Surfactant from diving aquatic mammals. *Journal of Applied Physiology*, **96**, 1626–1632.

Spragg, R. G., Lewis, J. F., Wurst, W., *et al.* 2003. Treatment of acute respiratory distress syndrome with recombinant surfactant protein C surfactant. *American Journal of Respiratory and Critical Care Medicine*, **167**, 1562–1566.

St. Aubin, D. J., Geraci, J. R., Smith, T. G., & Smith, V. I. 1978. Blood volume determination in the ringed seal, *Phoca hispida*. *Canadian Journal of Zoology*, **56**, 1885–1887.

Stacey, P. J. & Hvenegaard, G. T. 2002. Habitat use and behaviour of Irrawaddy dolphins (*Orcaellaa brevirostris*) in the Mekong River of Laos. *Aquatic Mammals*, **28**, 1–13.

Stahel, C. D. & Nicol, S. C. 1982. Temperature regulation in the little penguin (*Eudyptula minor*) in air and water. *Journal of Comparative Physiology B: Biochemical, Systemic, and Environmental Physiology*, **148**, 93–100.

Stahl, W. R. 1967. Scaling of respiratory variables in mammals. *Journal of Applied Physiology*, **22**, 453–460.

Steinfurth, A., Vargas, F. H., Wilson, R. P., Spindler, M., & Macdonald, D. W. 2008. Space use by foraging Galapagos penguins during chick rearing. *Endangered Species Research*, **4**, 105–112.

Stelle, L. L., Blake, R. W., & Trites, A. W. 2000. Hydrodynamic drag in Steller sea lions (*Eumetopias jubatus*). *Journal of Experimental Biology*, **203**, 1915–1923.

Stelle, L. L., Megill, W. M., & Kinzel, M. R. 2008. Activity budget and diving behavior of gray whales (*Eschrichtius robustus*) in feeding grounds off coastal British Columbia. *Marine Mammal Science*, **24**, 462–478.

Stephenson, R. 1993. The contributions of body tissues, respiratory system, and plumage to buoyancy in waterfowl. *Canadian Journal of Zoology*, **71**, 1521–1529.

Stephenson, R. 1994. Diving energetics in the lesser scaup (*Aythya affinis*, Eyton). *Journal of Experimental Biology*, **190**, 155–178.

Stephenson, R. 1995. Respiratory and plumage gas volumes in unrestrained diving ducks (*Aythya affinis*). *Respiration Physiology*, **100**, 129–137.

Stephenson, R. 2005. Physiological control of diving behaviour in the Weddell seal *Leptonychotes weddellii*: a model based on cardiorespiratory control theory. *Journal of Experimental Biology*, **208**, 1971–1991.

Stephenson, R. & Andrews, C. A. 1997. The effect of water surface tension on feather wettability in aquatic birds. *Canadian Journal of Zoology*, **75**, 228–294.

Stephenson, R., Butler, P., Dunstone, N., & Woakes, A. 1988. Heart rate and gas exchange in freely diving American mink (*Mustela vision*). *Journal of Experimental Biology*, **134**, 435–442.

Stephenson, R., Butler, P. J., & Woakes, A. J. 1986. Diving behaviour and heart rate in tufted ducks (*Aythya fuligula*). *Journal of Experimental Biology*, **126**, 341–359.

Stephenson, R., Hedrick, M. S., & Jones, D. R. 1992. Cardiovascular responses to diving and involuntary submergence in the rhinoceros auklet (*Cerorhinca monocerata* Pallas). *Canadian Journal of Zoology*, **70**, 2303–2310.

Stephenson, R. & Jones, D. R. 1992a. Blood flow distribution in submerged and surface-swimming ducks. *Journal of Experimental Biology*, **166**, 285–296.

Stephenson, R. & Jones, D. R. 1992b. Metabolic responses to forced dives in Pekin duck measured by indirect calorimetry and ^{31}P-MRS. *American Journal of Physiology*, **263**, R1309–R1317.

Stephenson, R., Jones, D. R., Kasserra, C. E., & Lemaire, C. 1997. The effects of skeletal muscle contractions and paralysis on the physiological responses to head immersion in Pekin duck. *Comparative Biochemistry and Physiology B*, **118**, 767–770.

Stephenson, R., Lovvorn, J. R., Heieis, M. A., Jones, D. R., & Blake, R. W. 1989a. A hydromechanical estimate of the power requirements of diving and surface swimming in lesser scaup (*Aythya affinis*). *Journal of Experimental Biology*, **147**, 507–518.

Stephenson, R., Turner, D. L., & Butler, P. J. 1989b. The relationship between diving activity and oxygen storage capacity in the tufted duck (*Aythya fuligula*). *Journal of Experimental Biology*, **141**, 265–275.

Sterling, J. & Ream, R. 2004. At-sea behavior of juvenile male northern fur seals (*Callorhinus ursinus*). *Canadian Journal of Zoology*, **82**, 1621–1637.

Stettenheim, P. R. 2000. The integumentary morphology of modern birds: an overview. *American Zoologist*, **40**, 461–477.

Stewart, B. S. & Delong, R. L. 1995. Double migrations of the northern elephant seal, *Mirounga angustirostris*. *Journal of Mammalogy*, **76**, 196–205.

Stewart, B. S., Harvey, J. T., & Yochem, P. K. 2001. Post-release monitoring and tracking of a rehabilitated California gray whale. *Aquatic Mammals*, **27**, 294–300.

Stewart, B. S., Petrov, E. A., Baranov, E. A., Timonin, A., & Ivanov, M. 1996. Seasonal movements and dive patterns of juvenile Baikal seals, *Phoca sibirica*. *Marine Mammal Science*, **12**, 528–542.

Stimpert, A. K., Deruiter, S. L., Southall, B. L., *et al.* 2014. Acoustic and foraging behavior of a Baird's beaked whale, Berardius bairdii, exposed to simulated sonar. *Scientific Reports*, **4**. DOI: 10.1038/srep07031.

Stirling, I. 1974. Midsummer observations on behavior of wild polar bears (*Ursus maritimus*). *Canadian Journal of Zoology*, **52**, 1191–1198.

Stirling, I. & van Meurs, R. 2015. Longest record dive by a polar bear. *Polar Biology*, **38**, 1301–1304.

Stockard, T. K., Heil, J., Meir, J. U., *et al.* 2005. Air sac P_{O2} and oxygen depletion during dives of emperor penguins. *Journal of Experimental Biology*, **208**, 2973–2981.

Stockard, T. K., Levenson, D. H., Berg, L., *et al.* 2007. Blood oxygen depletion during rest-associated apneas of northern elephant seals (*Mirounga angustirostris*). *Journal of Experimental Biology*, **210**, 2607–2617.

Stockin, K. A., Fairbairns, R. S., Parsons, E. C. M., & Sims, D. W. 2001. Effect of diel and seasonal cycles on the dive duration of the minke whale. *Journal of the Marine Biological Association*, **81**, 189–190.

Storey, K. & Hochachka, P. W. 1974. Glycolytic enzymes in muscle of the Pacific dolphin: role of pyruvate kinase in aerobic–anaerobic transitions during diving. *Comparative Biochemistry and Physiology B*, **49**, 119–128.

Strickler, T. L. 1980. The axial musculature of *Pontoporia blainvillei*, with comments on the organization of this system and its effect on fluke-stroke dynamics in the Cetacea. *American Journal of Anatomy*, **157**, 49–59.

Suburo, A. M. & Scolaro, J. A. 1990. The eye of the magellanic penguin (*Spheniscus magellanicus*): structure of the anterior segment. *American Journal of Anatomy*, **189**, 245–252.

Sumich, J. L. 2001. Direct and indirect measures of oxygen extraction, tidal lung volumes and respiratory rates in a rehabilitating gray whale calf. *Aquatic Biology*, **27**, 279–283.

Suzuki, A. T., Tsuchiya, T., Takahashi, Y., & Tamate, H. 1983. Histochemical properties of myofibers in longissimus muscle of common dolphins (*Delphinus delphis*). *Acta Histochemistry and Cytochemistry*, **16**, 223–231.

Suzuki, I., Naito, Y., Folkow, L., Miyazaki, N., & Blix, A. 2009. Validation of a device for accurate timing of feeding events in marine animals. *Polar Biology*, **32**, 667–671.

Suzuki, T. & Imai, K. 1998. Evolution of myoglobin. *Cellular and Molecular Life Sciences*, **54**, 979–1004.

Takahashi, A., Dunn, M. J., Trathan, P. N., *et al.* 2004a. Krill-feeding behaviour in a chinstrap penguin *Pygoscelis antarctica* compared with fish-eating in Magellanic penguins *Speniscus magellanicus*: a pilot study. *Marine Ornithology*, **32**, 47–54.

Takahashi, A., Dunn, M. J., Trathan, P. N., *et al.* 2003. Foraging strategies of chinstrap penguins at Signy Island, Antarctica: importance of benthic feeding on Antarctic krill. *Marine Ecology Progress Series*, **250**, 279–289.

Takahashi, A., Kokubun, N., Mori, K., & Shin, H. C. 2008. Krill-feeding behaviour of gentoo penguins as shown by animal-borne camera loggers. *Polar Biology*, **31**, 1291–1294.

Takahashi, A., Sato, K., Naito, Y., *et al.* 2004b. Penguin-mounted cameras glimpse underwater group behaviour. *Proceedings of the Royal Society London B, Supplement*, **251**.5: S281–S282.

Tamburrini, M., Condo, S. G., Di Prisco, G., & Giardina, B. 1994. Adaptation to extreme environments: structure–function relationships in emperor penguin hemoglobin. *Journal of Molecular Biology*, **237**, 615–621.

Tamburrini, M., Romano, M., Giardina, B., & Di Prisco, G. 1999. The myoglobin of emperor penguin (*Aptenodytes forsteri*): amino acid sequence and functional adaptation to extreme conditions. *Comparative Biochemistry and Physiology B*, **122**, 235–240.

Tarasoff, F. J. 1972. Comparative aspects of the hind limbs of the river otter, sea otter, and seals. In: Harrison, R. J. (ed.), *Functional Anatomy of Marine Mammals*. London: Academic Press.

Tarasoff, F. J. & Kooyman, G. L. 1973. Observations on the anatomy of the respiratory system of the river otter, sea otter, and harp seal: II. The trachea and bronchial tree. *Canadian Journal of Zoology*, **51**, 171–177.

Tawara, T. 1950. On the respiratory pigments of whale (studies on whale blood II). *Scientific Reports Whales Research Institute*, **3**, 95–101.

Taylor, B., Barlow, J., Pitman, R. L., *et al.* 2004. A call for research to assess risk of acoustic impact on beaked whale populations. Paper SC/56/E36, presented to the IWC Scientific Committee, July 2004, Sorrento, Italy.

Taylor, C. R., Heglund, N. C., & Maloiy, G. M. 1982. Energetics and mechanics of terrestrial locomotion: I. Metabolic energy consumption as a function of speed and body size in birds and mammals. *Journal of Experimental Biology*, **97**, 1–21.

Taylor, C. R., Karas, R. H., Weibel, E. R., & Hoppeler, H. 1987. Adaptive variation in the mammalian respiratory system in relation to energetic demand: II. Reaching the limits to oxygen flow. *Respiration Physiology*, **69**, 7–26.

Terrados, N., Jansson, E., Sylven, C., & Kaliser, L. 1990. Is hypoxia a stimulus for synthesis of oxidative enzymes and myoglobin? *Journal of Applied Physiology*, **68**, 2369–2372.

Tevis, Jr., L. 1950. Summer behavior of a family of beavers in New York State. *Journal of Mammalogy*, **31**, 40–65.

Thewissen, J. G. M., George, J., Rosa, C., & Kishida, T. 2011. Olfaction and brain size in the bowhead whale (*Balaena mysticetus*). *Marine Mammal Science*, **27**, 282–294.

Thomas, D. B. & Fordyce, R. E. 2007. The heterothermic loophole exploited by penguins. *Australian Journal of Zoology*, **55**, 317–321.

Thomas, D. B. & Fordyce, R. E. 2012. Biological plasticity in penguin heat-retention structures. *The Anatomical Record: Advances in Integrative Anatomy and Evolutionary Biology*, **295**, 249–256.

Thometz, N. M. 2014. Ontogeny of energetic demand and diving ability in the southern sea otter (Enhydra lutris nereis) and implications on diving and foraging behavior. PhD thesis, University of California Santa Cruz.

Thometz, N. M., Murray, M., & Williams, T. M. 2015. Ontogeny of oxygen storage capacity and diving ability in the southern sea otter (*Enhydra lutris nereis*): costs and benefits of large lungs. *Physiological and Biochemical Zoology*, **88**, 311–327.

Thompson, D., Duck, C. D., McConnell, B. J., & Garrett, J. 1998. Foraging behaviour and diet of lactating female southern sea lions (*Oaaria flavescens*) in the Falkland Islands. *Journal of Zoology London*, **246**, 135–146.

Thompson, D. & Fedak, M. A. 1993. Cardiac responses of grey seals during diving at sea. *Journal of Experimental Biology*, **174**, 139–164.

Thompson, D., Hammond, P. S., Nicholas, K. S., & Fedak, M. A. 1991. Movements, diving and foraging behaviour of grey seals (*Halichoerus grypus*). *Journal of Zoology London*, **224**, 223–232.

Thompson, P. M., Fedak, M. A., McConnell, B. J., & Nicholas, K. S. 1989. Seasonal and sex-related variation in the activity patterns of common seals (*Phoca vitulina*). *Journal of Applied Ecology*, **26**, 521–535.

Thoresen, A. C. 1989. Diving times and behavior of pigeon guillemots and marbled murrelets off Rosario, Washington. *Western Birds*, **20**, 33–37.

Thornton, S. J., Hochachka, P. W., Crocker, D. E., *et al.* 2005. Stroke volume and cardiac output in juvenile elephant seals during forced submersions. *Journal of Experimental Biology*, **208**, 3637–3643.

Thornton, S. J., Pelc, N. J., Spielman, J. R., *et al.* 1997. Vascular flow dynamics in a diving elephant seal (*Mirounga angustirostris*). *International Society of Magnetic Resonance Medicine*, **2**, 823.

Thornton, S. J., Spielman, D. M., Pelc, N. J., *et al.* 2001. Effects of forced diving on the spleen and hepatic sinus in northern elephant seal pups. *Proceedings of the National Academy of Sciences*, **98**, 9413–9418.

Thorson, P. H. 1993. Development of diving in the northern elephant seal. PhD thesis, University of California, Santa Cruz.

Thorson, P. H. & Le Boeuf, B. J. 1994. Developmental aspects of diving in northern elephant seal pups. In: Le Boeuf, B. J. & Laws, R. M. (eds.), *Elephant Seals: Population Ecology, Behavior, and Physiology*. Berkeley, CA: University of California Press.

Tibbs, R., Elghetany, M. T., Tran, L., *et al.* 2005. Characterization of the coagulation system in healthy dolphins: the coagulation factors, natural anticoagulants, and fibrinolytic products. *Comparative Clinical Pathology*, **14**, 95–98.

Tift, M. S., Ponganis, P. J., & Crocker, D. E. 2014. Elevated carboxyhemoglobin in a marine mammal, the northern elephant seal. *Journal of Experimental Biology*, **217**, 1752–1757.

Tift, M. S., Ranalli, E. C., Houser, D. S., Ortiz, R. M., & Crocker, D. E. 2013. Development enhances hypometabolism in northern elephant seal pups (Mirounga angustirostris). *Functional Ecology*. DOI: 10.1111/1365-2435.12111

Tikuisis, P. & Gerth, W. 2003. Decompression theory. In: Brubakk, A. O. & Neuman, T. S. (eds.), *Bennett and Elliott's Physiology and Medicine of Diving*. Edinburgh: Saunders.

Tinker, M. T., Costa, D. P., Estes, J. A., & Wieringa, N. 2007. Individual dietary specialization and dive behaviour in the California sea otter: using archival time–depth data to detect alternative foraging strategies. *Deep Sea Research Part II: Topical Studies in Oceanography*, **54**, 330–342.

Tipler, T. D., Edwards, Y. H., & Hopkinson, D. A. 1978. Developmental changes in the protein profiles of human cardiac and skeletal muscle. *Annals of Human Genetics*, **41**, 409–418.

Torday, H. S. & Rehan, V. K. 2002. Stretch-induced surfactant synthesis is coordinated by the paracrine actions of PTHrP and leptin. *American Journal of Physiology*, **283**, L130–L135.

Tran, T.-K., Sailasuta, N., Kreutzer, U., *et al.* 1999. Comparative analysis of NMR and NIRS measurements of intracellular P_{O2} in human skeletal muscle. *American Journal of Physiology*, **276**, R1682–R1690.

Trawa, G. 1970. Note preliminaire sur la vascularisation des membres des spheniscides. *L'Oiseau et RFO*, **40**, 142–156.

Tremblay, Y. & Cherel, Y. 2000. Benthic and pelagic dives: a new foraging behaviour in rockhopper penguins. *Marine Ecology Progress Series*, **204**, 257–267.

Tremblay, Y. & Cherel, Y. 2003. Geographic variation in the foraging behaviour, diet and chick growth of rockhopper penguins. *Marine Ecology Progress Series*, **251**, 279–297.

Tremblay, Y., Cherel, Y., Oremus, M., Tveraa, T., & Chastel, O. 2003. Unconventional ventral attachment of time–depth recorders as a new method for investigating time budget and diving behaviour of seabirds. *Journal of Experimental Biology*, **206**, 1929–1940.

Tremblay, Y., Cook, T. R., & Cherel, Y. 2005. Time budget and diving behaviour of chick-rearing Crozet shags. *Canadian Journal of Zoology*, **83**, 971–982.

Tremblay, Y., Guinard, E., & Cherel, Y. 1997. Maximum diving depths of northern rockhopper penguins (*Eudyptes chrysocome moseleyi*) at Amsterdam Island. *Polar Biology*, **17**, 119–122.

Trillmich, F., Kooyman, G. L., Majluf, P., & Sanchez-Griñan, M. 1986. Attendance and diving behavior of South American fur seals during El Niño in 1983. In: Gentry, R. L. & Kooyman, G. L. (eds.), *Fur Seals: Maternal Strategies on Land and at Sea*. Princeton, NJ: Princeton University Press.

Truex, R. C., Nolan, F. G., Schneider, H. P., & Perlmutter, H. I. 1961. Anatomy and pathology of the whale heart with special reference to the coronary circulation. *The Anatomical Record*, **141**, 325–353.

Tsuchimochi, H., Hayes, S. G., McCord, J. L., & Kaufman, M. P. 2009. Both central command and exercise pressor reflex activate cardiac sympathetic nerve activity in decerebrate cats. *American Journal of Physiology: Heart and Circulatory Physiology*, **296**, H1157–H1163.

Tsuchimochi, H., Matsukawa, K., Komine, H., & Murata, J. 2002. Direct measurement of cardiac sympathetic efferent nerve activity during dynamic exercise. *American Journal of Physiology: Heart and Circulatory Physiology*, **283**, H1896–H1906.

Tulsi, R. 1975. Observations on the structure of the dorsal muscle in the bottle-nose dolphin (*Tursiops truncatus*). *Journal of Anatomy*, **119**, 39–48.

Turner, A. W. & Hodgetts, V. E. 1959. The dynamic red cell storage function of the spleen in sheep. I. Relationship to fluctuations of jugular hematocrit. *Australian Journal of Experimental Biology*, **37**, 399–420.

Turner, D. L. & Butler, P. J. 1988. The aerobic capacity of locomotory muscles in the tufted duck, *Aythya fuligula*. *Journal of Experimental Biology*, **135**, 445–460.

Tyack, P. L., Johnson, M., Aguilar De Soto, N., Sturlese, A., & Madsen, P. T. 2006. Extreme diving of beaked whales. *Journal of Experimental Biology*, **209**, 4238–4253.

Tyack, P. L., Zimmer, W. M. X., Moretti, D., *et al.* 2011. Beaked whales respond to simulated and actual navy sonar. *PloS ONE*, **6**, e17009.

Ulenaers, P., Van Vessem, J., & Dhondt, A. A. 1992. Foraging of the great crested grebe in relation to food. *Journal of Animal Ecology*, **61**, 659–667.

Underwood, L. E. & Williams, R. S. 1987. Pretranslational regulation of myoglobin gene expression. *American Journal of Physiology*, **252**, C450–C453.

Upton, R. N. 2008. Organ weights and blood flows of sheep and pig for physiological pharmacokinetic modelling. *Journal of Pharmacological and Toxicological Methods*, **58**, 198–205.

US Department of Commerce & US Navy 2001. *US Dept Commerce & US Navy Joint Interim Report: Bahamas Marine Mammal Stranding Event of 15–16 March 2000*. www.nmfs.noaa.gov/prot_res/overview/Interim_Bahamas_Report.pdf.

Valtin, H. 1973. *Renal Function: Mechanisms Preserving Fluid and Solute Balance in Health*. Boston, MA: Little, Brown & Co.

Van Citters, R. L. & Franklin, D. L. 1969. Cardiovascular performance of Alaska sled dogs during exercise. *Circulation Research*, **24**, 33–42.

Van Citters, R. L., Franklin, D. L., Smith, Jr., O. A., Watson, N. W., & Elsner, R. W. 1965. Cardiovascular adaptations to diving in the northern elephant seal *Mirounga angustirostris*. *Comparative Biochemistry and Physiology*, **16**, 267–276.

Van Dam, R. P., Ponganis, P. J., Ponganis, K. V., Levenson, D. H., & Marshall, G. 2002. Stroke frequencies of emperor penguins diving under sea ice. *Journal of Experimental Biology*, **205**, 3769–3774.

Van Den Hoff, J. & Newbery, K. 2006. Southern giant petrels *Macronectes giganteus* diving on submerged carrion. *Marine Ornithology*, **34**, 61–64.

Van Nie, C. J. & Van Der Kamp, J. S. 1988. The closure of the foramen ovale and the ductus arteriosus in the common seal (*Phoca vitulina*, L. 1758) (a morphological approach). *Aquatic Mammals*, **14**, 82–85.

Van Opzeeland, I., Van Parijs, S., Bornemann, H., *et al.* 2010. Acoustic ecology of Antarctic pinnipeds. *Marine Ecology Progress Series*, **414**, 267–291.

Vázquez-Medina, J. P., Zenteno-Savín, T., & Elsner, R. 2006. Antioxidant enzymes in ringed seal tissues: potential protection against dive-associated ischemia/reperfusion. *Comparative Biochemistry and Physiology Part C: Toxicology & Pharmacology*, **142**, 198–204.

Vázquez-Medina, J. P., Zenteno-Savín, T., & Elsner, R. 2007. Glutathione protection against dive-associated ischemia/reperfusion in ringed seal tissues. *Journal of Experimental Marine Biology and Ecology*, **345**, 110–118.

Velten, B. P., Dillaman, R. M., Kinsey, S. T., McLellan, W. A., & Pabst, D. A. 2013. Novel locomotor muscle design in extreme deep-diving whales. *Journal of Experimental Biology*, **216**, 1862–1871.

Venditti, C. C., Casselman, R., Murphy, M. S. Q., *et al.* 2013. Chronic carbon monoxide inhalation during pregnancy augments uterine artery blood flow and uteroplacental vascular growth in mice. *American Journal of Physiology: Regulatory, Integrative and Comparative Physiology*, **305**, R939–R948.

Viamonte, M., Morgane, P., Galliano, R., Nagel, E., & McFarland, W. 1968. Angiography in the living dolphin and observations on blood supply to the brain. *American Journal of Physiology: Legacy Content*, **214**, 1225–1249.

Viles, P. H. & Shepherd, J. T. 1968. Relationship between pH, Po_2, and Pco_2 on the pulmonary vascular bed of the cat. *American Journal of Physiology*, **215**, 1170–1176.

Villegas-Amtmann, S. & Costa, D. P. 2010. Oxygen stores plasticity linked to foraging behaviour and pregnancy in a diving predator, the Galapagos sea lion. *Functional Ecology*, **24**, 785–795.

Villegas-Amtmann, S., Costa, D. P., Tremblay, Y., Salazar, S., & Aurioles-Gambola, D. 2008. Multiple foraging strategies in a marine apex predator, the Galapagos sea lion *Zalophus wollebaeki*. *Marine Ecology Progress Series*, **363**, 299–309.

Villouta, G., Hargreaves, R., & Rtveros, V. 1997. Haematological and clinical biochemistry findings in captive Humboldt penguins (*Spheniscus humboldti*). *Avian Pathology*, **26**, 851–858.

Vogl, A. W. & Fisher, H. D. 1982. Arterial retia related to supply of the central nervous system in two small toothed whales: narwhal (*Monodon monoceros*) and beluga (*Delphinapterus leucas*). *Journal of Morphology*, **174**, 41–56.

Vogl, A. W., Todd, M. E., & Fisher, H. D. 1981. An ultrastructural and fluorescence histochemical investigation of the innervation of retial arteries in *Monodon monoceros*. *Journal of Morphology*, **168**, 109–119.

Vogt, M., Puntschart, A., Geiser, J., *et al.* 2001. Molecular adaptations in human skeletal muscle to endurance training under simulated hypoxic conditions. *Journal of Applied Physiology*, **91**, 173–182.

Wagner, P. D. 1996. Determinants of maximal oxygen transport and utilization. *Annual Review of Physiology*, **58**, 21–50.

Wahrenbrock, E. A., Maruschak, G. F., Elsner, R., & Kenney, D. W. 1974. Respiration and metabolism in two baleen whale calves. *Marine Fisheries Review*, **36**, 3–9.

Waldrop, T. G., Bauer, R. M., & Iwamoto, G. A. 1988. Microinjection of GABA antagonists into the posterior hypothalamus elicits locomotor activity and a cardiorespiratory activation. *Brain Research*, **444**, 84–94.

Waldrop, T. G., Eldridge, F. L., Iwamoto, G. A., & Mitchell, J. H. 1996. Central neural control of locomotion, respiration, and circulation during exercise. In: *Handbook of Physiology*. Bethesda, MD: American Physiological Society.

Waldrop, T. G. & Iwamoto, G. A. 2006. Point:counterpoint. Supraspinal locomotor centers do/do not contribute significantly to the hyperpnea of dynamic exercise. *Journal of Applied Physiology*, **100**, 1077–1083.

Waldrop, T. G. & Stremel, R. W. 1989. Muscular contraction stimulates posterior hypothalamic neurons. *American Journal of Physiology: Regulatory, Integrative and Comparative Physiology*, **256**, R348–R356.

Walker, B. G. & Boersma, P. D. 2003. Diving behavior of Magellanic penguins (*Spheniscus magellanicus*) at Punta Tombo, Argentina. *Canadian Journal of Zoology*, **81**, 1471–1483.

Walls, G. 1942. *The Vertebrate Eye and its Adaptive Radiation*. New York: Hafner Publishing Co.

Wanless, S., Corfield, T., Harris, M. P., Buckland, S., & Morris, J. 1993. Diving behaviour of the shag Phalacrocorax aristotelis (Aves: Pelecaniformes) in relation to water depth and prey size. *Journal of Zoology London*, **231**, 11–25.

Wanless, S., Finney, S. K., Harris, M. P., & Mccafferty, D. J. 1999. Effect of the diel light cycle on the diving behaviour of two bottom feeding marine birds: the blue-eyed shag *Phalacrocorax atriceps* and the European shag *P. aristotelis*. *Marine Ecology Progress Series*, **188**, 219–224.

Wanless, S. & Harris, M. P. 1993. Use of mutually exclusive foraging areas by adjacent colonies of blue-eyed shags (*Phalacrocorax atriceps*) at South Georgia. *Colonial Waterbirds*, **16**, 176–182.

Wanless, S., Harris, M. P., Burger, A. E., & Buckland, S. T. 1997. Use of time-at-depth recorders for estimating depth and diving performance of European shags. *Journal of Field Ornithology*, **68**, 547–561.

Wanless, S., Harris, M. P., & Morris, J. 1992. Diving behaviour and diet of the blue-eyed shag at South Georgia. *Polar Biology*, **12**, 713–719.

Wanless, S., Morris, J. A., & Harris, M. P. 1988. Diving behaviour of guillemot *Uria aalge*, puffin *Fratercula arctica* and razorbill *Alca torda* as shown by radio-telemetry. *Journal of Zoology, London*, **216**, 73–81.

Warltier, D. C., Pagel, P. S., & Kersten, J. R. 2000. Approaches to the prevention of perioperative myocardial ischemia. *Anesthesiology*, **92**, 253.

Watanabe, S., Sato, K., & Ponganis, P. J. 2012. Activity time budget during foraging trips of emperor penguins. *PloS ONE*, **7**, e50357.

Watanabe, Y., Baranov, E. A., Sato, K., Naito, Y., & Miyazaki, N. 2004. Foraging tactics of Baikal seals differ between day and night. *Marine Ecology Progress Series*, **279**, 283–289.

Watanabe, Y., Baranov, E. A., Sato, K., Naito, Y., & Miyazaki, N. 2006. Body density affects stroke patterns in Baikal seals. *Journal of Experimental Biology*, **209**, 3269–3280.

Watanabe, Y. Y. & Takahashi, A. 2013. Linking animal-borne video to accelerometers reveals prey capture variability. *Proceedings of the National Academy of Sciences*, **110**, 2199–2204.

Watanabe, Y. Y., Takahashi, A., Sato, K., Viviant, M., & Bost, C.-A. 2011. Poor flight performance in deep-diving cormorants. *Journal of Experimental Biology*, **214**, 412–421.

Watanuki, Y., Daunt, F., Takahashi, A., *et al.* 2008. Microhabitat use and prey capture of a bottom-feeding top predator, the European shag, shown by camera loggers. *Marine Ecology Progress Series*, **356**, 283–293.

Watanuki, Y., Kato, A., & Naito, Y. 1996. Diving performance of male and female Japanese cormorants. *Canadian Journal of Zoology*, **74**, 1098–1109.

Watanuki, Y., Kato, A., Naito, Y., Robertson, G., & Robinson, S. 1997. Diving and foraging behaviour of Adelie penguins in areas with and without fast sea-ice. *Polar Biology*, **17**, 296–304.

Watanuki, Y., Niizuma, Y., Gabrielsen, G. W., Sato, K., & Naito, Y. 2003. Stroke and glide of wing-propelled divers: deep diving seabirds adjust surge frequency to buoyancy change with depth. *Proceedings of the Royal Society London B*, **270**, 483–488.

Watanuki, Y. & Sato, K. 2008. Dive angle, swim speed and wing stroke during shallow and deep dives in common murres and rhinoceros auklets. *Ornithological Science*, **7**, 15–28.

Watanuki, Y., Takahashi, A., Daunt, F., *et al.* 2005. Regulation of stroke and glide in a foot-propelled avian diver. *Journal of Experimental Biology*, **208**, 2207–2216.

Watanuki, Y., Wanless, S., Harris, M., *et al.* 2006. Swim speeds and stroke patterns in wing-propelled divers: a comparison among alcids and a penguin. *Journal of Experimental Biology*, **209**, 1217–1230.

Watkins, W. A. 1978. A radio tag for big whales. *Oceanus*, **21**, 48–54.

Watkins, W. A. 1979. A projectile point for penetrating whale blubber. *Deep Sea Research Part A: Oceanographic Research Papers*, **26**, 1301–1308.

Watkins, W. A., Daher, M. A., Dimarzio, N. A., *et al.* 1999. Sperm whale surface activity from tracking by radio and satellite tags. *Marine Mammal Science*, **15**, 1158–1180.

Watkins, W. A., Daher, M. A., Fristrup, K. M., Howald, T. J., & Notarbartolo Di Sciara, G. 1993. Sperm whales tagged with transponders and tracked underwater by sonar. *Marine Mammal Science*, **9**, 55–67.

Watkins, W. A. & Schevill, W. E. 1977. *The Development and Testing of a Radio Whale Tag*. Woods Hole: Woods Hole Oceanographic Institution.

Watkins, W. A. & Tyack, P. 1991. Reaction of sperm whales (*Physeter catodon*) to tagging with implanted sonar transponder and radio tags. *Marine Mammal Science*, **7**, 409–413.

Watkins, W. A. & Wartzok, D. 1985. Sensory biophysics of marine mammals. *Marine Mammal Science*, **1**, 219–260.

Watson, R. R., Fu, Z., & West, J. B. 2008. Minimal distensibility of pulmonary capillaries in avian lungs compared with mammalian lungs. *Respiratory Physiology & Neurobiology*, **160**, 208–214.

Watson, R. R., Kanatous, S. B., Cowan, D. F., *et al.* 2007. Volume density and distribution of mitochondria in harbor seal (*Phoca vitulina*) skeletal muscle. *Journal of Comparative Physiology B*, **177**, 89–98.

Watson, R. R., Miller, T. A., & Davis, R. W. 2003. Immunohistochemical fiber typing of harbor seal skeletal muscle. *Journal of Experimental Biology*, **206**, 4105–4111.

Watwood, S. L., Miller, P. J. O., Johnson, M., Madsen, P. T., & Tyack, P. L. 2006. Deep-diving foraging behaviour of sperm whales (*Physeter macrocephalus*). *Journal of Animal Ecology*, **75**, 814–825.

Webb, P. M., Andrews, R. D., Costa, D. P., & Le Boeuf, B. J. 1998. Heart rate and oxygen consumption of northern elephant seals during diving in the laboratory. *Physiological Zoology*, **71**, 116–125.

Webb, P. W. 1988. Simple physical principles and vertebrate aquatic locomotion. *American Zoologist*, **28**, 709–725.

Weber, R. E. 2007. High-altitude adaptations in vertebrate hemoglobins. *Respiratory Physiology & Neurobiology*, **158**, 132.

Weber, R. E. & Fago, A. 2004. Functional adapation and its molecular basis in vertebrate hemoglobins, neuroglobins, and cytoglobins. *Respiration Physiology and Neurobiology*, **144**, 141–159.

Weber, R. E., Hemmingsen, E. A., & Johansen, K. 1974. Functional and biochemical studies of penguin myoglobin. *Comparative Biochemistry and Physiology*, **49B**, 197–214.

Weidman, J. L., Shook, D. C., & Hilberath, J. N. 2014. Cardiac resuscitation and coagulation. *Anesthesiology*, **120**, 1009–1014.

Weimerskirch, H. & Cherel, Y. 1998. Feeding ecology of short-tailed shearwaters: breeding in Tasmania and foraging in the Antarctic? *Marine Ecology Progress Series*, **167**, 261–274.

Weimerskirch, H., Lecorre, M., Ropert-Coudert, Y., Katok, A., & Marsac, F. 2005. The three-dimensional flight of red-footed boobies: adapations to foraging in a tropical environment? *Proceedings of the Royal Society B*, **272**, 53–61.

Weimerskirch, H. & Sagar, P. M. 1995. Diving depths of sooty shearwaters *Puffinus griseus. Ibis*, **138**, 768–788.

Weise, M. J. & Costa, D. P. 2007. Total body oxygen stores and physiological diving capacity of California sea lions as a function of sex and age. *Journal of Experimental Biology*, **210**, 278–289.

Welch, Jr., K. C. & Altshuler, D. L. 2009. Fibertype homogeneity of the flight musculature in small birds. *Comparative Biochemistry and Physiology B*, **152**, 324–331.

Weller, P., Jeffreys, A. J., Wilson, V., & Blanchetot, A. 1984. Organization of the human myoglobin gene. *EMBO Journal*, **3**, 439–446.

Weller, P. A., Price, M., Isenberg, H., Edwards, Y. H., & Jeffreys, A. J. 1986. Myoglobin expression: early induction and subsequent modulation of myoglobin and myoglobin mRNA during myogenesis. *Molecular and Cellular Biology*, **6**, 4539–4547.

Wells, R. S., Manire, C. A., Byrd, L., *et al.* 2009. Movements and dive patterns of a rehabilitated Risso's dolphin, *Grampus griseus*, in the Gulf of Mexico and Atlantic Ocean. *Marine Mammal Science*, **25**, 420–429.

Welsch, U. & Aschauer, B. 1986. Ultrastructural observations on the lung of the emperor penguin (*Aptenodytes forsteri*). *Cell Tissue Research*, **243**, 137–144.

Welsch, U. & Riedel-Sheimer, B. 1997. Histophysiological observations on the external auditory meatus, middle, and inner of the Weddell seal (*Leptonychotes weddelli*). *Journal of Morphology*, **234**, 25–36.

Werner, R. & Campagna, C. 1995. Diving behaviour of lactating southern sea lions (*Otaria flavescens*) in Patagonia. *Canadian Journal of Zoology*, **73**, 1975–1982.

Werth, A. J. 2007. Adaptations of the cetacean hyolingual apparatus for aquatic feeding and thermoregulation. *The Anatomical Record: Advances in Integrative Anatomy and Evolutionary Biology*, **290**, 546–568.

West, J. B. 1972. *Respiratory Physiology: The Essentials*. Baltimore, MD: Williams & Wilkins.

West, J. B. 2007. *Pulmonary Physiology and Pathophysiology*. Philadelphia, PA: Lippincott, Williams & Wilkins.

West, J. B., Fu, X., Deerinck, T. J., *et al.* 2010. Structure–function studies of blood and air capillaries in chicken lung using 3D electron microscopy. *Respiratory Physiology & Neurobiology*, **170**, 202–209.

West, J. B., Hackett, P. H., Maret, K. H., *et al.* 1983. Pulmonary gas exchange on the summit of Mt. Everest. *Journal of Applied Physiology: Respiratory, Environmental and Exercise Physiology*, **55**, 678–687.

West, N. H., Bamford, O. S., & Jones, D. R. 1977. A scanning electron microscope study of the microvasculature of the avian lung. *Cell and Tissue Research*, **176**, 553–564.

West, N. H. & Van Vliet, B. N. 1986. Factors influencing the onset and maintenance of bradycardia in mink. *Physiological Zoology*, **59**, 451–463.

Westgate, A. J., Berggren, P., Koopman, H. N., & Gaskin, D. E. 1995. Diving behavior of harbour porpoises, *Phocoena phocoena*. *Canadian Journal of Fisheries and Aquatic Science*, **52**, 1064–1073.

Westgate, A. J., McLellan, W. A., Wells, R. S., *et al.* 2007. A new device to remotely measure heat flux and skin temperature from free-swimming dolphins. *Journal of Experimental Marine Biology and Ecology*, **346**, 45–59.

Wheaton, W. W. & Chandel, N. S. 2011. Hypoxia 2: hypoxia regulates cellular metabolism. *American Journal of Physiology: Cell Physiology*, **300**, C385–C393.

White, F. N., Ikeda, M., & Elsner, R. W. 1973. Andrenergic innervation of large arteries in the seal. *Comparative and General Physiology*, **4**, 271–276.

White, F. N. & Ross, G. 1966. Circulatory changes during experimental diving in the turtle. *American Journal of Physiology*, **211**, 5–18.

White, J. R., Harkness, D. R., Isaacks, R. E., & Duffield, D. A. 1976. Some studies on blood of the Florida manatee, *Trichechus manutus latirostris*. *Comparative Biochemistry and Physiology A*, **55**, 413–417.

Wickham, L. L., Bauersachs, R. M., Wenby, R. B., *et al.* 1990a. Red cell aggregation and viscoelasticity from blood of seals, swine and man. *Biorheology*, **27**, 191–204.

Wickham, L. L., Costa, D. P., & Elsner, R. 1990b. Blood rheology of captive and free-ranging northern elephant seals and sea otters. *Canadian Journal of Zoology*, **68**, 375–380.

Wickham, L. L., Elsner, R., White, F. C., & Cornell, L. H. 1989. Blood viscosity in phocid seals: possible adaptations to diving. *Journal of Comparative Physiology B*, **159**, 153–158.

Wiedeman, M. P. 1963. Dimensions of blood vessels from distributing artery to collecting vein. *Circulation Research*, **12**, 375–378.

Wienecke, B., Robertson, G., Kirkwood, R., & Lawton, K. 2007. Extreme dives by free-ranging emperor penguins. *Polar Biology*, **30**, 133–142.

Wieskotten, S., Dehnhardt, G., Mauck, B., Miersch, L., & Hanke, W. 2011a. The impact of glide phases on the trackability of hydrodynamic trails in harbour seals (*Phoca vitulina*). *Journal of Experimental Biology*, **213**, 3734–3740.

Wieskotten, S., Mauck, B., Miersch, L., Dehnhardt, G., & Hanke, W. 2011b. Hydrodynamic discrimination of wakes caused by objects of different size or shape in a harbour seal (*Phoca vitulina*). *Journal of Experimental Biology*, **214**, 1922–1930.

Wiig, O., Gjertz, I., Griffith, D., & Lydersen, C. 1993. Diving patterns of an Atlantic walrus *Odobenus rosmarus rosmarus* near Svalbard. *Polar Biology*, **13**, 71–72.

Wilber, C. G. 1960. Cardiac responses of *Alligator mississippiensis* to diving. *Comparative Biochemistry and Physiology*, **1**, 164–166.

Williams, C. L., Meir, J. U., & Ponganis, P. J. 2011a. What triggers the aerobic dive limit? Muscle oxygen depletion during dives of emperor penguins. *Journal of Experimental Biology*, **214**, 1801–1812.

Williams, C. L., Sato, K., Shiomi, K., & Ponganis, P. J. 2012. Muscle energy stores and stroke rates of emperor penguins: implications for muscle metabolism and dive performance. *Physiological and Biochemical Zoology*, **85**, 120–133.

Williams, E. E., Stewart, B. S., Beauchat, C. A., Somero, G. N., & Hazel, J. R. 2001a. Hydrostatic pressure and temperature effects on the molecular order of erythrocyte membranes from deep-, shallow-, and non-diving mammals. *Canadian Journal of Zoology*, **79**, 888–894.

Williams, L. R. & Leggett, R. W. 1989. Reference values for resting blood flow to organs of man. *Clinical Physics and Physiological Measurement* **10**, 187–217.

Williams, R. & Bryden, M. M. 1993. Observations of blood values, heart rate and respiratory rate of leopard seals (*Hydrurga leptonyx*) (Carnivora: Phocidae). *Australian Journal of Zoology*, **41**, 433–439.

Williams, T. D., Allen, D. D., Groff, J. M., & Glass, R. L. 1992a. An analysis of California sea otter (*Enhydra lutris*) pelage and integument. *Marine Mammal Science*, **8**, 1–18.

Williams, T. D., Briggs, D. R., Croxall, J. P., Naito, Y., & Kato, A. 1992b. Diving pattern and performance in relation to foraging ecology in the gentoo penguin, *Pygoscelis papua*. *Journal of Zoology, London*, **227**, 211–230.

Williams, T. M. 1989. Swimming by sea otters: adaptations for low energetic cost locomotion. *Journal of Comparative Physiology A*, **164**, 815–824.

Williams, T. M. 1999. The evolution of cost efficient swimming in marine mammals: limits to energetic optimization. *Philosophical Transactions of the Royal Society of London Series B: Biological Sciences*, **354**, 193–201.

Williams, T. M. 2001. Intermittent swimming by mammals: a strategy for increasing energetic efficiency during diving. *American Zoologist*, **41**, 166–176.

Williams, T. M., Davis, R. W., Fuiman, L. A., *et al.* 2000. Sink or swim: strategies for cost-efficient diving by marine mammals. *Science*, **288**, 133–136.

Williams, T. M., Friedl, W. A., Fong, M. L., *et al.* 1992c. Travel at low energetic cost by swimming and wave-riding bottlenose dolphins. *Nature*, **355**, 821–823.

Williams, T. M., Friedl, W. A., & Haun, J. E. 1993. The physiology of bottlenose dolphins (*Tursiops truncatus*): heart rate, metabolic rate and plasma lactate concentration during exercise. *Journal of Experimental Biology*, **179**, 31–46.

Williams, T. M., Fuiman, L. A., Horning, M., & Davis, R. W. 2004. The cost of foraging by a marine predator, the Weddell seal *Leptonychotes weddellii*: pricing by the stroke. *Journal of Experimental Biology*, **207**, 973–982.

Williams, T. M., Fuiman, L. A., Kendall, T., *et al.* 2015. Exercise at depth alters bradycardia and incidence of cardiac anomalies in deep-diving marine mammals. *Nature Communications*, **6.** DOI: 10.1038/ncomms7055.

Williams, T. M., Haun, J., Davis, R. W., Fuiman, L. A., & Kohin, S. 2001b. A killer appetite: metabolic consequences of carnivory in marine mammals. *Comparative Biochemistry and Physiology Part A: Molecular & Integrative Physiology*, **129**, 785–796.

Williams, T. M., Haun, J. E., & Friedl, W. A. 1999. The diving physiology of bottlenose dolphins (*Tursiops truncatus*): I. Balancing the demands of exercise for energy conservation at depth. *Journal of Experimental Biology*, **202**, 2739–2748.

Williams, T. M. & Kooyman, G. L. 1985. Swimming performance and hydrodynamic characteristics of harbor seals *Phoca vitulina*. *Physiological Zoology*, **58**, 576–589.

Williams, T. M., Kooyman, G. L., & Croll, D. A. 1991. The effects of submergence on heart rate and oxygen consumption of swimming seals and sea lions. *Journal of Comparative Physiology B*, **160**, 637–644.

Williams, T. M., Noren, S. R., & Glenn, M. 2011b. Extreme physiological adaptations as predictors of climate-change sensitivity in the narwhal, *Monodon monoceros*. *Marine Mammal Science*, **27**, 334–349.

Williams, T. M., Zavanelli, M., Miller, M. A., *et al.* 2008. Running, swimming, and diving modifies neuroprotecting globins in the mammalian brain. *Proceedings of the Royal Society B*, **275**, 751–758.

Willis, K., Horning, M., Rosen, D. A. S., & Trites, A. W. 2005. Spatial variation in heat flux in Steller sealions: evidence for consistent avenues of heat exchange along the body trunk. *Journal of Experimental Marine Biology and Ecology*, **315**, 163–175.

Wilson, R. P. 1985. The jackass penguin (*Spheniscus demersus*) as a pelagic predator. *Marine Ecology Progress Series*, **25**, 219–227.

Wilson, R. P., Bost, C. A., Putz, K., *et al.* 1997. Southern rockhopper penguin *Eudyptes chrysocome chrysocome* foraging at Possession Island. *Polar Biology*, **17**, 323–329.

Wilson, R. P., Cooper, J., & Plotz, J. 1992a. Can we determine when marine endotherms feed? A case study with seabirds. *Journal of Experimental Biology*, **167**, 267–275.

Wilson, R. P. & Gremillet, D. 1996. Body temperatures of free-living African penguins (*Spheniscus demersus*) and bank cormorants (*Phalacrocorax neglectus*). *Journal of Experimental Biology*, **199**, 2215–2223.

Wilson, R. P., Hustler, K., Ryan, P. G., Burger, A. E., & Noldeke, E. C. 1992b. Diving birds in cold water: do Archimedes and Boyle determine energetic costs? *The American Naturalist*, **140**, 179–200.

Wilson, R. P. & Peters, G. 1999. Foraging behaviour of the chinstrap penguin *Pygoscelis antarctica* at Ardley Island, Antarctica. *Marine Ornithology*, **27**, 85–95.

Wilson, R. P., Putz, K., Gremillet, D., *et al.* 1995. Reliability of stomach temperature changes in determining feeding characteristics of seabirds. *Journal of Experimental Biology*, **198**, 1115–1135.

Wilson, R. P., Shepard, E. L. C., Gomez Laich, A., & Quintana, F. 2010. Pedalling downhill and freewheeling up: a penguin perspective on foraging. *Aquatic Biology*, **8**, 193–202.

Wilson, R. P., Simeone, A., Luna-Jorquera, G., *et al.* 2003. Patterns of respiration in diving penguins: is the last gasp an inspired tactic? *Journal of Experimental Biology*, **206**, 1751–1763.

Wilson, R. P., Steinfurth, A., Ropert-Coudert, Y., Kato, A., & Kurita, M. 2002. Lip-reading in remote subjects: an attempt to quantify and separate ingestion, breathing, and vocalisation in free-living animals using penguins as a model. *Marine Biology*, **140**, 17–27.

Wilson, R. P., Vargas, F. H., Steinfurth, A., *et al.* 2008. What grounds some birds for life? Movement and diving in the sexually dimorphic Galapagos cormorant. *Ecological Monographs*, **78**, 633–652.

Wilson, R. P. & Wilson, M.-P. T. 1990. Foraging ecology of breeding *Spheniscus* penguins. In: Davis, L. S. & Darby, J. T. (eds.), *Penguin Biology*. San Diego, CA: Academic Press.

Wilson, R. W. & Culik, B. M. 1991. The cost of a hot meal: facultative specific dynamic action may ensure temperature homeostasis in post-digestive endotherms. *Comparative Biochemistry and Physiology*, **100A**, 151–154.

Wilson, R. W. & Liebsch, N. L. 2003. Up-beat motion in swinging limbs: new insights into assessing movement in free-living aquatic vertebrates. *Marine Biology*, **142**, 537–547.

Wislocki, G. B. 1942. The lungs of the Cetacea, with special reference to the harbor porpoise (*Phocaena phocoena*, Linnaeus). *The Anatomical Record*, **84**, 117–123.

Withers, P. C. 2001. Design, calibration and calculation for flow-through respirometry systems. *Australian Journal of Zoology*, **49**, 445–461.

Wittenberg, B. A. 2009. Both hypoxia and work are required to enhance expression of myoglobin in skeletal muscle: focus on "Hypoxia reprograms calcium signaling and regulates myoglobin expression". *American Journal of Physiology: Cell Physiology*, **296**, C390–C392.

Wittenberg, B. A. & Wittenberg, J. B. 1989. Transport of oxygen in muscle. *Annual Review of Physiology*, **51**, 857–878.

Wittenberg, J. B. 1959. Oxygen transport: a new function proposed for myoglobin. *Biological Bulletin*, 117, 402.

Wittenberg, J. B. 1970. Myoglobin-facilitated oxygen diffusion: role of myoglobin in oxygen supply into muscle. *Physiological Reviews*, **50**, 559–636.

Wittenberg, J. B. & Wittenberg, B. A. 2003. Myoglobin function reassessed. *Journal of Experimental Biology*, **206**, 2011–2020.

Witteveen, B. H., Foy, R. J., Wynne, K. M., & Tremblay, Y. 2008. Investigation of foraging habits and prey selection by humpback whales (*Megaptera novaeangliae*) using acoustic tags and concurrent fish surveys. *Marine Mammal Science*, **24**, 516–534.

Woakes, A. J. 1988. Metabolism in diving birds: studies in the laboratory and the field. *Canadian Journal of Zoology*, **66**, 138–141.

Woakes, A. J. & Butler, P. J. 1983. Swimming and diving in tufted ducks, *Aythya fuligula*, with particular reference to heart rate and gas exchange. *Journal of Experimental Biology*, **107**, 311–329.

Woakes, A. J., Butler, P. J., & Bevan, R. M. 1995. Implantable data logging system for heart rate and temperature: its application to estimation of field metabolic rates in Antarctic predators. *Medical and Biological Engineering and Computing*, **33**, 145–151.

Woakes, A. J., Butler, P. J., Bevan, R. M., & Boyd, I. L. 1992. The metabolic rate of free-ranging Antarctic animals. *Biotelemetry*, **12**, 213–223.

Woodin, M. & Stephenson, R. 1998. Circadian rhythms in diving behavior and ventilatory response to asphyxia in canvasback ducks. *American Journal of Physiology: Regulatory, Integrative and Comparative Physiology*, **274**, R686–R693.

Woodward, B. L. & Winn, J. P. 2006. Apparent lateralized behavior in gray whales feeding off the central British Columbia coast. *Marine Mammal Science*, **22**, 64–73.

Woodward, J. D. & Maina, J. N. 2008. Study of the structure of the air and blood capillaries of the gas exchanges tissue of the avian lung by serial section three-dimensional reconstruction. *Journal of Microscopy*, **230**, 84–93.

Worthy, G. A. J. & Edwards, E. F. 1990. Morphometric and biochemical factors affecting heat loss in a small temperate cetacean (*Phocoena phocoena*) and a small tropical cetacean (*Stenella attenuata*). *Physiological Zoology*, **63**, 432–442.

Wozniak, B., & Dera, J. 2007. *Light Absorption in Sea Water*. New York: Springer.

Wright, A. K., Ponganis, K. V., McDonald, B. I., & Ponganis, P. J. 2014. Heart rates of emperor penguins diving at sea: implications for oxygen store management. *Marine Ecology Progress Series*, **496**, 85–98.

Wright, K. L. B., Pichegru, L., & Ryan, P. G. 2011. Penguins are attracted to dimethyl sulphide at sea. *Journal of Experimental Biology*, **214**, 2509–2511.

Wright, T. J. & Davis, R. W. 2015. Myoglobin oxygen affinity in aquatic and terrestrial birds and mammals. *Journal of Experimental Biology*, in press.

Wu, L. & Wang, R. 2005. Carbon monoxide: endogenous production, physiological functions, and pharmacological applications. *Pharmacological Reviews*, **57**, 585–630.

Würsig, B., Wells, R. S., & Croll, D. A. 1986. Behavior of gray whales summering near St. Lawrence Island, Bering Sea. *Canadian Journal of Zoology*, **64**, 611–621.

Wursig, B. & Wursig, M. 1980. Behavior and ecology of the dusky dolphin, *Lagenorhynchus obscurus*, in the South Atlantic.. *Fishery Bulletin*, **77**, 871–890.

Yamamoto, M., Kato, A., Niizuma, Y., Watanuki, Y., & Naito, Y. 2011. Oxygen store and diving capacity of rhinoceros auklet *Cerorhinca monocerata*. *Ornithological Science*, **10**, 27–34.

Yamato, M., Ketten, D. R., Arruda, J., Cramer, S., & Moore, K. 2012. The auditory anatomy of the minke whale (*Balaenoptera acutorostrata*): a potential fatty sound reception pathway in a baleen whale. *The Anatomical Record: Advances in Integrative Anatomy and Evolutionary Biology*, **295**, 991–998.

Ydesen, K. S., Wisniewska, D. M., Hansen, J. D., *et al.* 2014. What a jerk: prey engulfment revealed by high-rate, super-cranial accelerometry on a harbour seal (*Phoca vitulina*). *Journal of Experimental Biology*, **217**, 2239–2243.

Yeates, L. C., Carlin, K. P., Baird, M., Venn-Watson, S., & Ridgway, S. 2014. Nitric oxide in the breath of bottlenose dolphins: effects of breath hold duration, feeding, and lung disease. *Marine Mammal Science*, **30**, 272–281.

Yeates, L. C., Williams, T. M., & Fink, T. L. 2007. Diving and foraging energetics of the smallest marine mammal, the sea otter (*Enhydra lutris*). *Journal of Experimental Biology*, **210**, 1960–1970.

Yenari, M. A. & Han, H. S. 2012. Neuroprotective mechanisms of hypothermia in brain ischaemia. *Nature Reviews Neuroscience*, **13**, 267–278.

Yumino, D. & Bradley, T. D. 2008. Central sleep apnea and Cheyne–Stokes respiration. *Proceedings of the American Thoracic Society*, **5**, 226–236.

Zapol, W. M. 1987. Diving adaptations of the Weddell seal. *Scientific American*, **256**, 100–105.

Zapol, W. M. 1996. Diving physiology of the Weddell seal. In: Fregly, M. J. & Blatteis, C. M. (eds.), *Handbook of Physiology*. Oxford: Oxford University Press.

Zapol, W. M., Liggins, G. C., Schneider, R. C., *et al.* 1979. Regional blood flow during simulated diving in the conscious Weddell seal. *Journal of Applied Physiology*, **47**, 968–973.

Zavalaga, C. B., Benvenuti, S., Dall'antonia, L., & Emslie, S. D. 2007. Diving behavior of blue-footed boobies *Sula nebouxii* in northern Peru in relation to sex, body size and prey type. *Marine Ecology Progress Series*, **336**, 291–303.

Zavlaga, C. B. & Jahncke, J. 1997. Maximum dive depths of the Peruvian diving-petrel. *The Condor*, **99**, 1002–1004.

Zenker, W. & Kubik, S. 1996. Brain cooling in humans: anatomical considerations. *Anatomy and Embryology*, **193**, 1–13.

Zenteno-Savín, T., Clayton-Hernández, E., & Elsner, R. 2002. Diving seals: are they a model for coping with oxidative stress? *Comparative Biochemistry and Physiology Part C: Toxicology & Pharmacology*, **133**, 527–536.

Zenteno-Savin, T., St. Leger, J., & Ponganis, P. J. 2010. Hypoxemic and ischemic tolerance in emperor penguins. *Comparative Biochemistry and Physiology Part C: Toxicology & Pharmacology*, **152**, 18–23.

Zenteno-Savín, T., Vasquez-Medina, J. P., Cantu-Medellin, N., Ponganis, P. J., & Elsner, R. 2011. Ischemia/reperfusion in diving birds and mammals: how they avoid oxidative damage. In: Zenteno-Savín, T., Vasquez-Medina, J. P., & Abele, D. (eds.), *Oxidative Stress in Aquatic Ecosystems*. Oxford: Wiley-Blackwell.

Zierath, J. R. & Hawley, J. A. 2004. Skeletal muscle fiber type: influence on contractile and metabolic properties. *PLoS Biology*, **2**.10, e348. DOI: 10.1371/journal.pbio.0020348.

Zimmer, W. M. X., Johnson, M. P., D'Amico, A., & Tyack, P. L. 2003. Combining data from a multisensor tag and passive sonar to determine the diving behavior of a sperm whale (*Physeter macrocephalus*). *IEEE Journal of Oceanic Engineering*, **28**, 13–28.

Zimmer, W. M. X. & Tyack, P. L. 2007. Repetitive shallow dives pose decompression risk in deep-diving beaked whales. *Marine Mammal Science*, **23**, 888–925.

Index

For EU product safety concerns, contact us at Calle de José Abascal, 56–1°,
28003 Madrid, Spain or eugpsr@cambridge.org.

www.ingramcontent.com/pod-product-compliance
Lightning Source LLC
LaVergne TN
LVHW062206200926
848484LV00006B/1333

* 9 7 8 0 5 2 1 7 6 5 5 5 8 *